THE
ENCYCLOPEDIA
OF
WORKING WITH
GLASS

THE
ENCYCLOPEDIA
OF
WORKING WITH
GLASS

by MILTON K. BERLYE

EVEREST HOUSE · PUBLISHERS · NEW YORK

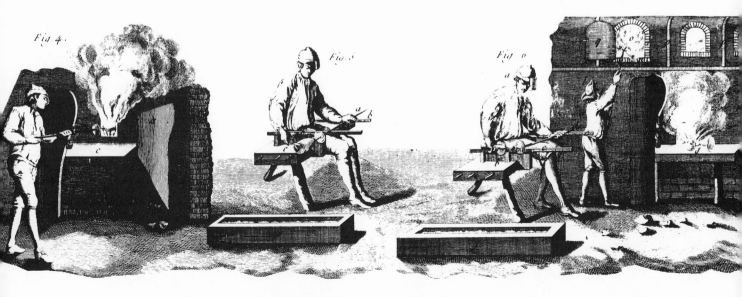

Library of Congress Cataloging in Publication Data

Berlye, Milton K.
    The encyclopedia of working with glass.

    Includes bibliographies and index.
    1. Glass blowing and working.   I. Title.
TP859.B5   1983   666'.1   83-1537
ISBN 0-89696-193-1

Contents

Preface, & Acknowledgements vii

1. THE HISTORY 1
   From the Beginning There was Glass/1
   Glass Comes to America/3
   Glass in Today's World/4
   THE PROCESS 8
   What is Glass?/8
   How is Glass Made?/8
   Glassmaking by Hand/8
   Glassmaking by Machine/9
   Finishing Glass/12
   Types of Glass/14
   Special Glasses/15
   Properties of Glass/17
   How is Glass Used?/19
   Products Made with Glass/20
2. SAFETY IN WORKING WITH GLASS 21
3. CUTTING STRAIGHT PIECES OF
   GLASS 25
   The Equipment Needed/25
   How to Cut Straight Pieces of Glass/30
   Causes and Cures of Glass Cutting
      Problems/32
4. CUTTING LAMINATED SAFETY GLASS 35
   Cutting Curved Windshields/37
   Cutting the Plastic/37
   Problems in Cutting Laminated Safety
      Glass—Causes and Cures/38
5. CUTTING GLASS DISKS 41
   Glass Disk Cutters/41
   Cutting Disks by the Customary
      Method/41
   Improvised Disk Cutters/42
   Cutting Glass Disks with Tin Snips/43
6. HOW TO MAKE UNUSUAL CUTS IN
   GLASS 45
7. CUTTING GLASS CYLINDERS 49
   Methods of Scoring for Cutting
      Cylinders/49
   Cutting Cylinders by Heating on Scored
      Lines/51
   Cutting Tubing/52
   Cutting Bottles, Jars and Jugs/52
   Polishing Cut Edges of Cylinders/53
8. CUTTING HOLES IN GLASS 55
   Drilling Holes up to ½ Inch/55
   Drilling Holes ½ Inch or Larger/56
   Glass Drilling Machines/58
   Cutting Odd-shaped Holes in Glass/58
9. FROSTING AND ETCHING GLASS 61
   Chemical Methods of Frosting/61
   Commercially Prepared Solutions/64
   Details on Procedure for Using Liquid
      Frosting Solution/64
   Physical Methods of Frosting/66
   Sandblasting/67
10. CHIPPING GLASS SURFACES 69

11. ENGRAVING GLASS 71
    Glass Engraving with Portable Hand
       Grinder/72
    Glass Engraving with a Sandblast Gun/73
    Copper Wheel Engraving/74
12. DECORATING GLASS 77
    Preparing the Design/77
    For Painting on Decorating Designs/79
    Scratch-designing/80
    Marbleizing/80
    The Rubber Stamp Method/81
    Dipping/81
    Spraying/81
    Silk Screen Process/82
    Decalcomanias/83
    Cleaning the Glass/83
    Semi-permanent Decorating Materials/83
    Permanent Decorating Materials/84
    Vitrified Decorating Materials/85
13. FINISHING GLASS EDGES 87
    Edging Glass in School and Small
       Shops/87
    Preparing the Edge on Safety Glass/90
    Beveling Procedure used in Glass Fabrica-
       tion Shops/91
14. REMOVING SCRATCHES FROM
    GLASS 93
    Limits of Polishing/93
    Removing Slight Scratches by Hand/93
    Removing Scratches with a Portable
       Polisher/93
15. FIRING GLASS 95
    Effect of Heat on Glass/95
    Separators for Glass/96
    Firing Devices/97
    Procedure for Firing Glass in a Kiln/101
16. GLASS SAGGING (Bending) 105
    Metal Molds/105
    Grog Mix Molds/106
    Fire Brick Molds/107
    Asbestos Molds/107
    Glass Sagging Operation/107
    Miscellaneous Glass Objects that Can be
       Sagged/108
    Glass Jewels/108
17. LABORATORY GLASS BLOWING 111
    Material/111
    Equipment/111
    Supplies/115
    Procedure/115
    Preparation of Glass/115
    Identification of Glasses/116
    Cutting Tubing—General Instructions/117
    Methods of Marking Tubing/117
    Properties of Soft Glass/118
    Properties of Pyrex Brand Glass
       No. 7740/118
    Procedure for Working Tubing/119

Round and Flat Bottom Tubes/119
Bends U-L/120
End Finish and Flaring/121
Straight Seal/121
T Seals/122
Ring Seals/122
Sealing Capillary Tubing/123
Joining Cane/124
Bulbs/124
Closed Circuits of Tubing/124
Coils from Tubing/125
Annealing/125
Furnace Annealing/126
Flame Annealing/126
Etching Pyrex Brand Glasses/126
Grinding Glass/126
Glass to Glass Seals (Graded Seals)/127
Glassworking Lathes/127

18. OFFHAND GLASS BLOWING 129
Glass used/131
Blending/131
Adding Cullet/131
Gathering/131
Systems used/133
The Servitor/135
The Gaffer/135
The Bit-gatherer/135
Cooling/136

19. JOINING GLASS TO GLASS AND OTHER
MATERIALS 137
Non-fired Glass Joints/137
Fired Glass Joints/139
Fired Glass-to-metal Joints/140
Metal Sealed into Glass/142
Bonding Metal to Outside of Glass
(Metalizing)/144
Mending Broken Glass/146

20. MAKING STAINED GLASS PANELS AND
WINDOWS 149
What is Stained Glass Work/149
The Glass Used/149
How a Stained Glass Panel is Made/150

21. SIMULATED STAINED GLASS 163
22. MAKING SLAB GLASS PANELS 171
Thin Glass Slab Panel/171
Thick Glass Slab Panels/174

23. FIBER GLASS CONSTRUCTION METHODS
AND MATERIALS 175
Four Principles in Using Fiber Glass
Reinforced Plastics/175
Fiber Glass Materials (Reinforcements)/177
Resins Used for Fiber Glass Reinforced
Plastics/180
Other Materials Used in FRP Resin
Mix/182
Selecting a Molding Process/183
Design Considerations/188
Suggested Procedures for School and
Home Workshop/190
Spray-up Method of Applying Fiber
Glass/199
Using Fiber Glass as a Permanent
Covering/199
Safety Suggestions/201
Flake Glass/201

24. FIBER GLASS PAINTING 203
25. FIBER GLASS SCULPTURE 205
Fiber Glass Sculpture/205
Solid Form Fiber Glass Sculpture/208

26. REPAIRING DAMAGED FIBER GLASS
PARTS 209
Small to Moderate Holes (Punctures)/209
Split, Torn or Cracked Parts/210
Repair of Larger Damaged Area/211
Repairing Metal Parts with Fiber
Glass/212
Suggestions for Better Fiber Glass
Repairs/213

27. GLASS MOSAIC PANELS AND
FURNITURE 215
Materials used/215
Procedure for making Glass Mosaics/217
Tiling Edges of Mosaic Panels and
Tables/218
Procedures for Different Types of
Design/219

28. CRUSHED GLASS ORNAMENTS 223
Ornaments without Supporting
Panels/224
Ornaments with Supporting Panels/224

29. ENAMELING WITH GLASS 227
Metals that can be Enameled/228
Tools and Equipment needed for
Enameling/228
Supplies needed for Enameling/229
Preparation for Enameling/229
Applying the Enamel/230
Enameling Procedure/230
Designing on the Enameled Piece/232
Finishing the Enameled Piece/236
Possible Defects in Enameling and
Corrections/237

30. MIRROR MAKING 239
The Mirror Making Process/239
Cold-table Hand Pouring Method of
Mirror Making/239
Mixing Formulas for Cold-table
Silvering/241
Peacock (Brashear Type) Cold-table
Silvering Formulas/242
Mirror Making Procedure/242
Hot-table Hand Pouring Method of
Mirror Making/244
Hot-table Mirror Making Procedure/245

31. GLAZING WINDOWS AND PANELS 247
General Conditions Governing
Glazing/247
Glasses used in Glazing/247
Glazing Materials/250
Preparation before Glazing/252
Glazing Details for Single Glass in
Wood and Metal Sash/253
Glazing Details for Insulating Glass in
Wood and Metal Sash/256
Suspended Glazing/258
Special Glazing Considerations/258

32. MISCELLANEOUS INFORMATION 261
The Painting of Display Windows/261
Why is Glass Transparent?/262
Polarized Light/262
Markings on Windows/262

SOURCES OF SUPPLY 264
GLOSSARY 266
RELATED READING 276
INDEX 277

Preface and Acknowledgments

If you can build objects of wood, or form shapes in metal, or mold designs in clay, or prepare sculptures and paints, or repair a car or a TV set, or cook a meal of which you can be proud, or if you can perform any of the skills we usually accept as commonplace, then you can also shape and form glass to your needs and desires.

This book strips away the unfounded fear of breakage and injury which haunts most people from their early experiences with glass. It shows glass for what it is—a substantial material that can be shaped, formed, moulded, and manipulated into useful and artistic shapes with a minimum of material and modest degree of skill.

The activities involving glass which are included in this book can mean many different things to many different people: To the home craftsperson it will mean hours of joyous experience and lead to the development of readily saleable items and skills. To the artist and the modern art teacher it is a new and exciting medium of expression ideally suited for fashioning art objects of infinite and lasting beauty. To the progressive shop teacher it presents the opportunity to expand his/her curriculum by including the interesting study of glass—one of the most important materials of our society. To the arts and crafts and rehabilitation instructors it is an opportunity to get away from "the-same-old-thing" by offering challenging activities for people of all ages. To the hobbyist, housewife, handyman, and all other individuals looking for "something-to-do" that is satisfying, enjoyable, and within their financial means, glass activities offer the ideal answer.

The reason glass activities are ideally suited to meet the needs of so many different people is to be found in the nature of these activities. The crafts described and explained in this book, when interpreted in their most elementary form are suited to the young; however, the same activity performed on a more exacting and elaborate level will prove challenging to the most demanding and skilled adult professional.

The Author wishes to express his thanks and appreciation to the many corporations, institutions and organizations which made available technical data, catalogs, illustrations and other material.

A. Ludwig Klein & Son
American Craftsmen's Council
American Crayon Company
American-Saint Gabain Corporation
Anderson Manufacturing Company
Arts and Crafts Distributors
Berton Plastics Corporation
Bethlehem Apparatus Company, Inc.
Bellco Glass, Inc.
B.F. Drakenfeld & Company, Inc.
Blenco Glass Company
Borden Chemical Company
Bull Dog Lock Company
Central Scientific Company
Chicago Wheel & Manufacturing Company
Cleveland Museum of Art
Corning Glass Works
Corning Museum of Glass
Doerr Glass Company
Dow Corning
E.I. Du Pont de Nemours & Company
Electric Hotpack Company
Empire Machine Company
Ermax Corporation
Etchall, Inc.
Flat Glass Jobbers Association
Fletcher-Terry Company
General Electric Company

Glass Industry Magazine
Glass Plastics Corporation
Glo-Glass Corporation
Immerman and Sons
John Royle and Sons
Kimble Glass Company (Division Owens, Illinois)
La Crosse Glass Company
Libby-Owens-Ford Glass Company
Litton Engineering Laboratories
Loire Imports, Inc.
L. Rensche & Company
Metalizing Industries, Inc.
Minnesota Mining and Manufacturing Company
O. Hommel Company
Owens-Corning Fiberglas Corporation
Paasche Airbrush Company
Pangborn Corporation
Payne-Spiers Studios
Peacock Laboratories, Inc.
Pittsburgh Plate Glass Company
Popular Ceramics Magazine
Pressure Blast Manufacturing Company, Inc.
Red Devil Tools
Reumelin Manufacturing Company
Salem County Vocational-Technical Institute
Sapolin Paints

Sauereiser Cement Company
Sommer and Maca Glass Machinery Company
S.S. White Company
Stain Glass Products
Standard Ceramic Supply Company
Stephens College
Steuben Glass
Stewart-Carey Glass Company, Inc.
Surface Combustion Division, Midland-Ross Corporation
Thiokol Chemical Corporation
Thomas C. Thompson Company
Toledo Museum of Art
Tysaman Machine Company
White Metal Rolling and Stamping Corporation
White Motor Corporation
Yachting Magazine

Chapter 1
The History

FROM THE BEGINNING THERE WAS GLASS

During the prehistoric ages, while man was still living in caves and learning to fashion and use stone implements, Nature herself was engaged in the making of glass. She mixed sand and ashes according to her own formula, heated them in volcanic fires and spewed them forth to cool into a shiny, black rock-like substance called Obsidian. In Yellowstone National Park there is today a mass of this material nine miles long, five miles wide, and rising to a height of about 250 feet above the adjoining terrain.

At other times and other places on the earth Nature prepared different forms of crystallized silica ("glass") as quartz, amethyst, onyx, jasper, rock crystal and agate. The heat when lightning strikes sometimes fuses sand into long, slender glass tubes called fulgurites.

Stone Age man discovered that these forms of natural glass could be broken readily into sharp elongated pieces for arrow and spear heads, and used as cutting implements, knife blades, and razors, but because of its abundance and his limited needs he was not interested in its reproduction.

Just when man discovered the process for making glass is not known. According to an oft-repeated legend, glass was first produced quite by accident. The legend is credited to Pliny the Elder, a Roman historian, who describes a scene on a sandy shore of a tidal river in

Fig. 1-1. NECKLACE OF GLAZED CERAMIC AND GLASS BEADS. Egypt—Probably before 1500 B.C. *(Courtesy: The Corning Museum of Glass)*

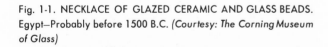

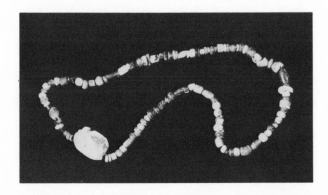

Fig. 1-2. AMPHORISKOS OF IVORY AND RED-BROWN GLASS. Eastern Mediterranean Area 6th-4thc. B.C. Ht. 4-7/8" *(Courtesy: The Corning Museum of Glass)*

Syria, with the men huddled around a fire and in the process of preparing food. Lumps of natron (soda), carried as ballast in their ship, were used to support the cooking pot. In the dying embers of that fire, which had fused the sand and soda, they discovered the first glass known to man.

Modern scientists contend that glass could not have been formed under the conditions described by Pliny. The legend does stand, however, as the earliest written indication of the existence of glass.

Archaeologists' discoveries indicate that the making of glass started about 6,000 years ago. Its earliest use was a crude coating or glaze for pottery. There are in existence specimens of glass bearing definite dates which

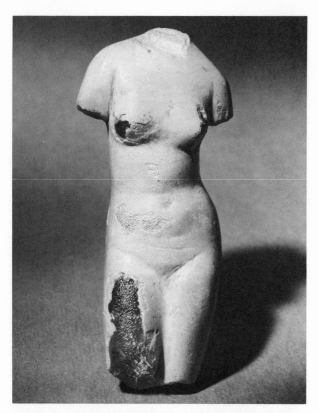

Fig. 1-3. TORSO OF APHRODITE. Probably Near East or Italy. Ca. 1st-2nd c. A.D. Ht. 3-3/4" (*Courtesy: The Corning Museum of Glass*)

Fig. 1-4. GIANT UNGUENTARIUM. Near East—Probably 4th-6th c. A.D. Ht. 15-1/4" (*Courtesy: The Corning Museum of Glass*)

prove that the manufacture of glass slowly emerged from prehistoric obscurity to become staple Egyptian and possibly Syrian industries at least 3,500 years ago. They would wind layer after layer of long thin strips of molten glass around molds formed of sand. When the sand was dumped out the hollow glass object remained.

Although the glass was crudely colored it was highly prized by nobility and royalty who demanded it for beads, cups and oil bottles.

The invention of the blowpipe, credited to a Sidonian artisan, in the third century B.C. made possible quantity production of glass articles in shapes and designs previously impossible to achieve. The blow pipe in essentially the same form was to remain for almost two thousand years as the productive instrument with which increasingly skilled artisans explored and expanded the true capabilities of glass. Blocks, cylinders, hollow vessels, vases, urns, jars and bottles became important terms of commerce.

It was about 200 years B.C. that Roman glassmakers next learned to "roll out" thin slabs of glass mosaics. These were first used only for decoration purposes. Then a dramatic change took place. It started to be used for the closing of window areas in buildings—despite the fact that this semi-opaque glass admitted only a very feeble light.

Transparent glass first appeared about 2,500 years ago and with the development of this property glass was on its way to fulfilling its ultimate destiny—to serve mankind as no other material has or could. Glass had already given some indication of other unusual properties which man was not yet able fully to appreciate or exploit, but none so momentous as transparency.

Although the affinity between light, sight and the transparency of glass was apparent, over a thousand years passed before man began to realize the potential value of this quality and to apply it. During this period the fate of glass alternately suffered and prospered. The scene shifted from Rome to Byzantium-Asia-Germany-France-England. Repeatedly when glass manufacture thrived as a profitable industry it was literally taxed out of existence. It suffered near extinction with the invasion of the barbaric tribes during the fourth and fifth centuries A.D.

During the sixth and seventh centuries magnificent stained glass windows appeared in churches in every land. By the eleventh century the art of glass making was again alive in all its glory.

It was the Crusaders, returning from the wars to the great seaport of Venice, who rekindled an interest in the craft of glass making. Being ideally located in terms of raw materials and trade opportunities, glass making flourished in Venice as it had never done before. Anxious to retain the thriving and lucrative industry for themselves, the Grand Council of Venice in 1290 virtually imprisoned their glass-maker artisans on the island of Murano.

In spite of all the precautions imposed by Venice, the art did spread to Bohemia and then to France and England. The craftsmen of these countries, after first copying Venice techniques, soon introduced new qualities and colors.

Roger Bacon (1214-1292), English philosopher and man of science, is credited with inventing the glass-lensed spectacles. A portrait painted in 1352 shows a man wearing spectacles. Glass—ordained to aid man's vision—was now embarked on a career of service to humanity—unrivalled in the annals of history.

The science of optics flourished and many adapted their talents to this profession. The accidental discovery of the telescope in 1590 by Zacharias Jansen of Middleberg, Holland opened a new dimension for man's eyes to an intimate knowledge of his universe. It was the Italian genius, Galileo, who constructed an improved refracting telescope, later manufacturing them in quantity. In 1627 Johann Kepler, a German astronomer and mathematician, further improved the telescope by using two convex lenses combined with a device for measuring distance.

It was left to Anthony Van Leeuwenhoek, a Dutch naturalist and expert workman in the art of glass lens-making, to reveal the tiny world of microbes and bacteria. In 1660 he perfected the single lens, short focus

Fig. 1-5. ENGRAVED BOWL, INSCRIBED "VITA BONA FRUAMUR FELICS" (We fortunates enjoy the good life). Roman Empire—Probably 3rdc. A.D. Ht. 3-1/4" (Courtesy: The Corning Museum of Glass)

Fig. 1-6. MOLDED AND CUT BOWL. Mesopotamia—CA. 3rd c. B.C. (Courtesy: The Corning Museum of Glass)

microscope. Glass had now made possible the establishment of basic sciences without which glass science would be blind.

It was in 1674 that George Ravenscraft, an Englishman, succeeded in developing lead oxide glass—a "crystali glass" of unusual clarity and brilliant reflectance. This was the glass the very skillful Dutch glass cutters and copper wheel engravers imported for their magnificent decorated pieces.

At about this same time flat glass was being improved and its use in windows spreading. One of the most important developments that took place in England was the substitution of coal for wood in the furnaces used to fuse glass. Unfortunately, however, both England and France retarded the increasing popularity of glass-enclosed windows by short-sighted and almost prohibitive taxes.

A method of casting invented in 1688 by Louis Lucas de Nehou gave glass a brilliance and character never before known. For the first time it permitted the production of large polished plates of flat clear glass of relatively uniform thickness from which it was possible to make very excellent mirrors. It started a fashion in mirrors that covered Europe and extended across the seas to America.

GLASS COMES TO AMERICA

It was natural that glass should be brought to the new world where there was plenty of the raw materials and fuel. The first factory in what is now the United States was built in 1608 in Jamestown, Virginia. It was a tiny place where only beads (for barter with the Indians) and green glass bottles were made. Eight Dutch and Polish glass workers had started the company, augmented in 1620 by six Italian artisans, supplemented thereafter at periodic intervals by many others of various nationalities. During the following hundred years many glass factories (or glass houses as they were called) making crude bottles, bowls and window panes came into existence, and most soon faded out. Glass was still a very rare item among the colonists.

The first important contribution from America was Benjamin Franklin's invention of bifocal glass spectacle lenses in 1760. This was but the forerunner of many astounding developments.

Glass making actually became a permanent industry in this country with the formation of Casper Wistar's company in Salem County, New Jersey, in 1739, which operated until 1781. The second great glass maker was Henry William Stiegel, who operated his company at Manheim, Pa., from 1764 to 1774. Though both companies were in existence a comparatively short time they made profound contributions to the industry. Stiegel gave distinct style to American glass work. Stiegel introduced art glass of outstanding design.

By the nineteenth century glass-making had become a stable part of the economy. In 1825 Deming Jarves founded the famous Boston and Sandwich Glass Co., which introduced the inexpensive "Sandwich" pressed glass—frequently called the "poor man's glass." This was formed by pressing molten glass into iron molds.

Fig. 1-8. CUT PLATE. United States T.G. Hawkes & Co.—Ca. 1900 D. 13-1/2" (Courtesy: The Corning Museum of Glass)

The introduction of this inexpensive glass gave great impetus to the entire trade. During these years some of the most successful and stable companies were founded. The Corning Glass Works was established in Corning, New York, by the Houghtons in 1869. The bulbs for Edison's electric light were produced here. In 1880 the first truly successful Plate Glass Company was established at Creighton, Pa. by John B. Ford, John Pitcoirn and associates. This became the famous Pittsburgh Plate Glass Co. In 1888, Edward Drummond Libby moved his New England Glass Works to Toledo, Ohio, which then became and has remained a center of the glass industry. It was here that Michael J. Owens, in 1809, conceived the fully automatic bottle machine which completely revolutionized glass production methods. The Mississippi Glass Company at St. Louis, Missouri, started the manufacture of rough, ribbed, and figured glass, and in 1893 produced the first wire glass.

GLASS IN TODAY'S WORLD

During the period of twenty years from 1880 to 1900 the United States experienced a great industrial expansion. The building of large railroad systems, manufacturing plants, skyscrapers, and other large industrial and commercial projects greatly increased the need for flat glass of all kinds.

Man was now demanding of glass the fulfillment of those services it had already proved so capable of rendering. His buildings needed windows to admit the natural daylight so that gainful occupation, rest and

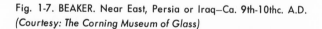

Fig. 1-7. BEAKER. Near East, Persia or Iraq—Ca. 9th-10thc. A.D. (Courtesy: The Corning Museum of Glass)

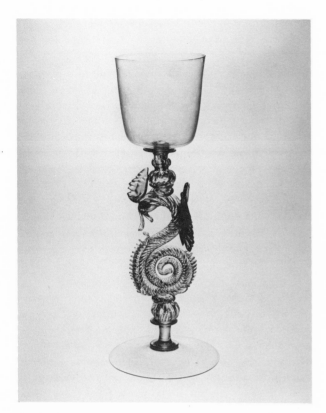

Fig. 1-9. DRAGON-STEM GOBLET. Venice—Late 16th c. Ht. 10-1/4"
(Courtesy: The Corning Museum of Glass)

scourge of many dread diseases abated or eliminated, his span of life increased because of the glass-lensed microscope. During the period 1860 to 1905 the magnifying power of glass-lensed microscopes was steadily increased until it was possible to view an object enlarged two thousand times its actual size. They brought to light the secrets of the origin of disease, the habits and characteristics of bacteria, the atomic structure of all matter: animal, vegetable, or mineral.

Practically every important development in modern science and industry owes its present state of existence to glass.

The huge electronic telescopes which magnify objects twenty-five thousand times their normal size and can further enlarge pictures to one hundred thousand diameters are no exception.

The simple magnifying glasses with their limited power of twenty diameters are not only aids to easy reading but invaluable to such important professions as watchmaking, engraving, die cutting, carving, and a host of other occupations.

Not yet have mortal eyes beheld all the sights that glass possesses power to reveal.

In 1788 William Herschel, noted English astronomer, constructed what was at that time the largest and most powerful telescope in the world. Its reflecting mirror

leisure could proceed unhampered by the vagaries of nature's weather. Only glass could adequately serve this purpose. His buildings needed glass for the display and protection of his wares in the marts of trade and commerce. All these needs and more did glass supply in constantly accelerating tempo. By 1900 there were one hundred twenty-four flat glass establishments in operation in this country catering to its ever-growing necessities.

The automobile, which had come into being in 1879, was joined by the airplane in 1903. These and other methods of travel subjected man to new and unaccustomed exposure from which he needed protection. In vehicles of transport, on land and sea, under water and in the air, glass forms an invisible barrier against the elements and dangerous missiles and other instruments of destruction, preserving his comfort, protecting his life. Protection and preservation were not new services for glass to perform.

Glass protects the products of man's labor in fields, orchards, laboratories, and workshops, protecting them from theft and corruption, from abuse through handling, and from atmospheric contact. It protects his sight, and health, and life, impartially and with equal efficiency.

Glass became the indispensable servant in man's continual battle for survival against the ravages of disease and infection. Man's health has been guarded, the

Fig. 1-10. CANTIR. Spain, Catalonia—18th c. Ht. 16-1/6" *(Courtesy: The Corning Museum of Glass)*

was forty-eight inches in diameter, its focal length forty feet. With this tremendous instrument he obtained a comprehension of the immensity and wonders of the universe never previously attained.

Remarkable as was this achievement, it fades into insignificance when compared with the mammoth twenty-ton glass mirror disc, two hundred inches in diameter and twenty-six inches thick, completed by Corning Glass Works. It is the largest telescope mirror ever conceived by man. The telescope for which it is the light gathering eye has a sixty-foot focal length.

Through its glass lenses man is able to peer one billion light years into space. It can magnify the heavenly bodies ten thousand times and increase thirty-fold the present known volume of the stellar universe. This largest piece of glass ever made, for any purpose, is undoubtedly the most spectacular achievement in the world's history.

The possibilities of its revelations actually stagger the imagination, but who can now value their importance to man, compared with the common, everyday services that glass lenses are performing. Binoculars, in reality two telescopes mounted together to serve both eyes simultaneously, scan great distances. The success in modern warfare is due to the efficiency of glass in periscopes, gunsights, bombsights, rangefinders, and crystals. As protective covers for instrument dials and gauges of every type and description does glass function effectively.

Controlled illumination is possible because of glass incandescent light whose glowing glass bulbs are now produced in excess of one billion a year. Flourescent light of varying intensity and quality is contained in glass tubes. Globes and shades and fixtures come in many different sizes, shapes, colors, and kinds of glass. Molded into very accurate shapes, glass becomes an efficient parabolic reflector from which is projected millions of candle power of all-revealing light, flooding the skies and earth alike with its brilliant radiance.

Glass serves to diffuse light, direct its beams, control its intensity, whether emanating from artificial sources in bulb or tube, the burning flame, or the heavenly bodies.

Our greatest scientific discoveries—in chemistry, physics and physiology—were actually born in glass. In our scientific laboratories where all the elements of nature are confined and almost constantly subjected to exhaustive research, as man patiently separates or combines them in hopes of making an important discovery, glass is his indispensable aid. Thermometers, test tubes, beakers, retorts, vials, bottles, flasks, hypodermic syringes, all of these are glass. For glass is impervious to prac-

Fig. 1-11. BEAKER WITH LOOPED PRUNTS. Germany—Probably 16th c. Ht. 7-15/16" *(Courtesy: The Corning Museum of Glass)*

Fig. 1-12. TRICK GLASS WITH A STAG. Netherlands—17th-Early 18th c. Ht. Ca. 10" *(Courtesy: The Corning Museum of Glass)*

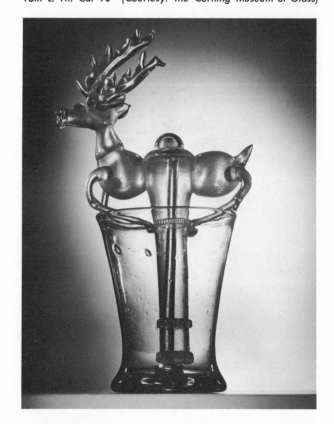

tically all chemicals, resistant to thermal shock as well as physical impact and has the property of transparency which permits full view of the action taking place.

Man's knowledge of invisible solar radiations was spawned in a glass test tube. The X-ray is entirely dependent on glass for its function—in all its applications—curative and exploratory. Violet, ultraviolet and infared rays are either transmitted or absorbed by various special types of glass. Less commonly known but equally important are the alpha, beta, gamma, cathode, and Roentgen rays, whose existence and benefits would still be unknown without glass.

Nor has glass failed in its mission to contribute to man's pleasure and happiness. The truth of that ancient Chinese proverb "A picture is worth ten thousand words" has long since been verified, for pictures have become the universal language by which men of all countries, civilized or not, arrive at a mutual understanding, regardless of differences in speech. The science of photography depends on glass for both still and moving pictures and again, on glass to project them. Today motion pictures so accurately portray the dramas, tragedies and comedies of human life as it exists all over the world, that they have become our principal source of entertainment and of instruction.

In this modern world of today glass contributes sub-

Fig. 1-13. DIAMOND-STIPPLE ENGRAVED GOBLET (signed "F. Greenwood") England and Netherlands—Dated 1746 Ht. 9-7/8" (Courtesy: The Corning Museum of Glass)

stantially to every phase of human endeavor. Electricity—the telephone—radio and now television depend on glass insulators for their efficacy.

Insulation against sound, heat, and cold—heat radiation—commercial refrigeration—air conditioning—flotation—dehydration—are but a few of the modern industries where glass, in a variety of forms such as double or multiple sheets of flat glass—glass blocks, cellulated glass slabs—fibre glass, lend their invaluable properties.

Not bread alone—but also glass—may truly be termed the staff of life. Our modern civilization enjoys the unparalleled advantages of glass walls, glass doors, glass floors, and glass roofs, glass furniture and fixtures, glass utensils, vessels and containers, glass tanks, glass enclosures, glass pipes and glass tubes, glass thread, glass fabrics, and glass apparel, glass armor, glass ornaments and other articles too numerous to recount. Glass—either transparent, translucent or opaque—developments of the twentieth century—examples of man's ingenuity and the versatility of the material.

Well may we share the wonder expressed two hundred years ago by Dr. Samuel Johnson, great English lexicographer and critic, who said:

"Who, when he first saw the sand and ashes by casual intenseness of heat melted into a metalline form, rugged with excrescences and clouded with impurities, would have imagined that in this shapeless mass lay concealed so many conveniences of life as would in time constitute a great part of the happiness of the world."

The history of glass began with the creation of the world—it has no ending. Ageless, indefatigable, incomparable servant of man and all his possessions, glass will survive for the duration of time. Its future, and the future of man because of it, is beyond the power of imagination.

THIS IS GLASS—THE MATERIAL

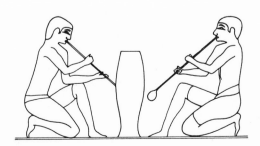

The Process

WHAT IS GLASS?

Man discovered how to make and use glass long before he began to fathom its exact nature. Even today, there are questions still to be answered about this useful substance.

Although it appears to be a solid, glass is technically a supercooled liquid made of fused inorganic materials. Unlike most materials, glass does not change from a liquid to a solid at some fixed temperature. It is always in a non-crystalline state, no matter how hard and brittle it seems to be.

Most of us think of glass as we use it. To the glassmaker, it is the material he obtains by heating such things as sand, lime, and soda until they fuse into a liquid which can be worked to form useful or decorative objects. To the housewife, it is a heat-resistant pie plate or a sparkling tumbler. To the chemist, it is a reagent bottle or a chemical flask. To the young girl, it is a shining mirror.

The large number of different glasses melted by a modern glass factory prevents us from easily defining the material from a viewpoint of its properties. Today we have glass products which are light as cork or almost as heavy as iron; strong as steel or fragile as eggshells; soft as cotton or hard as precious stones.

The variety of possible properties points to the great advantage of modern glass. That is, that through engineering skill and scientific knowledge, the properties of glass can be controlled to a remarkable degree.

We can therefore define glass as a material which is obtained by cooling without crystallization, an inorganic mixture that has been fused at a high temperature.

HOW IS GLASS MADE?

The furnaces of the first glassmakers were tiny pots of clay suspended over wood fires. After hours of patient fire tending, the raw materials melted, fused together, and formed into a glassy mass.

As the efficiency and size of the melting units increased, the pot furnaces took on the appearance of a beehive with a number of clay pots, each holding a few hundred pounds of glass, surrounding a central source of heat. Eventually, coal was substituted for wood as glass-melting fuel, to be replaced in turn by oil, gas, and to some extent, electricity.

Modern glass furnaces are giant tanks holding up to hundreds of tons of glass. A careful mixture of raw materials, called batch, is fed continuously into the rear of the furnace while the molten mixture flows from the other end directly into forming machines.

The object of glass melting is to convert the raw materials into a well-mixed viscous mass. Added to the necessary sand and other compounds in the batch is a quantity of cullet-waste glass of the same type as that being melted. The amount of cullet used varies but it aids melting and is almost always included.

Both pot furnaces and modern continuous tanks are made of ceramic materials known as refractories. The pot furnace is a single refractory unit; the tank is constructed of refractory blocks.

A number of materials, such as silica, alumina, and zircon, are used as refractories. These must be chosen with care, for molten glass is highly corrosive. Refractory materials must resist this corrosion as much as possible. At the same time the refractory dissolved by the glass must not contaminate the batch.

GLASSMAKING BY HAND

Glassmaking requires unique manufacturing techniques for two related reasons: the material must be worked while it is an extremely hot syrup; and the forming operation must be completed in a very short time.

From antiquity until the last quarter of the 19th century, all glass was formed by hand. The first major method of turning molten glass into usable form was off-hand blowing. Eventually, molds to guide the shape of the blown mass were brought into use. Then, in the nineteenth century, methods of hand pressing and drawing were developed.

Hand processing is limited today to special items for which there is a limited demand, and for high quality art and tableware. A knowledge of these hand methods, some of which use tools designed in medieval times,

Fig. 1-14. HAND PROCESSES. When hand-forming glass, the molten material is manually gathered at the end of a tank or pot furnace. It is then formed, usually with the aid of simple mechanical devices. Finishing operations are often identical with those carried out on mass-production lines. *(Courtesy: Corning Glass Works)*

however, aids in the understanding of the giant high-speed machines of the modern glass factory.

Hand Blowing

Hand blowing is usually carried out by a group of men under the direction of a gaffer. The gaffer is responsible for the quality of the finished product and does the more difficult operations himself.

Through an opening in the furnace a gob of molten glass is gathered on the end of an iron blowpipe. Blowing through the pipe forms the hot glass into a hollow ball. The size, shape, and wall thickness of this hollow mass can be controlled by the air the glassblower forces into it, the angle at which he holds the pipe, and the degree to which he allows the glass to cool.

During forming, the glass is handled much like taffy and can be cut, compressed, and stretched with simple tools. Additional small gathers of glass are sometimes added to the original mass to form handles, feet, and stems. To keep the partly-finished piece pliable, it may be reheated a number of times in a small furnace.

For the past few hundred years, the majority of blown glass objects have been formed in molds. In this process, the hot bulb of glass is formed and then placed in an open mold. The mold is closed and the bulb is blown against the wall of the mold, which determines the shape and dimensions of the blown piece.

Casting

Casting is a difficult and usually unreliable method of forming glass. In the early days of glass-making, casting was accomplished by heating pieces of glass in a clay mold until they melted and flowed into the mold shape. Later, the glass was melted separately and then poured into the mold. It is difficult to fill an intricate mold by this method, so casting is restricted to large, simple pieces. However, the largest piece of glass in the world—the 200 inch telescope disk for the Hale Telescope—was successfully made in this manner.

Hand Pressing

Considered a revolution in glass forming 125 years ago, pressing is an outgrowth of the blow mold. Molten glass is gathered on a solid steel rod, or "punty iron." The white-hot gob is then held over the mold and the gaffer cuts off the exact amount required with a pair of hand shears. A plunger is forced into the mold, causing the glass to flow into the mold shape. Different types of mold equipment known as block, spit, and font, are used to make differently-shaped pieces.

Hand Drawing

Molten glass may be drawn into rods or tubing. When hand-drawing tubing, a large gather of glass is made on a blowing iron and rolled into a partially conical shape. A small bubble of air is forced into the gather. The unsupported end of the glass is then attached to another iron in the hands of a helper. The gaffer, keeping the bore open by occasional puffs of air, stretches the glass by walking backward away from his helper. At the same time, a fanner cools the tubing and checks its diameter. When the tubing has cooled it is cut in desired lengths. In this same fashion, glass rods can be made from a solid gather.

GLASSMAKING BY MACHINE

The first machines used in the glass industry were designed to assist hand workers. An example still used today is the thermometer tubing updraw machine. A gather of glass is formed by hand and put into place at the bottom of a tower 185 feet high. The glass is then mechanically drawn to the top of the tower to form tubing with precise inside and outside dimensions.

While most of the first machines were essentially imitations of hand processes, as the demands for glassware increased this approach was abandoned. The most modern glass machines disregard precedent to achieve high speed production by simpler, and sometimes revolutionary methods.

The development of many specialized methods of forming glass has also contributed a great deal to the wide expansion of glass applications. The origin of some of these processes can be seen in the old hand methods of forming glass, but a number of them are purely original results of the work of glass factory engineers and scientists. Examples of unusual glass manufacturing methods include Multiform processing, the manufacture of fused silica and 96%-silica glasses, foam glass processing, fiber glass forming, and centrifugal casting.

Fig. 1-15. MACHINE PROCESSES. *(Courtesy: Corning Glass Works)*

FORMING PROCESSES

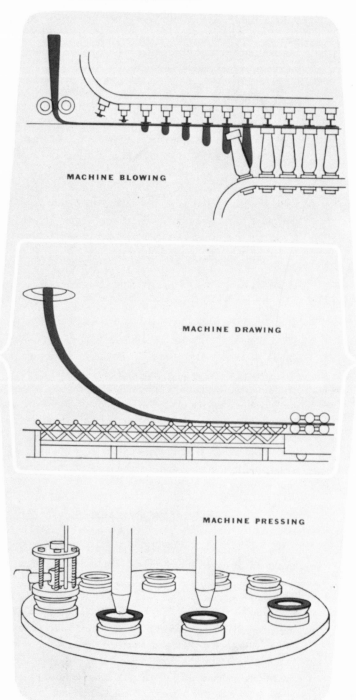

MACHINE BLOWING

MACHINE DRAWING

MACHINE PRESSING

FINISHING PROCESSES

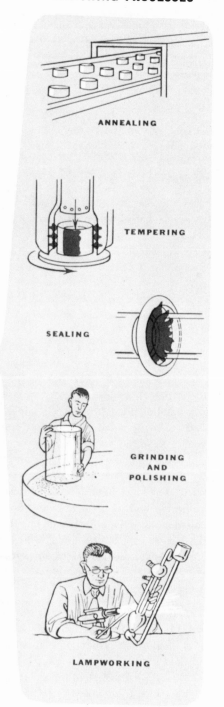

ANNEALING

TEMPERING

SEALING

GRINDING AND POLISHING

LAMPWORKING

Machine Blowing

Glassblowing machines are used to manufacture a wide variety of products, including bottles, jars, laboratory beakers and flasks, coffee makers, signal lights, Christmas ornaments, and enclosures of incandescent lamps and radio tubes.

A number of different types of machines have been developed, among them the turret-chain machine for production of large size glassware, and press-and-blow and blow-and-blow machines for container production. The latter are two-operation machines which first form the neck of the container, either by pressing or blowing, and then complete the forming operation by blowing.

The most remarkable glassblowing machine is the Corning ribbon machine. One of the fastest production machines in the world, it is capable of turning out enclosures for incandescent lamp bulbs at speeds up to 2,000 a minute.

The Corning machine is an outstanding example of the latest stage in glass machinery design—discarding completely the practice of imitating the human glass blower. Molten glass flows continually from a melting tank through a pair of rollers which form it into a rapidly-moving ribbon. Traveling along a steel track, the ribbon sags through holes as air-blowing plungers come down from above to fill out the gobs. Almost simultaneously, molds come up from below to clamp around the molten glass. More puffs of air are forced into the glass, filling it out into the mold shape. The molds separate and pull away revealing a moving chain of finished bulbs which are knocked off onto conveyor belts.

Fig. 1-16. BLOWING. A continuous stream of glass from a forehearth is carried on a perforated track. As air from above, blowing through the holes, creates glass bubbles, molds from below form the bubbles to shape. (Courtesy: Corning Glass Works)

Machine Pressing

Fully automatic presses are used today in the manufacture of table and oven ware, fuse plugs, insulators, automobile head lamps, glass blocks, and a myriad of other products. The pressing process is completely mechanized. From a large tank, gobs of glass are fed continuously into the molds of a revolving press. The mold, containing the hot gob, is moved beneath a plunger which forces the glass into final shape. At the same time, another gob is being dripped into the following mold. After pressing, the hot formed piece of glass remains in the mold while it passes under cooling streams of air. Then, only seconds after the molten glass has left the furnace, the slightly cooled, finished piece is automatically transferred to a moving conveyor belt.

Fig. 1-17. PRESSING. Molten glass from the second forehearth is mechanically sheared off in "gobs" by scissor-like blades, then fed into the molds of an automatic press. Plungers from above, pressing into the molds, squeeze and shape the glass. (Courtesy: Corning Glass Works)

Machine Drawing

Glass is now drawn automatically at speeds approaching 40 miles per hour to form tubing and rod of all sizes for all types of industrial and home use, including glass piping and fluorescent and neon tubing. A great variety of tube sizes and diameters are made automatically on the Danner, Vello, and Updraw machines—the three major methods of drawing glass tubing. The principle of each machine is basically the same: molten glass, flowing directly from the furnace, passes around a ceramic cone, called a mandrel. The glass is then pulled rapidly along a series of pulleys. Air blowing through the center of a mandrel helps maintain the glass as a continuous tube, while drawing speed and glass temperature, along with air pressure, control the dimensions.

Specialized drawing methods are also used for other forms of glass. Sheet glass is made by a drawing process, as is paper-thin ribbon glass and some fiber glass.

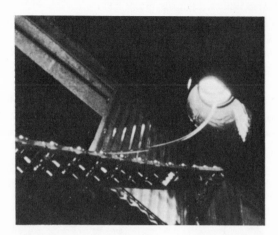

Fig. 1-18. DRAWING. From the forehearth, glass in a thin, steady stream is drawn over a spindle and carried over pulleys at speeds over twenty miles an hour, until it is hardened. The tubing is then cut and finished. *(Courtesy: Corning Glass Works)*

FINISHING GLASS

One of the advantages of glass over other materials is the superiority of the surface obtained in primary forming operations. While metal surfaces usually require grinding or polishing, glass surfaces rarely require additional treatment. Because of the peculiarities of this super-cooled liquid, the surface of a piece of pressed glass is often superior to the surface of the mold in which it was made.

However, a number of glass objects are not complete when they are first formed. Most require some type of secondary or finishing operation. These operations may be conveniently classified as thermal, mechanical, and chemical.

Thermal Finishing

There are a number of glass finishing processes using heat. These include annealing, sealing, tempering, flame cutting, as well as forming pouring lips, fire-polishing edges, backing on enamels and lacquers, and hole forming by localized heating followed by the use of a punch.

Annealing

Almost all glassware is annealed immediately after it has been formed. This process is a controlled heating-cooling cycle which is carried out in an annealing oven, or lehr. The conventional lehr has the appearance of a long tunnel through which conveyor belts carry the newly-formed glass. In the lehr, the glass is first heated, then slowly cooled. This controlled process helps remove the strains in the glass caused by uneven cooling when the piece is formed.

Sealing

A number of glass objects are the product of more than one basic forming method. The enclosure of a television picture tube is an example: the face plate is made by pressing; the funnel is spun; and the neck is drawn. The neck is sealed to the funnel by heating the joining edges of both in a gas flame until they begin to melt. The two pieces are then brought together and allowed to cool, forming a seal as strong as any other part of the funnel or neck. The funnel and face plate are also joined in a similar method—the heat, however, is usually derived from both gas and electricity.

Trimming

Surplus glass is often trimmed from a piece of glassware by means of a flame cut-off. This process or a similar mechanical method is almost always used when finishing hand or machine-blown ware.

Fig. 1-19. MACHINE FINISHING. Finishing operations transform glass components into articles ready for use. Most glassware is annealed to relieve internal stresses; or tempered, a heat treatment in which the glass is reheated, then quickly cooled to strengthen its surfaces. Another finishing method is sealing. (left) The complex parts of a TV picture tube are shown being sealed together with electricity. A large lens is shown (right) being smoothly ground and polished. *(Courtesy: Corning Glass Works)*

Tempering

Certain types of glassware are subjected to heat treatment known as tempering, which appreciably increases their working mechanical strength. After forming, the glass is heated close to the softening point. It is then removed from the heat and quickly chilled, which places the outside glass surfaces under a high compressive stress. Since glass fractures only in tension and cannot be broken in compression, this rapid surface chilling can effectively treble practical glass strength.

Glass-to-Metal Sealing

Glass can be successfully sealed to most metals. The principal problem is satisfactorily matching the coefficients of expansion of the glass and the metal. The wide variety of expansion characteristics available in glass aids this work, as do a number of special alloy metals developed principally for sealing to glass. A number of special techniques have also been devised which include the use of graded glass seals. In this method, glasses with slightly increasing or decreasing expansion coefficients are sealed one to another, until the characteristics of the final glass are sufficiently close to those of the metal.

Fig. 1-21. LAMPWORKING. (Courtesy: Corning Glass Works)

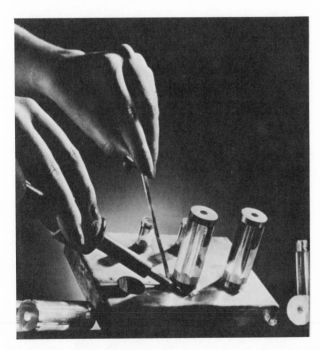

Fig. 1-20. GLASS-TO-METAL SEALING. (Courtesy: Corning Glass Works)

Lampworking

Complicated pieces of laboratory and scientific ware are fashioned by dexterous glass fabricators known as lampworkers. Starting with assorted pieces of glass rod,

tubing and parts, these men fashion with the heat of a gas burner, or lamp, an imposing variety of glass laboratory items.

Metallizing Glass

Glass surfaces can be metallized in various ways to form a base for sealing, for decoration, and to provide a current carrier in the manufacture of electrical devices. Two of the most common techniques are spraying the metal on a cold or hot glass surface and, especially in the case of noble metals, firing the metal on at high temperatures. Meter windows, for example, are sprayed with brass and soldered to a metal fitting to make a moisture-proof seal. Fired-on coatings are used in the manufacture of such products as instrument windows, encapsulating tubes, electronic components, and glass decorated with metallic coating.

Chemical Finishing

Glass is chemically treated during finishing operations to permanently mark it with graduations; to achieve the frosting effect familiar on incandescent lamp bulbs; and as a method of polishing and decorating. Hydrofluoric acid is used for polishing and etching graduations and decorative effects on the glass surface. This is done by first covering the glass with an acid-resisting wax. Lines are cut through the wax with a stylus. The glass is then dipped in acid, which etches the glass where the wax has been cut. Hydrofluoric acid is also used in frosting machines, which lightly etch the interior surface of lamp bulbs.

Glass can also be stained by copper of silver. Copper stains the glass red, while the use of silver results in an amber color. The red color on a number of therapeutic heat lamps is the result of copper staining.

Mechanical Finishing

Mechanical finishing of glass includes grinding, polishing, and cutting. Actually glass is rarely "cut." A diamond or some other hard material is used to score the glass. Slight mechanical force is then applied near the score mark, breaking the glass at the desired point.

Grinding and Polishing

Closely related, both grinding and polishing are used extensively in the glass industry. Edges are ground to make them square, joints are ground for vacuum-fitting scientific apparatus, and plate glass, television picture tube face plates and optical glass are polished to give them a smooth, accurately-finished surface. A number of abrasive materials, such as diamond grains and silicon carbide, are suitable for grinding glass. Polishing is done with finely powdered materials, such as ferric oxide rouge or ceric oxide.

TYPES OF GLASS

Modern glasses are classified into a number of general groups or types. These classifications indicate either the distinguishing materials in the glass, such as lead, or a particular characteristic, such as photosensitivity.

Fig. 1-22. Soda-Lime glass objects. (Courtesy: Corning Glass Works)

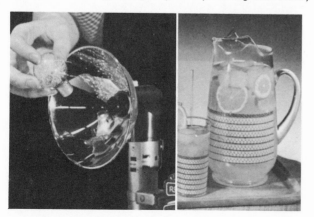

Soda-Lime Glass

Soda-lime glass is made from inexpensive materials and is relatively easy to melt and shape. As a result, this glass generally costs less than any other type. The earliest glass made, soda-lime accounts for nearly 90% of the great tonnage of glass produced in the world each year. Used where heat resistance and chemical durability are not required, the products made from it are infinite—window and plate glass, glass blocks, bottles, tableware, incandescent and fluourescent lamps, and inexpensive novelties.

Fig. 1-23. Lead glass objects. (Courtesy: Corning Glass Works)

Lead Glass

Lead glass is used in the manufacture of fine crystal because of its ease of working and exceptional sparkle and luster. For example, famed Steuben crystal manufactured by Corning is one type of lead glass. The characteristic brilliance of the glass is due to its high index of refraction, which also accounts for its use in many optical lenses and prisms. This glass also has good electrical properties, making it desirable for certain types of electronic applications. Nearly all neon tubing is made of lead glass.

Fig. 1-24. Borosilicate glass objects. (Courtesy: Corning Glass Works)

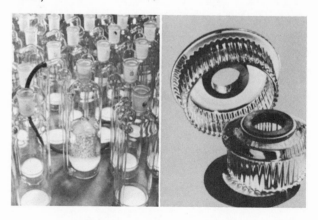

Borosilicate Glass

Borosilicate glass has a high resistance to heat, excellent chemical durability, and good electrical characteristics. Because of its higher softening temperature and lower coefficient of expansion, as well as its great resistance to the corrosive effects of acids, borosilicate glass is used for glass ovenware as well as for many industrial and technical application. These include laboratory glassware, glass piping, boiler gauge glass, and high temperature thermometers. One noteworthy use is the 200-inch mirror disk for the Hale Telescope.

Fig. 1-25. 96% - Silica glass objects. *(Courtesy: Corning Glass Works)*

96%-Silica Glass

96%-silica glass is made by an unusual process which removes almost all the elements except silica from a borosilicate glass after it has been formed by conventional glassworking. Characterized by thermal endurance, chemical resistance, and excellent electrical properties, this glass can be heated cherry-red and then plunged into ice water without breaking. Products made from it include Vycor brand chemical glassware and home appliance parts. Certain 96%-silical glasses transmit a high amount of ultra-violet light and are used for germicidal, photochemical and sun lamps.

Fig. 1-26. Photosensitive glass objects. *(Courtesy: Corning Glass Works)*

Photosensitive Glass

Starting in 1947 with a glass in which a photograph could be permanently formed, a family of photosensitive glass with varying properties has been developed. Included in this group are Fotoform glass, which can be chemically machined with photographic accuracy, and Fotoceram, a ceramic made from glass with exceptional mechanical and thermal properties. These glasses are used for a number of mechanical and electrical parts which require precise machining, and hole forming, as well as lighting panels, name plates, and decorative tiles.

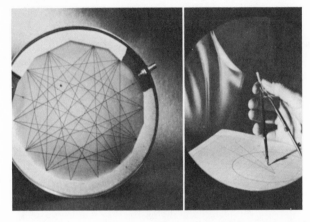

Fig. 1-27. Fused Silica objects. *(Courtesy: Corning Glass Works)*

Fused Silica

Fused silica is the simplest glass chemically and physically and is made directly from silica without other constituents. The melting temperature, however, is extremely high and beyond the range of commercial furnaces. This glass is difficult to work and forming operations are restricted. It has a high melting point and outstanding physical and chemical properties. Because of high manufacturing costs, applications of fused silica are limited today to specialized electronic items such as ultrasonic delay lines and certain optical products, including camera windows on aircraft.

Note: Many glasses of the same general type are melted. For instance, Corning markets some 150 borosilicate glasses. Thus, a wide range of different properties can be found in each classification. Some of the general statements made on these pages apply only to the more common formulations.

SPECIAL GLASSES

Most glasses identified by an obvious property, such as optical glass or colored glass, are special variations of one of the types outlined on the preceding pages. Slight changes are often made in formulation to emphasize certain properties, or some specialized method of procedure is used.

Colored Glass

Colored glasses are made by adding very small amounts of different metal oxides to clear glass formulas: nickel or cobalt for purples; cobalt with copper for blues; copper or chromium for green; iron or carbon for yellows; and gold, copper or selenium for reds. While these metals do not affect the glass properties, the color obtained can vary with each type of glass. Nickel oxide, for example, produces colors ranging from yellow to purple, depending on the glass composition. Almost any type of glass can be colored, but a great deal of skill is needed to insure uniformity. Colored glasses are used in signalware for automotive, railroad, aviation and marine transportation, light filters, colored lights; and sunglasses.

Electrically-Conducting Glass

Electrically-conducting glass is manufactured by permanently bonding a metallic oxide coating to the surface of a borosilicate glass. The radiating and electrical insulating qualities of glass are thus effectively combined with the conductivity of metal. The size and shape of the glass may be varied; and the electrical resistance of the coating may be controlled for various voltages and powers. This glass is used in heating panels for industrial heating and drying, and home heating. In rod and tube form, it is used in the construction of electronic resistors of varying sizes.

Multiform Glass

Multiform glass was developed to produce difficult shapes not readily made by conventional glass fabrication methods. It is formed by first powdering any of a number of glass compositions. The powdered glass is mixed with a binder, pressed into the desired shape, and then heated, or sintered. A translucent glass, Multiform has properties essentially the same as those of the glass from which it is made. Multiform glasses are used to make irregularly shaped chemical apparatus such as bubble plates, pump seal rings, and filters, as well as other unusually shaped mechanical and electrical parts for a variety of applications.

Fiber Glass

Fiber glass can be drawn from a number of glass compositions. Formed by any of the three different processes, each glass fiber is a solid thread with properties generally similar to the glass from which it is made. Heat and acid resistant, fiber glass is noncombustible and an excellent electrical insulator. Fiber glass has extensive applications in insulating structures and appliances; as air filter material; in fabric form for general textile uses; and, in combination with other materials, in the fabrication of such diverse products as chairs and fishing rods.

Opal Glass

Opal glasses have very small particles in otherwise transparent glass which disperse the light passing through the glass. This results in a translucent and milky white appearance. The particles themselves are colorless, but because their index of refraction is different from the base glass, opalescence results. Fluorides are among the ingredients necessary to produce these glasses, and special manufacturing conditions are required. Opal glasses are used for lighting fixtures, tableware, ovenware, and decorative building panels.

Radiation-Absorbing Glass

Radiation-absorbing glasses, some of which are almost as heavy as iron, are made from both soda-lime and lead glasses. Stabilized by adding materials such as cerium oxide to the glass, they do not darken like ordinary glass under bombardment. Normally, these glasses are cast in massive slabs as much as ten inches thick. The slabs are used as radiation-shielding windows mounted in nuclear laboratories to protect scientists observing atomic experiments.

Cellular Glass

Foam or cellular glass is made by heating a mixture of pulverized glass and finely divided carbon in a closed container. The mixture expands into a mass of black foam and then solidifies. The resulting material is almost as light as cork and can be shaped with conventional woodworking tools. Noncombustible and unaffected by dampness and condensation, it is an excellent insulating material. Both side walls and roofs of buildings, as well as refrigerator equipment and piping, can be insulated with this material.

Radiation-Sensitive Glass

Radiation-sensitive glasses are used as atomic radiation indicators. While all glasses are probably affected in some way by nuclear bombardment, certain glasses glow brightly under ultraviolet light after radiation exposure. The degree of fluorescence is directly related to the amount of radiation to which the glass has been exposed. Other radiation-sensitive glasses discolor visibly under bombardment. Both these types of glass can be used as exposure-indicating dosimeters. Photosensitive glasses may also be considered radiation-sensitive.

Optical Glass

Optical glasses are made of soda-lime, lead, borosilicate, and other compositions, as well as some that do not contain silica. More than 100 different types are melted each year. Requirements for optical glass include certain characteristics, such as indices of refraction and dispersion as well as special quality standards. To obtain the required optical properties, it is sometimes necessary to limit other desirable properties, such as resistance to scratching and chemical durability. Optical glasses are used for lenses, prisms, and mirrors in microscopes, cameras, binoculars, range-finders, and other scientific and photographic equipment.

PROPERTIES OF GLASS

The most important characteristic of modern glass is the remarkable degree to which its properties can be varied and controlled.

In fact, the discovery that careful choice of any of a wide number of constituents would give the glassmaker an exceptional variety of glass properties marks the beginning of the modern era in glassmaking and the birth of glass as a basic engineering material.

At the same time, glassmakers have learned to control glass not only by varying the basic formula, but by modifying, improving, and developing entirely new methods of glass manufacture. Added to this extensive engineering and scientific skill has been the realization that the apparent properties of glass products can be greatly modified by their final shape.

The properties of any one glass product are thus the combined result of the knowledge, talent, and activities of the scientist, engineer, and designer.

Mechanical Properties

The mechanical characteristics of glass are among the most important of all its properties. Although glass was thought of for thousands of years as an essentially fragile substance, a rod of glass with perfect surfaces can be stronger than steel. Other valuable mechanical properties are smoothness, resistance to scratching, high ratio of strength to weight, and permanence of shape.

Glass strength is directly related to the condition of the glass surface. Flexural strengths of ordinary glasses with normal surface scratches can range from 5,000 to 15,000 pounds per square inch. Composition has little practical effect on glass strength, but glasses that resist scratching give better service.

Tempering, or heat treating, contributes to the practical strength of glass. Even with an adequate safety factor, tempered borosilicate glass products are produced with working stresses of 4,000 psi—approximately the strength of ordinary grey cast iron.

For practical purposes, glass can be considered a perfectly elastic material. It is elastic up to the point of fracture. No matter how much force is applied, as long as this force is not sufficient to cause fracturing, the glass will return to its original shape when the force is removed.

The hardness of glass cannot be measured by conventional methods. In the glass industry, the terms "soft" and "hard" are used to indicate low and high softening temperatures, rather than actual mechanical hardness.

Mechanical hardness represents an entirely different property-resistance to surface abrasion, or scratch hardness.

Fig. 1-28. MECHANICAL. For centuries the most fragile substance known to man, some forms of glass can now be strong enough to hammer nails into wood—can have strengths up to 400,000 pounds per square inch. *(Courtesy: Corning Glass Works)*

Hardness usually varies with glass composition, but most glasses will scratch mica, mild steel, copper, aluminum, and marble. Common materials hard enough to scratch glass include sand, hard steel, and emery.

Chemical Properties

Most glasses have an outstanding chemical characteristic—corrosion resistance. This property alone tremendously increases the industrial usefulness of the material. Glass can be made that will not dissolve and contaminate materials contacting it; that will not be eaten away by normal acid solutions; and that is easily cleaned and sterilized. These are the basic reasons glass is so widely used in medicine, in all types of laboratories, and in chemical and food processing.

Practically, glass is affected only by hydrofluoric acid, hot concentrated phosphoric acid, alkaline solutions and water at high temperatures.

While glasses possess chemical durability superior to most materials, they have this property in widely varying degrees. Borosilicate and 96%-silica glasses are normally used where acid attack is a problem. However, certain soda-lime glasses are more resistant to alkalis than the high-silica content types. It is also possible to control appreciably chemical attack on certain glasses by chemical or heat treatment.

Thermal Properties

There are a number of measurable characteristics in glass which can be classified as thermal properties. Most important to the user are those concerned with thermal endurance-resistance to the effect of temperature changes. Low-expansion "heat-resisting" glasses are among the most versatile members of the glass family, and are widely used in the laboratory, factory, and kitchen.

A second thermal property, absorption of heat, is of particular interest to the housewife. While glass is not a good conductor, food bakes faster in glass than metal because of the radiant heat striking its surface.

The thermal endurance of any glass is directly related to the coefficient of expansion—the amount a material expands per degree of temperature. There is considerable variation between the expansion coefficients of glasses. Those with a high silica content have exceptionally low expansion coefficients, as do certain glasses containing boric oxide. The coefficient of a commercial soda-lime glass, for example, is fifteen times higher than the coefficient of fused silica.

Radiation Effects

The effect on glass of bombardment by high-energy or nuclear radiation is a matter of extreme importance to scientists concerned with harnessing the power of the atom. There is still a great deal to be learned in this field, but progress in developing glasses for specific atomic energy applications is being made. Since 1951, special extremely heavy glasses have been used in radiation-shielding windows. Other glasses, especially sensitive to gamma radiation, have been developed which can be used as radiation dosage indicators.

Radiation-shielding glasses are made from extremely dense lead glasses and from other types of glasses stabilized to prevent discoloration or darkening as a result of bombardment. Unless specially compounded for these applications, the majority of glasses darken when subjected to X-rays and gamma-rays until they lose their transparency.

Glasses used for radiation dosage indicators are usually of a non-silicate type. The one type which has been used to the greatest extent is a silver-phosphate glass. After exposure to radiation these glasses will fluoresce under ultra-violet light in direct relationship to the amount of radiation they have received.

Electrical Properties

Glass is essential today for such varied electrical uses as insulators, radio tubes, electronic components, X-ray tubes, and television picture tubes. Probably its primary value in these cases is its property of electrical insulation.

The resistance glass offers to the passage of electricity depends on the type of glass, the temperature of the glass, and the surface conditions. However, glasses are among the best insulators; even at high voltages, only small amounts of current leak through or over glass.

Among the other measurable electrical properties, the dielectric constant of glass (its ability to store up electricity) is one of the most important.

Glasses with a low dielectric constant are required for high-frequency and high-voltage insulation, while glasses with a high dielectric constant are needed to construct electronic capacitors. A unique quality of glass is the wide range of dielectric constants available.

Optical Properties

The optical properties of glass are of the greatest value in applications where transmission, refraction, or absorption, of light rays are essential. Today we have glasses which transmit and control radiations through a broad part of the visible and invisible light spectrum, as well as all other forms of energy radiations. Optical glass is highly transparent and contains materials which absorb very little light. It must be of such perfect uniformity that a light ray going through the glass is not distorted.

The optical properties of glass have been more extensively investigated than any other characteristic.

There are hundreds of different optical glasses with appearances varying from crystal clear to deeply colored. Each of these glasses has been formulated for a specific purpose and has varying characteristics and varying relationships between separate characteristics. Some of the many optical properties of glass which can be controlled are **refraction**, the bending of light rays passing through a piece of glass; **dispersion**, differences in refraction of various length rays; **reflection**, the amount of light returned from a glass surface; and **absorption** and **transmission**, both of which control the amount of light that passes through glass.

HOW IS GLASS USED?

There are few modern activities in which glass does not play a substantial role. From the dairy barn to the dress shop, from the living room to the laboratory, glass serves in some useful fashion.

During the last fifty years, glass has emerged as a basic engineering material. Thousands of glass articles, scientifically designed and mass-produced with consistently high quality, perform predictably and faithfully in homes, industries, and laboratories.

For the manufacture of glass products—useful, unique, or decorative—several hundred different glasses are melted each year. Each of these glasses is carefully tailored for the proper combination of specific, optical, thermal, mechanical, electrical, and chemical characteristics.

Even the most imaginative person would have difficulty describing a world without glass. Much of what we see is seen through glass. Optical glasses help us investigate the stars and discover bacteria. Heat-resistant glasses allow us to bake a cherry pie and transport hot sulphuric acids. Other glasses enable us to harness heat, electricity, and light.

We have only begun to recognize the infinite variety and fundamental usefulness of glass. For our health, our pleasure, and our comfort, continuing research discovers each year new glasses and new uses for glass.

Lighting

All forms of efficient artificial light require glass. Incandescent and vapor-discharge lights such as fluorescent lamps are made of glass, and even the kerosene lantern uses a glass lamp chimney.

The most common artificial light source is the incandescent bulb—millions are made each year in nearly 1,000 shapes and sizes. Regardless of the use of the lamp—from desk light to street light—the glass envelope holds and insulates a glowing filament in a vacuum or an atmosphere of inert gases while absorbing only a small percentage of the light passing through it. The smallest lamp made, the size of a wooden match, is used in surgical instruments. The largest, hand-blown by Corning for the 75th anniversary of light, is 42 inches high and 20 inches in diameter.

Another major man-made light source is the vapor-discharge lamp, in which gas molecules emit radiation under electron bombardment. The most common is the fluorescent light. Made by coating the inside surface of glass tubing with fluorescent powder, which transforms ultra-violet rays emitted by the gas into visible light, this new lighting method requires hundreds of miles of low-cost, precisely-drawn tubing each year. Other vapor-discharge lights are mercury and sodium lamps, as well as neon tubes.

Glass is not only an integral part of most light sources. It is used to control the direction, intensity and diffusion of natural and artificial light through the use of engineered lightingware in the form of shades, lenses, panels, and reflectors.

Science and Medicine

Glass has immeasurably accelerated the evolution of modern science and medicine. In research laboratories, it is an indispensable material insuring reliable and accurate experimental results. In scientific instruments, such as microscopes, telescopes, spectroscopes, and cameras, it is an irreplaceable part.

The work of the physician, the dentist and the ophthalmologist would be seriously handicapped without glass—whether as hypodermic syringe parts, spectacle lenses, thermometer tubing, medicine bottles, or microscope slide covers. Operating rooms are more comfortable, thanks to glass engineered to cut off the intense heat of thousands of watts of light. Chemical glassware plays a leading role in blood transfusions, because its composition prevents contamination and makes possible repeated sterilization. And special glasses are used in the construction of barrels of "electron guns" capable of developing cancer-killing rays with twice the power of those given off by all the medical radium in the world.

In the scientific search for information, whether it be the evolution of the universe or the construction of the atom, glass has an essential role.

Manufacturing

Many industrial glass applications demand combinations of diverse glass properties. For example, the widely used gauge glass—a vital part of steam power plants—must have mechanical strength, to resist vibration shock; thermal resistance, to withstand the high temperatures; chemical resistance, because of the corrosive effects of steam; and transparency.

Different compositions and manufacturing methods make it possible to produce glasses with predetermined characteristics to solve many specific industrial problems, such as corrosion, heat, cleanliness, and abrasion. In almost every manufacturing plant, glass, as well as special glass-industry developed refractories, is used in a variety of forms to do many important jobs.

Glass heating panels are now used in industrial drying and baking operations. Made of electrically-conducting glass, these panels efficiently radiate a controlled, even source of heat for rapid heating. Other manufacturing applications of glass include godet wheels, which are used to pull threads for synthetic fibers from an acid bath, mercury switch enclosures, jewel bearings, blueprint machine cylinders, and even protecting units for thermocouples used for temperature measurement inside steel furnaces.

PRODUCTS MADE WITH GLASSES

For the Home

art pieces
ash trays
baking ware
basters
beverage sets
binoculars
bottles
bowels
broiler trays
cameras
candle warmers
carafes
Christmas ornaments
clock faces
coffee makers
dinnerware
dishes
drinking glasses
dryer windows
electric lamps
eyeglasses
flash bulbs
fluorescent lamps
food blenders
fuses

ice tub liners
jars
juice extractors
lamps
lamp chimneys
light bulbs
lighting panels
measuring cups
mirrors
mixers
mixing bowls
nursing bottles
oil lamps
packaging
picture framing
refrigerator dishes
serving trays
snack servers
space heaters
stemware
sunglasses
table decorations
vacuum bottles
washer windows
windows

For Science

ampules
atomic windows
barometers
beakers
bell jars
burettes

graduated cylinders
jars
laboratory flasks
lamp reflectors
lenses
level switches

cathode ray tubes
cloud chambers
color filters
cover glasses
crucibles
culture tubes
desiccators
dental mirrors
diodes
distilling apparatus
distilling receivers
dosimeters
extraction apparatus
extractors
filters
fine screens
fritted glass
funnels
gamma ray filters
gas dispersion bottles

manometers
microscopes
opthalmic ware
pharmaceutical bottles
pipettes
precision bore tubing
prisms
radio tubes
raegent bottles
spectroscopes
stirrers
stopcocks
stoppers
telescopes
television tubes
test tubes
thermometers
transistors
x-ray equipment

For Industry

airport beacons
airport runway lights
attenuator plates
aviation filters
aviation fresnels
battery jars
blueprint cylinders
buoy lights
bushings
calcining trays
candle coolers
capacitors
cascade coolers
coil forms
condensers
delay lines
distillation columns
drying bulbs
electrical insulators
explosion globes
explosive containers
fire extinguisher bulbs
flares
flood lights
flow meters
fresnel lenses
furnace sight glasses
fuses

glascast molds
godet wheels
gauge glasses
heat exchangers
inductances
industrial dryers
institutional dinnerware
instrument windows
insulators
lightning arrestors
meter windows
neon signs
oil cups
piping
pumps
radar bulbs
radiant panels
railroad signal chimneys
rangefinders
reflectors
rotameters
roundels
scrubbing columns
sealed beam lamps
street lights
thermocouples
traffic lights
urn liners

Chapter 2
Safety in working with glass

Few materials are as unjustly maligned as is glass as concerns its safety. When the word glass is mentioned many people have visions of sharp edges, cut fingers, blood, and other gory details. The fact is, most people go through a whole lifetime of using glass everyday—drinking glasses, eye glasses, window panes, light bulbs, mirrors, watch crystals, etc.—and never experience a single injury as a result of the association. Glass, when handled with a minimum of care, is no more dangerous than dozens of other materials we use daily.

The only time glass becomes a potential danger is when it is involved in a mishap. Such mishaps are rare and are almost always a result of negligence that could have been avoided. The fear of glass is a characteristic of the raw novice alone. Anyone who has ever had a small experience in handling glass soon loses any fear of the material.

Following are listed some of the suggested precautions that should be observed when working with glass:

1. Maintain Good Housekeeping Practices.

Most glasses will break when hit or pressed against hard objects. Keep the area where you are going to work on glass cleared of unnecessary hard objects and materials. Store glass against a smooth, even surface, avoiding edges and protrusions that might touch the glass in one small area. A table that is to be used for glass work on frequent occasion should be covered with an old carpet or thick felt. On a temporary basis the table could be covered with a pad of newspapers. In any case, the table should be kept clean of all accumulation of any kind; this includes glass chips. It is a good practice to brush off the table with a counter brush after each time glass has been cut on it.

2. Avoid Raw Edges.

There are two times when there are sharp "raw" (not ground or polished) edges on glass; first, when it is cut; secondly, when it is broken. The bare skin should not be pressed or rubbed against such edges at any time. When carrying a small sheet of glass with raw edges, grasp it at the top of the sheet with the thumb and first finger making sure the raw edge does not touch the skin between the fingers. When a heavy sheet of glass is being moved about and must be held at the bottom, protect the hand by wearing a glove with a rubber coated palm or use a "Non-Slip" rubber hand grip which is a piece of rubber 7" by 7" by 1/8" thick

Fig. 2-1. A glass sling for carrying large sections of glass. Requires two men. (Courtesy: Sommer and Maca Glass Machinery Co.)

Fig. 2-2. The "Third-Man" sling enables one man to manage a large section of glass. (Courtesy: Sommer and Maca Glass Machinery Co.)

Fig. 2-3. Vacuum cup holder for lifting glass. *(Courtesy: Red Devil Tools)*

Fig. 2-4. The glass lifting tool enables a man to conveniently lift plate glass. *(Courtesy: Sommer and Maca Glass Machinery Co.)*

that is folded over the raw edge. When moving a large piece of glass that you estimate to be beyond your capability of handling alone, secure some assistance. Never take a chance and move it by yourself as the results can be very destructive.

There are a few different devices available that can be used to carry about large pieces of glass. One is the **Glass Sling** (Fig. 2-1) which consists of a strong belt about 50 inches long by 4 inches wide with handles on both ends. Two people hold the handles on the ends of the sling while the glass is cradled on the belt between them. A device called a **"Third-Man Sling"** (Fig. 2-2) is a unique type of sling that enables man to carry safely a maximum sheet of glass. The **Vacuum Cup Holders** (Fig. 2-3) are safe lifting devices, positively capable of holding as much weight as any strong man can lift. They can be secured in "single, twin, and triple grips." The vacuum holder is easy to operate; simply place it on the clean, dry surface of the glass and push down the lever. The **Glass Lift Tool** (Fig. 2-4) enables a man to lift plate glass from ground level directly into any moderate height bulkhead and onto a setting block without the use of fingers, straps, or vacuum cups. The entire device is rubber coated except for the handle.

3. Handle Broken Glass With Care.

Broken glass requires care in its removal, as the edges are usually knife-sharp. It is best if gloves can be worn for such an operation. Proceed by removing

the largest pieces first, then work on down to the smallest pieces. The smaller pieces can be picked up with a pair of tweezers. These small pieces are sometimes hard to find. Resist the temptation of feeling around for these pieces with the bare hand. It is far better to use a brush or broom to sweep a large area around where the glass was dropped.

4. Avoid Touching Fired Glass.

Heated glass does not change in color or shape until it has been subjected to very high temperatures. There-

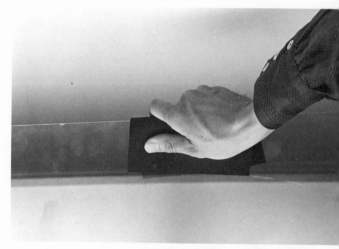

Fig. 2-5. Non-slip rubber hand grips enable a person to hold glass firmly while moving it about. *(Courtesy: C.R. Laurence Co., Inc.)*

fore, a fired (heated) piece of glass can look perfectly normal and still have sufficient heat in it to inflict a disturbing burn. When firing glass it is best to use spatulas, tweezers, and similar devices for handling the glass. It is also good to wear asbestos or even regular work gloves as there is always a temptation to pick up and inspect a fired piece before it has sufficiently cooled.

5. Protective Devices For Glassworkers.

There are a number of safety devices available for those persons who handle large pieces of glass regularly. These devices include:

Wrist Protectors, usually made of heavy-duty leather or mesh nylon that covers the wrist and arm a distance of between 7 and 9 inches.

Protective Sleeves, usually made of strong, reenforced white cotton jean with staggered grommets to protect against glass puncture and to give cool ventilation. They cover the entire arm and shoulder.

Leg Protectors are leather forms that strap over the front of the legs all the way down to the shoes.

Shin and Foot Protectors consist of leggings made of leather and reinforced with a steel frame. The foot guard is a 1/8-inch piece of rubber attached to the leggings and cut to follow the foot contour.

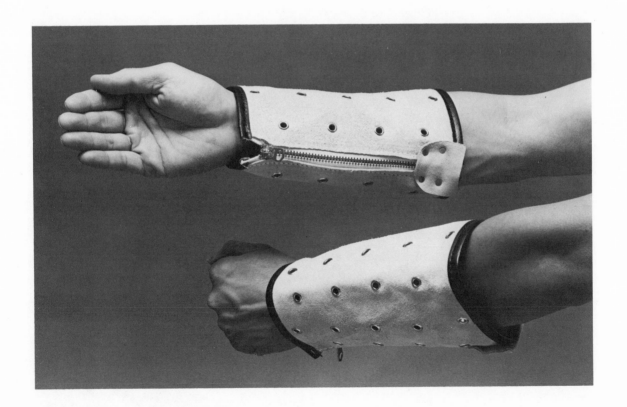

Fig. 2-6. Glassworker's leather wrist protector. Lighter versions are made of nylon. (Courtesy: C.R. Laurence Co., Inc.)

Fig. 2-7. Glassworker's protective sleeves. *(Courtesy: C.R. Laurence Co., Inc.)*

Fig. 2-8. Glassworker's leg protectors. *(Courtesy: C.R. Laurence Co., Inc.)*

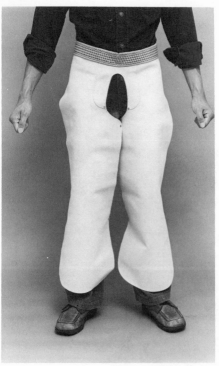

Fig. 2-9. Glassworker's shin and foot protectors. *(Courtesy: C.R. Laurence Co., Inc.)*

Chapter 3
Cutting straight pieces of glass

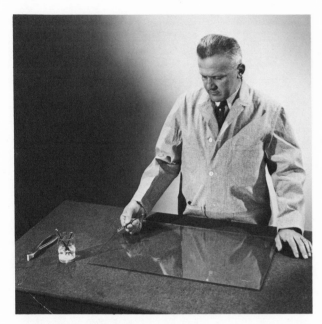

Fig. 3-1. The tools needed for cutting glass are simple and few. *(Courtesy: Libby-Owens-Ford Glass Co.)*

In the Middle Ages glass was cut with a tool which was nothing more than a sharply pointed rod of iron, heated to a high temperature. The red hot point was drawn along the moistened surface of the glass causing it to snap apart. The fracture was not very accurate and the rough piece had to be chipped or grozed down to the exact shape with the help of a hooked tool called a grozing iron.

The present day Steel Wheel Cutter, which is almost universally used, was invented in 1869 by Samuel Monce in Bristol, Connecticut.

The term "glass cutting" is rather misleading as the glass cutter does not actually cut the glass. It merely fractures the surface, upsetting the molecular structure, which causes the glass to part along the scored line.

The whole trick in cutting glass is the making of a clean, sharp score to a uniform width and depth throughout the length of the cut. This is the single requirement for the perfect glass cut. To accomplish this, careful attention must be given to a few simple details.

THE EQUIPMENT NEEDED

The Cutting Table

Choose a firm, level, flat-topped surface for the cutting table. Thirty-four inches above the floor is a good working height for the average student. Make certain that the lighting is good, and placed well above the line of sight.

Cover the work table with soft felt or a piece of deep-piled carpeting. A covering of this type prevents the tiny glass chips from previous cuttings from marring or scratching the glass as it is slid around while making the cut. A hard surface holds the chips tight against the glass, causing scratches. In an emergency or for occasional cutting, a flat wad of newspaper could be used.

The soft felt or carpeted table top has the extra advantage of giving slightly under pressure. This allows the glass to bend when pressed down on after making a cut, helping to "run" the cut.

A whisk broom should be kept handy to sweep the table top clean and free of chips.

A commercial table is shown in figure 3-2. This table is made of steel and can be raised, lowered and tilted for convenient handling of glass.

Fig. 3-2. Tilt-top cutting table in flat position. *(Courtesy: Sommer and Maca Glass Machinery Co.)*

The Straight Edge

A good straight edge is needed as a guide in cutting straight lines, and it will speed up the operation. For this purpose glazier's rules (Fig. 3-3) as well as the Glass cutters "L" and "T" squares (Fig. 3-4) are used.

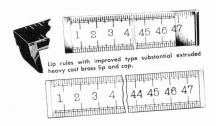

Fig. 3-3. Glazier's rules. *(Courtesy: Sommer and Maca Glass Machinery Co.)*

These are usually made of selected hard maple, thoroughly kiln dried and graduated in 1/8 inches.

For cutting window glass accurately and uniformly to size, a glass cutting board complete with adjustable straight edge (Fig. 3-5) may be used. These are made in sizes from 24" x 36" to 48" x 72", of narrow strips of well seasoned lumber and ruled in inches both ways.

Where the above are not available, any good yard stick or straight and true piece of wood or metal will do as a guide.

Should the student have trouble with the straight edge slipping while the cut is being made, he can overcome the trouble by dampening or placing strips of friction tape on the bottom side.

The Glass Cutter (Fig. 3-6)

Use a good cutter! The best is inexpensive and a good job cannot be done with a poor one.

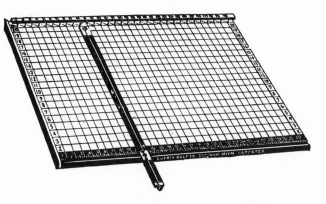

Fig. 3-5. Glass cutting board. *(Courtesy: Sommer and Maca Glass Machinery Co.)*

Cutters employed everywhere today are the **hardened-steel wheel type,** made by a number of manufacturers. Figure 3-7, A through H shows many of the popular types. Cutters A and B are the types used for general cutting. Cutter B has the ball knob that is used for tapping the bottom of the glass after it has been scored. Cutter C is the type wherein the wheel can be replaced when it gets dull. Both D and E have turret heads with six separate cutters. As each wheel gets dull a new one can be rotated and locked in place. When all six are dull they can be replaced with a new turret head. Cutters E and F both have a wooden handle which is preferred by many glass cutters because of its larger grip. The magazine refill wheel glass cutter shown in G carries

Fig. 3-4. Glass cutter's "L" and "T" squares. *(Courtesy: Sommer and Maca Machinery Co.)*

Fig. 3-6. A sharp cutter is of prime importance. *(Courtesy: Libby-Owens-Ford Glass Co.)*

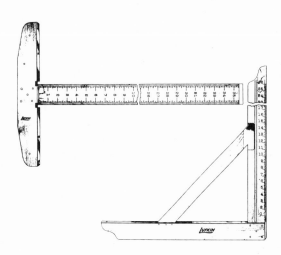

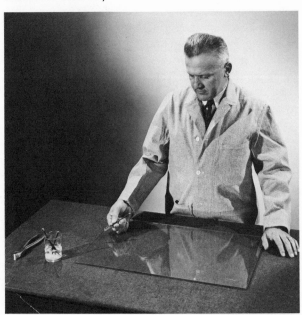

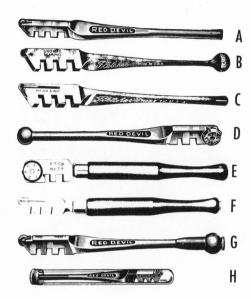

Fig. 3-7. Various types of glass cutters.

six extra wheels in reserve in an extra heavy ball head; when a wheel gets dull it is easily slipped out and a new one slipped in. In H is shown how many of the better glass cutters are packaged in a plastic tube. This tube or "Caddie" is kept as a protective constant storage for the cutter when not in use.

The Raven Glass Cutter is a new product that has recently been placed on the market. It allows the user to apply pressure with the palm of the hand and to control direction with the fingers. Both the cutter's tip and the palm rest are adjustable to meet virtually everyone's needs. The Raven Glass Cutter gives the user the greatest freedom to exercise the most intricate design. The special design gives greater comfort, more precision and a wider range of action. It comes in right and left handed models, it permits the interchanging of five different tips for different glasses and applications, it operates when pushed or pulled, and it has a tapered tip for better cutting visibility.

The wheel of the glass cutter is the all important part. The ordinary type is made of very high grade tool steel, carefully ground, balanced and axled. When

properly lubricated and maintained it can give a long period of satisfactory service. Within the past decade carbide wheels have come into rather popular acceptance. Utilizing tungsten carbide, long recognized as one of the "hardest metal known to man", these wheels permit a relatively longer life than the standard steel wheel and offer certain advantages in use—greater penetration in cutting and greater ease in parting the glass after use of the glass cutter.

At the time the first steel wheel glass cutter was invented in 1869, glass manufacturers produced mostly plain flat glass which was used primarily for windows. Great progress has taken place over the years until now

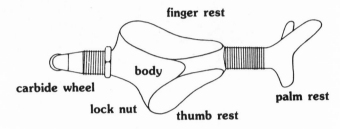

Fig. 3-8a. The Raven Glass Cutter

we have hard glass, soft glass, heat-resisting glass, thick glass, thin glass, decorative glass and countless other varieties. With these many variations it becomes necessary to change the penetration action of the wheel to compensate for the difference in the structural composition of the glass. This is accomplished by varying the degree of the bevel which produces a progressively sharper or duller "cut." A soft glass such as plate glass requires a dull cutting while the harder glass requires a sharper wheel.

There is also a reason for the difference in cutter wheel diameters. The great majority of wheels are 7/32" in diameter. The pressure of the wheel as it rolls over the glass has a bearing on the penetration, and there is a ratio between the pressure and the arc of the wheel that is important. With average hand

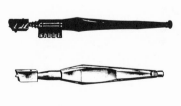

Fig. 3-8. Diamond glass cutters. (Courtesy: Sommer and Maca Machinery Co.)

Fig. 3-9. Diamond glass cutter particularly good for novice.

pressure the 7/32'' diameter wheel gives best results. For a duller wheel on soft glass a slightly different pressure is required. This is compensated for with a larger wheel, 1/4'' in diameter, making it unnecessary for the operator to change the hand pressure. The smaller wheel, 1/8'' in diameter, is used for cutting patterns, and for cutting circles. Here the wheel with a small arc is necessary to follow curved lines without dragging.

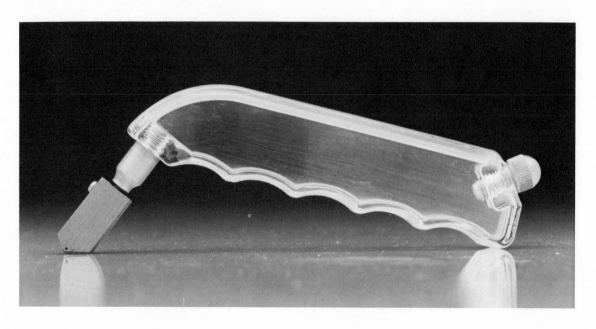

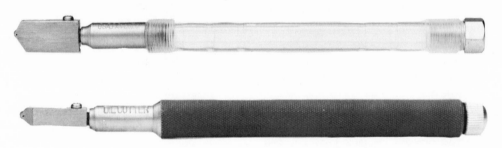

Fig. 3-10. Oil type glass cutters. Every time the blade is depressed a drop of oil is fed to the blade. The head of the cutter is replaceable. (Courtesy: C.R. Laurence Co., Inc.)

Fig. 3-11. Red Devil stained glass cutter. This cutter's revolutionary design lets you cut comfortably with less fatigue than with ordinary cutter. (Courtesy: C.R. Laurence Co., Inc.)

Care of Wheel Cutter

The wheel cutter should be kept in good condition by immersing it in oil when not in use; a mixture of one part light oil and one part kerosene is recommended. This prevents rusting, keeps the cutter clean, and makes certain that the wheel will turn freely on its shaft at all times. Pocket and table containers can be secured that are made especially for this purpose. A clean 1/32 gallon paint can with cotton or rag in the bottom and just enough oil and kerosene mixture in the bottom to cover the cutter wheel makes a very fine permanent holder for the cutter.

Treat the cutter with respect; it should not be thrown down and forgotten until the time comes to use it again. Use a sharp cutter. There is no economy in trying to work with a dull one. When a non-renewable cutter has become too dull for further use, destroy it.

Diamond Glass Cutter (Fig. 3-8)

Diamond glass cutters are used to a very limited degree. These cutters depend on a hard commercial diamond embedded in the head to do the cutting. The diamonds are ground to steeper or broader angles for cutting hard window glass or soft plate glass. The dia-

Fig. 3-12. Glass cutting machine used where large quantities of fast, accurately cut glass is needed. *(Courtesy: Fletcher-Terry Co.)*

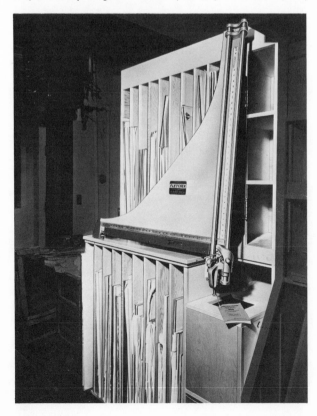

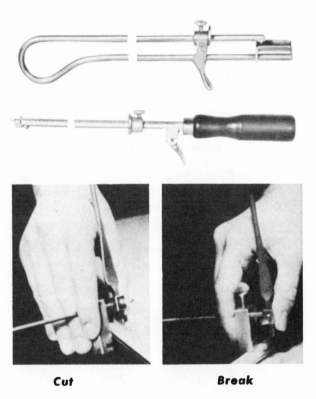

Cut   **Break**

Fig. 3-13. Glass cutting gauges.

mond glass cutters are preferred by some experienced glass cutters because of their long useful life span and the ease with which they trace odd shapes. They are, however, extremely expensive and hard to control and therefore not recommended for the inexperienced. In figure 3-9 is shown a type of diamond glass cutter that can be used by the novice with a degree of success. The proper angle of the diamond is set in this tool, and it is only necessary to hold it in the manner shown to obtain the best results.

The Glass Cutting Machine

Where a great deal of straight glass cutting is to be done, a cutting machine can prove to be most worthwhile (Fig. 3-12). These machines are designed to as-

(Cont. p.32, col.2)

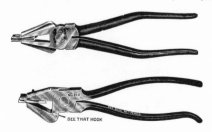

Fig. 3-14. Glass pliers.

HOW TO CUT STRAIGHT PIECES OF GLASS

Fig. 3-15.

1. Rest the glass on the cutting table or layers of newspaper and clean it with a clean rag. Get rid of all dirt or film which would prevent the cutter from making a uniform cut. Many glass workers "wipe" kerosene across the line to be cut with a soft brush. This eliminates much of the flaking from the cut and reduces the tendency for the cutter to slip. The kerosene should be wiped off of the glass after making the cut, as it hides the cut mark.

Fig. 3-15. *(Courtesy: Red Devil Tools)*

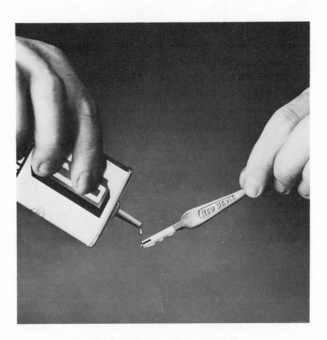

Fig. 3-16. *(Courtesy: Red Devil Tools)*

Fig. 3-16.

2. Lubricate the wheel of the cutter, using any light household oil such as 3-in-1. Dipping it in a mixture of one part light motor oil and one part kerosene is ideal.

Fig. 3-17. *(Courtesy: Red Devil Tools)*

Fig. 3-17.

3. Place the straight edge along line to be cut to guide the cutter. If the bottom side of the straight edge is dampened it will stay in position without slipping.

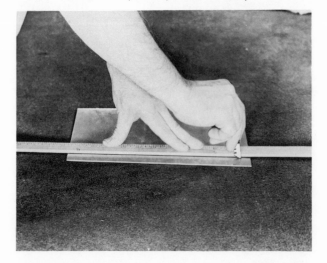

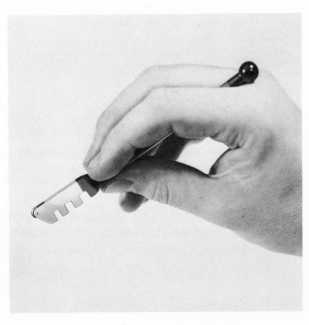

Fig. 3-18A. *(Courtesy: Red Devil Tools)*

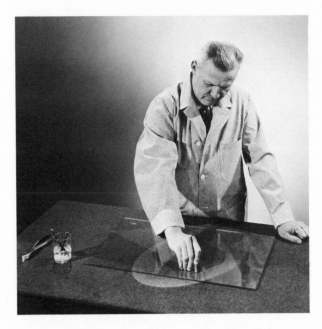

Fig. 3-18B. *(Courtesy: Libby-Owens-Ford Glass Co.)*

Fig. 3-18.
A & B

4. Hold the cutter correctly, between the first and second fingers with the thumb on the underside of the handle, as shown in photograph. Do not hold it too firmly. Press only enough to make a fine hair line. Let the last three fingers of the same hand rest on the glass being cut, or on the surface of the cutting table. This steadies the hand and makes the work less tiring.

Fig. 3-19. *(Courtesy: Red Devil Tools)*

Fig. 3-19.

5. Gently but firmly press the cutter to the glass, holding it upright. Start the cut about 1/8" from the farthest edge. Do not hold the cutter on an angle—it cuts poorly in this position. Make a straight, even, continuous stroke across the whole surface and off the edge of the glass. Practice on pieces of scrap to get the feel.

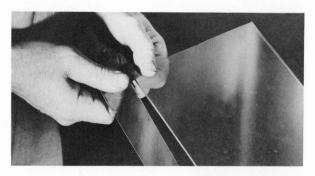

Fig. 3-20. *(Courtesy: Red Devil Tools)*

Fig. 3-20.

6. To break the glass, hold it firmly on both sides of and close to the cut. Then give a quick bending motion away from the cut. Always break right after cutting, so the cut does not get "cold." The break can also be made by placing a wooden matchstick or similar object on the flat surface under the cut near one edge and pressing down on both sides.

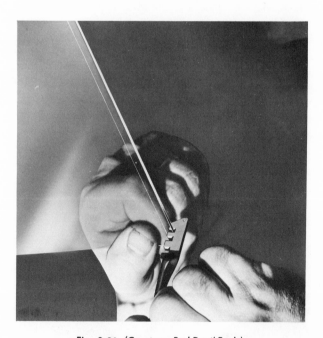

Fig. 3-21. *(Courtesy: Red Devil Tools)*

Fig. 3-21.

7. To break off narrow strips, use the slots in the cutter. Hold the glass in one hand, the cutter in the other. A firm movement away from the cut will separate the glass. (Tapping the underside of the glass right after making the cut may make it separate more easily).

sure fast, proper performance for all types of glass cutting to close tolerances.

Glass Cutting Gauge

When several narrow pieces of glass are to be cut parallel with the edge, a glass cutting gauge may be used (Fig. 3-13). The gauge can be quickly adjusted from 3/16" to 3". It also includes two breaker openings for thin and thick glass.

Glass Pliers

Glass pliers (Fig. 3-14) are used for breaking off narrow strips of glass. They are also used for nibbling away unwanted small pieces of glass that have remained from an improperly executed break.

Cutting Patterned Glass

Patterned glass should be cut on the smoothest side with an even (not heavy) pressure. (Be sure the cutter is lubricated with kerosene and oil mixture mentioned above.) Thus on a pattern like Doublex (Fig. 3-22), the cut should be on the side where the pattern is rounded; on textured Randex (Fig. 3-23), it should be on the pebbled, not the grooved side. Further, do not cut along a groove; the cutting wheel may get stuck if the groove is uneven.

For a smooth pattern, a sharp cutting wheel should be used, as on clear glass; for a rough pattern, on the other hand, a dull carborundum wheel should be used.

Cutting Bent Glass

All bent glass must be cut on the concave or inside surface of the glass. Never remove more than a 2 inch strip at a time when cutting bent glass.

CAUSES AND CURES OF GLASS CUTTING PROBLEMS

If any difficulty is experienced in making cuts, opening or running cuts, or cutting plastic in shatter-proof glass, the operator should go over the following very carefully.

1. **Poor, Ragged, and Chipped Cuts:**

   Skips or blank sections in the mark made by the cutting wheel.
   (a) **Cause**—Poor or dull cutting wheel. Wheels occasionally become flat on part of the edge. **Cure**—Try a new wheel.
   (b) **Cause**—Uneven pressure on wheel in making cut.

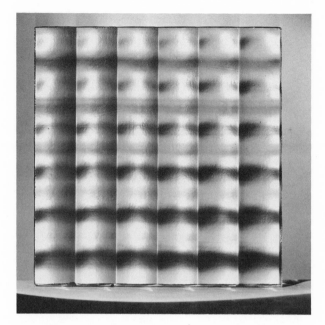

Fig. 3-22. Daublex patterned glass. *(Courtesy: American-Saint Gobain Corp.)*

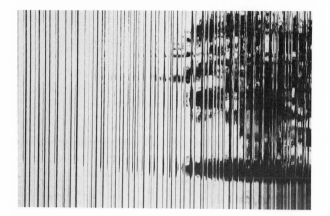

Fig. 3-23. Randex patterned glass. *(Courtesy: American-Saint Gobain Corp.)*

Cure—Use an even pressure, practice on scrap glass. (Too much pressure on a new wheel may be bad.)

(c) **Cause**—Frozen wheel—wheel not revolving as cut is made.

Cure—Free the wheel or try new stem and wheel. Wheel usually can be freed with light oil. To avoid flaking, coat the surface of the plate where the cut is to be made with the lubricant. It is suggested that cutting wheels be kept immersed in a small, shallow can-lid partially filled with kerosene and light oil mixture when not in use. This is also convenient for dipping cutter prior to making cuts on glass.

(d) **Cause**—Too much pressure.

Cure—Press just enough to mark glass clearly.

(e) **Cause**—Glass not lying entirely on level table.

Cure—See that there is no overhang.

Note: A firm even pressure will make the best cut. A good cut is clean and even, with a shine when wiped dry. Flakes of glass, rough lines, chips, wide marks and continued chipping indicate the need of further practice on the part of the operator, or that the plate surface needs oil. Never attempt to cut through drops of water.

2. Cracking of Glass Away From the Line of Cut:

(a) **Cause**—Poor cutting. See above causes.

Cure—New wheel. More practice.

(b) **Cause**—Failure to hold glass cutter square with rule, or pattern.

Cure—If cutter wheel drags sideways slightly, the cut will appear "snowy," somewhat the same as with the application of too much pressure. Be sure to hold the glass cutter so that the side is flat against the rule or pattern. If the edge of the rule or pattern causes the cutter to drag, apply oil or grease to the edge. Never allow cutter to be out of line with the direction of the cut.

(c) **Cause**—Cold air drafts on the glass.

Cure—Be sure you are cutting in a normally warm room, 60 to 75 degrees F. Glass must be at 75 degrees F. or above when cutting. A cold piece of glass brought into a warm room has internal strains in it because the outside surface is warm and the inside is colder. Allow plate to warm up thoroughly before cutting.

Fig. 3-24. Wire saw used for cutting thick glass. *(Courtesy: Tysaman Machine Co.)*

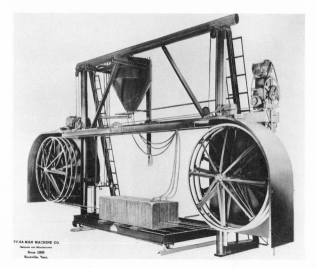

(d) **Cause**—Failure to run the crack before the cut gets cold. A cut will heal in one or two minutes, appearing the same but losing its ability to guide the crack.

Cure—Do not allow much time to elapse before running the crack. As each cut is made, it should be opened.

(e) **Cause**—Too much pressure in running crack.
Cure—Apply just enough pressure to crack clear across the plate by one operation. Never apply pressure ahead of the crack.

(f) **Cause**—Especially on LAMINATED GLASS—Carelessness on the part of the cutter in failing to make the second cut directly over the first.
Cure—Good lighting conditions, careful cutting and experience. The offset formed by uneven cuts subjects the glass to a severe strain when the cracking operation takes place.

Note: A new cutting wheel usually works best on sheet glass and a slightly used cutting wheel on plate glass. If the wheel is too sharp, causing excessive flaking, oil the glass and use wheel with no more than ordinary pressure.

Chapter 4
Cutting laminated safety glass

Laminated safety glass consists of two plates of glass (or lights) with inter-layers of tough transparent plastics that are bonded tightly together with the aid of heat and pressure. It is made in this fashion so that even though the laminated plate is broken, practically no part of the glass will be exposed to inflict injury. This type of glass is available in many thicknesses to satisfy the requirements of a multiplicity of uses. One of its most common uses of laminated safety glass is in automobiles, and so the following cutting instructions will be centered around this type of operation.

The same procedure as for plate glass and window glass is followed in cutting laminated safety glass, with several additional conditions to be met.

Preparing the Glass

In working with laminated safety glass, make certain that the glass is kept in a warm room, or put it in a warm place, before cutting. Glass is much easier to cut, and less likely to break, when warm than when cold.

If glass is taken from a cold room, place it near a radiator for 5 to 10 minutes before it is cut. This will allow the glass to warm up all the way through and prevent internal strains which might cause breakage in cutting.

Use of the Pattern

Laminated safety glass is frequently cut to odd shapes such as the windows of cars. To make sure the glass will be properly cut, a full size pattern is first drawn on heavy paper. The block-size safety glass (which refers to the size of the glass as it arrives in the shop) is placed over the pattern and the design traced on the glass with the cutter. In shops where a great deal of car window repair work is done, prepared professional patterns are purchased for this use.

On some occasions the cracked or broken original glass is used as a pattern. The original glass is placed on top of the block-size safety glass and traced. Should the novice have trouble with the glass pattern slipping while it is being traced, the trouble can be eliminated by placing a few pieces of friction tape between the glass pattern and the block-size glass.

> **Note:** Before starting, note the type of edgework used on the original glass. If ground or polished edges are to be furnished, the cut should be made the exact size; if not, cut 1/32″ undersize.

Fig. 4-2. Pushing the cutter makes it easier to follow a guide line. *(Courtesy: Libby-Owens-Ford Glass Co.)*

Fig. 4-1. Body light pattern on block size glass showing cutting marks. *(Courtesy: Pittsburgh Plate Glass)*

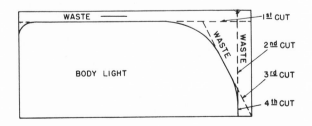

Fig. 4-3. Thumb pressure can be used to run the cut. (Courtesy: Libby-Owens-Ford Glass Co.)

This will make the glass easier to install and will not require forcing.

Place block-size safety glass over the pattern, making sure that the block-size allows enough margin for cutoff. If the cut is less than one inch from edge, use glass pliers to carefully run the cut. The same results can be achieved by tapping.

Making the Cut

Do not press hard on the cutter. For best results, exert just enough pressure to make a fine, light cut. Too much pressure causes flaking, which may destroy the cut. Then, too, these chips are often actually minor runs which may become cracks in the glass later on.

Pull or push the cutter. In following a straight edge or a pattern thick enough to serve as a guide for the cutter, you will probably pull the cutter toward you. However, where you must guide the cutter by watching a pattern placed under the glass, you will probably find it easier to push the cutter (Fig. 4-2). This enables you to follow the pattern without having the cutter itself hiding it. This applies, too, in making the second cut on a piece of safety glass, where the one must be directly over the cut on the opposite light. (Having one cut directly above the other avoids a lot of grinding later on and makes it much easier to cut the plastic.)

Hints on Cutting

In cutting the combination safety glass (one light single strength and the other double strength) it is best

to cut the heavy glass first, making certain that you run it clear through its thickness. Then cut and run the thinner glass.

Do not try to hurry. Remember the time you save by rushing a job will be lost if you break the glass, and you have lost the price of the block of glass.

Running the Cut

Using thumb pressure on glass, when it is standing on edge, as demonstrated in the picture is practical (Fig. 4-3). However, laying the glass flat and using the tapping method to run the cut proves more successful.

Here is where the "give" of your felt table top comes in handy; press on the glass with your thumb and you'll find it bends enough to run the cut. In doing this, however, keep the pressure behind the run. Avoid excess pressure, as that often causes chipped edges or it may vent the glass. If the cut is less than an inch from the edge, use glass pliers to run it.

Good results are also obtained by the tapping method; there is less chance of chipping (Figure 4-4). Corners may be run by flexing or tapping (Fig. 4-5).

Regardless of which method is used, make certain that the run goes clear through each light of glass.

After making the cut on one side of the safety glass, run that cut out before turning the glass over and making the second cut. If this is not done, a vent will invariably occur along the first cut.

Fig. 4-4. Best results are obtained by the tapping method. (Courtesy: Libby-Owens-Ford Glass Co.)

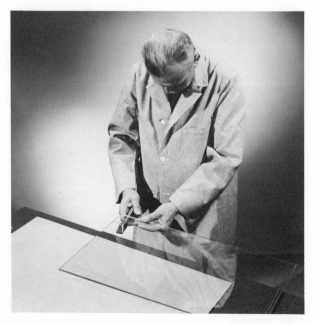

Fig. 4-5. Corners are run by flexing and tapping. *(Courtesy: Libby-Owens-Ford Glass Co.)*

CUTTING CURVED WINDSHIELDS

The outboard light should always be cut first. To cut the inboard light turn the curved windshield over with the outboard light facing the table. Before cutting, support both sides of the curved windshield to prevent sagging.

A windshield body opening improperly re-aligned after an accident necessitates special care in trimming glass for replacement.

After making the score or cut, use glass nippers with several thicknesses of cloth between the jaws. This will distribute the pressure to conform to the curvature of the glass and, with a little practice, you will find that a curved windshield is as easy to cut as flat safety glass.

Caution must be observed when cutting shaded windshields so the proper shaded area will be retained. Match the fadeout lines of the shaded portions and run the cut around the periphery. It is suggested that you take an old curved windshield and practice a few cuts before attempting to cut a good windshield.

Fig. 4-6. Locator.

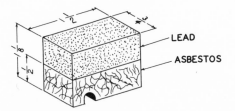

LEAD

ASBESTOS

CUTTING THE PLASTIC

After the safety glass is cut and flexed on both sides, the waste glass is held to the remainder of the light by the flexible plastic only. Extreme care should be taken to avoid chipping of the edges of the glass or overstretching the plastic.

If glass is cold (below 75), it is advisable to heat the plastic along the cut line. Trying to cut the plastic while it is cold may cause chipping of the glass and subsequent breaks.

A heating coil, or other device, applied to the cut for a few moments will soften the plastic sufficiently. Care should be used, however, to prevent overheating, as too much heat will cause the plastic to bubble. If a heating coil is used, do not let the wire get hotter than cherry red.

In figure 4-6 are shown Electric Heating Coil Locators that can be made to hold the heating coil for softening the plastic. The nichrome wire heating coil fits into the groove of the asbestos while the weight of the lead holds the coil in position to conform to the cut. The hot wire should be placed on the **waste** side of the glass 1/16 to 3/16 inches from the cut.

In place of the hot wire method, the stretching and cutting of the plastic may be facilitated if the glass is immersed in water at a temperature of 160° F. to 170° F. for a short period not to exceed 60 seconds. It is important that the whole sheet be immersed. Then, immediately upon removing the glass from the bath, the

Fig. 4-7. Cutting the plastic. *(Courtesy: Libby-Owens-Ford Glass Co.)*

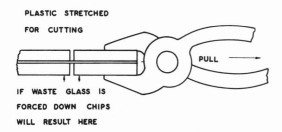

PLASTIC STRETCHED
FOR CUTTING

PULL

IF WASTE GLASS IS
FORCED DOWN CHIPS
WILL RESULT HERE

Fig. 4-8. *(Courtesy: Pittsburgh Plate Glass)*

Fig. 4-10. Laminated glass cutting and heating table. *(Courtesy: Sommer and Maca Glass Machinery Co.)*

plastic membrane is cut.

Use a sharp razor blade of the single edge type to cut the plastic (Fig. 4-7). Stretch the plastic only enough to allow for the free entrance of the razor blade. Do not let the cut-off glass sag, as it may stretch the plastic (Fig. 4-8 & 9).

> **Note:** The razor blade should have its temper removed by heating. This makes it more flexible, so that it follows the contours of the cuts better and reduces blade breakage.

There are special laminated glass cutting and heating tables available (Fig. 4-10). They are made with a regular cutting table top that is hinged so it can be lifted out of the way to expose an asbestos-top table which includes a variable hot wire facility. Figure 4-11 shows a safety glass cutting board. A specially designed tank used for softening the plastic in safety glass is pictured in figure 4-12.

Preparation of Glass for Grinding

Be sure that all excess plastic is cut off flush with the glass. Otherwise, the extended plastic will cause bumping and possibly chipping, of the glass while on the grinding wheel.

"Be a good housekeeper" applies particularly to your glass grinding equipment. Keep your wheels in first-class condition, properly dressed and clean. Recommendations on types of grinding wheels for various operations in your shop, how to use them and their care, are given in pages 50-51.

PROBLEMS IN CUTTING LAMINATED SAFETY GLASS— CAUSES AND CURES

1. **Difficulty In Stretching Plastic**

   (a) **Cause**—Not enough heat.
   Cure—Use the hot wire for a longer time.

   (b) **Cause**—Failure to cut plastic quickly enough after heat application.
   Cure—Cut plastic as soon as possible. It must be cut before the plate cools.

   (c) **Cause**—Cut not open at some point; this is often the case when the waste strip is narrow.
   Cure—Rerun the cut.

2. **Bubbling In Center Of Plastic**

   (a) **Cause**—Heating too long with hot wire. The plastic bubbles when burned.
   Cure—Shorten heating period or lengthen wire. if too hot. The wire when heated should be about a Cherry Red. If glowing brightly, increase its length. Heating wire should always be on waste side of cut.

3. **Excess Separation Along Edge Of Cut**

   (a) **Cause**—Bending the waste strip down too far, thereby stretching the plastic until it becomes thin.
   Cure—Pull straight out with glass pliers.

Fig. 4-9. *(Courtesy: Pittsburgh Plate Glass)*

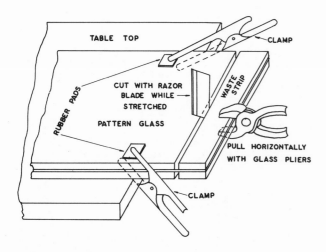

TABLE TOP

CLAMP

RUBBER PADS

CUT WITH RAZOR
BLADE WHILE
STRETCHED

WASTE STRIP

PATTERN GLASS

PULL HORIZONTALLY
WITH GLASS PLIERS

CLAMP

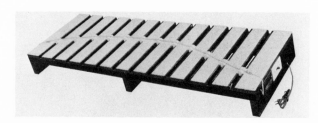

Fig. 4-11. (Courtesy: Sommer and Maca Glass Machinery Co.)

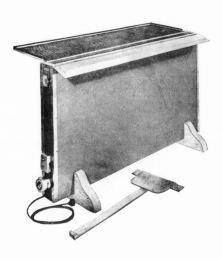

Fig. 4-12. Tank for softening plastic in safety glass. (Courtesy: Sommer and Maca Glass Machinery Co.)

4. Don'ts

(a) Do not exert too much pressure in cracking and stretching plastic.

(b) Do not jar or bump a plate while hot.

(c) Do not make a fresh cut on a hot plate.

(d) Do not wet a hot plate.

(e) Do not apply hot wire to plate colder than room temperature.

(f) Do not apply heat too long.

(g) Do not place hot wire on pattern side of cut.

(h) Do not stop cuts before they have been run to the edge of the light.

(i) Do not try to remove too much waste at one time.

(j) Do not allow any overhang beyond the table when making cut.

Chapter 5
Cutting glass disks

Glass disks of various sizes may be needed for such things as table tops, picture frames, replacement of glass for clocks or flashlights, etc. All that is needed to produce a disk of glass is to make a continuous clean light cut on glass equal to the exact size of the circle desired. It makes no apparent difference whether the glass is held stable and the cutter rotated on the glass or the cutter held stable and the glass rotated below the cutting wheel. The former procedure is the one customarily used; however, the latter is superior for cutting small disks.

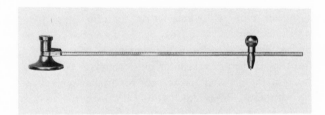

Fig. 5-1. Circle cutter. *(Courtesy: Fletcher-Terry Co.)*

GLASS DISK CUTTERS

In figure 5-1 is shown a glass disk cutter that is customarily used for cutting circles. It consists of a large suction cup in the center with a moveable rod sticking out above it. On the rod is a device for adjusting the cutter to different radii. The cutter illustrated will cut circles from 2 inches to 24 inches in diameter. Larger ones are available.

The disk-cutting machine shown in figure 5-2 cuts much faster and more accurately than the cutter mentioned above. It is, of course, much more expensive than the above and is only recommended for use where quantity cutting of glass disks is a consideration. The machine shown in figure 5-3 cuts ovals as well as circles.

CUTTING DISKS BY THE CUSTOMARY METHOD

Note: In these instructions, the simple disk cutter as illustrated in figure 5-4 is used to score the desired circle in preparation for the cut.

1. Pull out the rod of the cutter to the proper diameter and lock it in this position with the thumb-screw provided.
2. Lubricate the cutter wheel with a drop of light oil or light oil and kerosene mixture.
3. Place the glass flat on the cutting table and press the suction cup down firmly on the spot that is to be the center of the disk.
4. Press firmly on the suction cup with the left hand while rotating the cutting wheel with the right hand (Fig. 5-4).
5. Scribe random radial lines from the circle outward, to the edge of the panel (Fig. 5-5).
6. With the ball end of the cutter start tapping on the underneath side of the circle. Watch for the crack to carry through below the cut. Continue to carry the crack all around the circle, then do the same on the underneath side of each radial line.
7. As the cracks are carried through each of the radial lines the waste glass will break away in segments (Fig. 5-6).

Note: Circles do not usually break as cleanly as do straight cuts. Large chips that continue to hang on after the rest of the waste has been removed can be nibbled off with glazer's pliers. Small irregularities can be polished away (See polishing).

Fig. 5-2. Circle cutting machine. *(Courtesy: Red Devil Tools)*

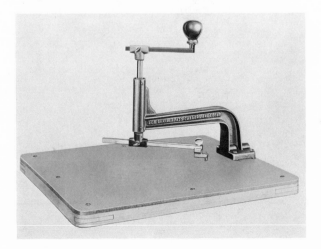

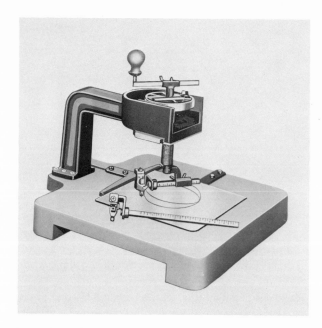

Fig. 5-3. Oval and circle cutting machine. *(Courtesy: Fletcher-Terry Co.)*

IMPROVISED DISK CUTTERS

When a commercial disk cutter, as described above, is not available, a cutter may be improvised from parts found in most shops. Following are a few suggested methods for scoring circles. Once the circle is scored the operation would be continued by starting with step 5 on the above list under "Cutting Disks by the Customary Method."

Method 1: Lock a regular glass cutter in the pencil leg of a carpenter's trammel-point set. Place a suction cup in the center of the circle and place the center point of the trammel on the suction cup. Holding the center point firmly on the suction cup, swing the cutter slowly about the circle.

Method 2: Pivot one end of a strip of iron or hard board on a screw-fitted suction cup (these can be purchased in auto-supply stores). Clamp a regular cutter to the strip at a distance from the center of the suction cup equal to the required radius, then score the circle as described above.

Method 3: In this method, the cutter is held stable and the glass is rotated. First devise a turntable: For cutting small pieces of glass one may be made by drilling a hole in the center of a piece of metal or hardboard and loosely screwing it to a base piece of wood so the metal or hardboard can be rotated. For large pieces of glass a turntable can be made by using the bearings that are used to make Lazy Susan table tops (these bearings can be secured in a few different sizes from most large hardware stores.)

The bottom plate of the bearings can be screwed or clamped to a bench top while a piece of wood connected to the top plate will rotate freely. Stick the glass, from which the disk is to be cut, to the turntable by using melted beeswax or with masking tape pressed along the edges. Locate a regular glass cutter at the radius desired, then clamp it securely to a fixer beam over the turntable. Press down firmly on the cutter as you rotate the glass on the turntable beneath it.

Method 4: When a machine lathe is available a disk

Fig. 5-4. *(Courtesy: Red Devil Tools)*

Fig. 5-5.

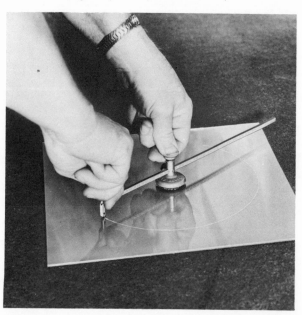

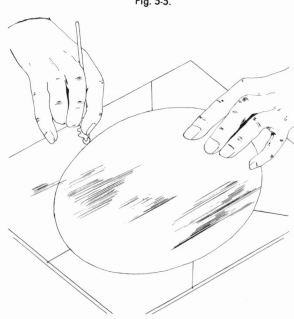

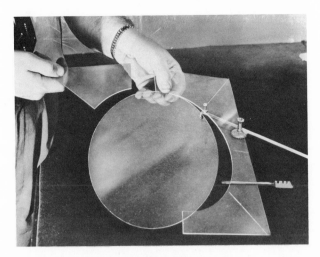

Fig. 5-6. (Courtesy: Red Devil Tools)

may be scored as follows: Mount a piece of wood on a
face plate and turn it up perfectly flat on the lathe.
Mount the glass to be cut to the wood on the face
plate, taping it securely with masking tape on all edges.
Place the faceplate back on the lathe. Mount the cutter
in the tool rest and run it up to the glass until the cutting wheel is pressing firmly against the glass at the
radius desired. The face plate can then be rotated by
hand for one revolution or the lathe can be set to the
slowest speed possible and turned on for one rotation
(Fig. 5-7).

Note: Equally satisfactory methods for cutting
disks can be devised from an old record-player
turntable or potter's wheel when available.

CUTTING GLASS DISKS WITH TIN SNIPS

An easy way to cut a thin glass disk that does not require an accurate and clean edge is to cut the glass
under water with a compound-lever tin snips having
serrated cutting edges. To do this it is important that
both snips and glass be submerged in the water. First
mark a circle on the glass with a grease pencil, then
nibble off the corners of the waste, gradually forming
a circle by taking bites not over 1/8" wide. When done
carefully it is possible to get right down to the marked
circle.

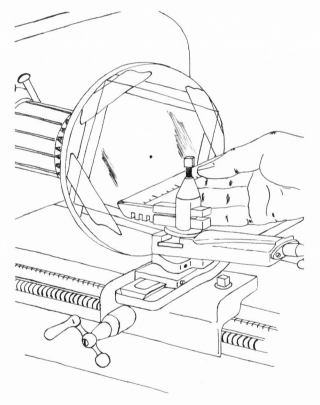

Fig. 5-7.

Chapter 6
How to make unusual cuts in glass

It is occasionally necessary to make cuts in glass that are other than straight lines and complete circles. When such cutting problems occur they must be given special consideration and cut with extreme care if an accurate glass segment is to result.

Before commencing to make an odd shape cut in glass, the shape required should be carefully considered in relation to the glass sheet, for much glass may be saved by the economical use of materials. If it is not easy to see how the shape can be cut from the sheet with economy, the shape may be traced off on to scrap paper. The template, cut out in paper, may be moved about on the glass until a suitable formation is discovered which will cause a minimum of waste. When the position is ascertained, the glass sheet should be placed conveniently on the cutting bench with the template beneath. The cutter should be slowly run around on top of the glass following the outline of the template. Should the novice have difficulty following the template in this free-hand manner, he may cut the template from thick cardboard or plywood, place the template on top of the glass and use it for guiding the cutter.

EXAMPLES OF UNUSUAL CUTS

Fig. 6-1. Proper methods for cutting re-entrance angles.

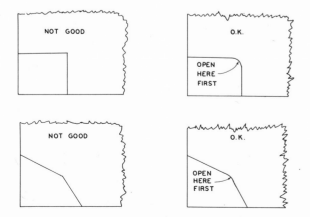

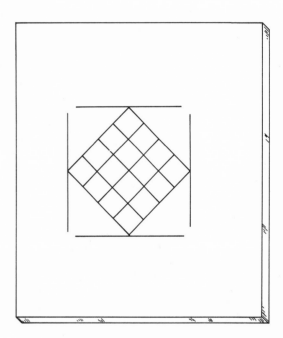

Fig. 6-2. To cut internal squares score as shown.

Re-entrance Angles in Figures 6-1 are sketches illustrating the proper methods for cutting re-entrance angles in a glass sheet. The sharp corners will vent or break—the arrowed curves show the correct method for making this type of cut. It must be remembered that when separating the cut segment from the sheet that the internal curved corner must be separated first.

Internal Squares. A square opening may be cut in glass if the following directions are observed: First, cut the outlines of the opening desired but note in Figure 6-2 that the cuts are not completed to the corners. Next, tap the underside of the glass sheet with the ball end of the cutter handle at the center of each cut. Continue to tap the glass at these points until a crack starts to develop. Following this, make diagonal cuts in two directions and equally spaced as indicated. Tap these cuts from the underside to start cracks, or fractures, in both directions. Then knock out the center square. If the cuts are complete and the breaks well started, it will drop out when tapped lightly with a mallet. Then break or nibble away the triangular corners that remain.

Arches. Arches as shown in figure 6-3 may be cut as follows: First, cut the semi-circular and straight cuts to one edge as shown. Then cross-hatch the waste with evenly spaced cuts. Carefully start to tap the waste glass on the underside of the diagonal cuts working from the edge in. As the cracks start to develop along the cut lines it is possible to remove the waste one piece at a time until the top of the arch is reached. Do not expect the cuts along curved lines to be as smooth as those which can be made on a straight cut. The curved edge will usually need a little "nibbling" with glass pliers and polishing.

Odd Shapes. It is impossible to anticipate all the different kinds of cuts that will confront the student as he becomes more involved in glass work. Following are a few illustrations that will suggest solutions to most of the problems that will arise.

To cut the piece of glass A-B-C-D in figure 6-4, from an irregular glass sheet, proceed as follows: The shape would be laid out on the glass sheet as shown to take advantage of the straight edge of the sheet for side C-D. The first incision should be made from X to Y. This will complete the side A-B as required and at the same time separate the desired piece from the main sheet. The two remaining pieces are removed by cutting along A-C and B-D.

Cutting concave and convex curves present a particular problem because they are breaks that run contrary

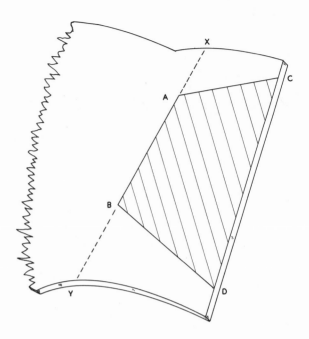

Fig. 6-4.

to the normal breaking pattern of glass. To cut out Curve X in figure 6-5 first make cut A-B then C-D and finally E-F. The nearer to a complete circle the curve desired happens to be, the more cuts will have to be carried to the outside edges as B and D. When all cuts are completed start tapping underneath the glass at the center of the curve and work in both directions. Carry the cracks out to B and D by tapping underneath and the cut segments will break away a piece at a time leaving piece X.

Should a piece such as Y in Figure 6-6 be desired, requiring the removal of an internal curve such as piece

Fig. 6-3. Internal arches are scored for cutting.

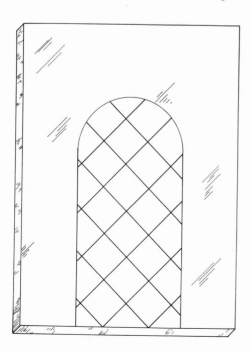

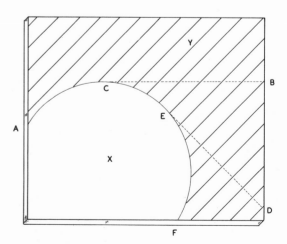

Fig. 6-5.

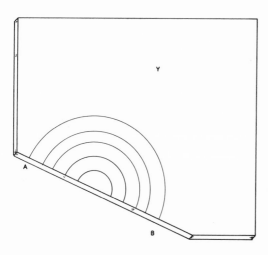

Fig. 6-6.

X, it can be accomplished as follows: First make cut A-B as shown in figure 6-6 the full size of the piece to be removed. Following this make a series of concentric incisions on the waste portion of the glass. Each cut should be taken out to the edge of the glass at intervals of about an eighth of an inch for normal work, though for large work more space may be allowed between the parallel incisions. The main consideration is to weaken the waste portion as much as possible. When all the incisions have been made the cuts may be tapped, beginning at both edges of the glass and working towards the center. Before all the cuts have been tapped the waste glass should fall away, but if it continues to hold after considerable tapping the unwanted portion may be nibbled away with the glazier's pliers.

Chapter 7
Cutting glass cylinders

Very beautiful, inexpensive objects such as fish bowls, terrariums, canister sets, tumblers, candlesticks, vases and the like can be made from nearly all kinds of cylindrical glass bottles and jugs. From cylindrical tubings of different sizes can be made laboratory equipment of wide use.

The cutting of all cylinders is accomplished by causing a hairline fracture to develop completely around the cylinder at a predetermined location. There are three methods that may be used to develop the fracture for separating cylinders:

In the first method, the cylinder is scored in a complete circle at the desired location, then pressure is applied at right angles to the scored line to cause the separation.

In the second method, the cylinder is scored as before, then heat is applied on the scored line causing it to part.

The third method is the same as the second with the exception that should the cylinder not part after the heat is applied, an application of cold is immediately applied. The method used is determined by the size, thickness and type of glass the cylinder to be cut is made of.

Following are described a number of methods and devices that might be used for the convenient cutting of cylinders.

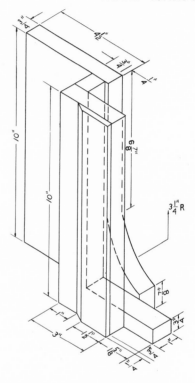

CYLINDER CUTTING JIG

Fig. 7-2.

Fig. 7-1. Glass cutting gauge. *(Courtesy: Red Devil Tools)*

METHODS OF SCORING FOR CUTTING CYLINDERS

1. An **ordinary three-cornered file** may be used to score a cyliner. It works quite well on small-diameter tubing. It is not satisfactory for continuous use as it soon wears away at the corner, and the scratch produced is rather wide.

2. A **glass knife** is more satisfactory than a three-cornered file for scoring tubing and other small cylinders. The glass knife should be designed to produce a sharp scratch, and since it must be forced through the surface of the glass it should be quite hard. Grinding the edges of an ordinary machinist's mill file forms corners which make satisfactory cutting edges, and they may be easily sharpened by re-grinding. Smooth steel knives are also available. Knives with tungsten carbide cutting edges are excellent and produce a very fine scratch. They are durable, and because of the hardness of the tungsten carbide do not wear or dull easily.

3. A **glass gauge** (Fig. 7-1) may be used for cutting tubing of larger diameter. The gauge is very accurate and efficient. However, it is limited in the length it

Fig. 7-3. Scoring a jar in preparation for cutting.

Fig. 7-5. Portable grinder used for scoring and engraving on glass. (Courtesy: Chicago Wheel & Mfg. Co.)

will cut by the shafts holding the cutter and backer.

4. An **ordinary glass cutter** can be used to score cylinders such as bottles, jars, and jugs. The cutter is usually clamped in a jig or vise in a manner that will permit the cylinder to be conveniently rotated against the cutting wheel with a firm even pressure. In figures 7-2 and 7-3 is shown a simple jig that can be made up in the shop. The cutter is clamped at the desired height. The cylinder is steadied against the board next to the cutter as it is rotated against the cutting wheel (Fig. 7-4).

> **Note:** Do not go over the scored line a second time as it tends to dull the cutter. Apply a drop of light oil to the cutting wheel of the cutter before each scoring. If the cutter does not give you an even, fine cut it is most likely dull. Replace it with a new one.

5. A **fine cut-off grinding wheel** of the mounted types (Fig. 7-4) may be used to score larger-size cylinders such as bottles and jugs. The cut-off wheel may be placed in the chuck of the drill press. The table of the drill press can be raised or lowered until the cut-off wheel is in line with the part of the bottle to be cut. Switch

the drill press on, bring the cylinder to the wheel, and keep it rotating to prevent cracking from heat caused by grinding. Score the circumference of the bottle six or seven times on the same line.

6. A **portable grinder** of the "Handee" type (Fig. 7-5) may be used instead of the drill press to score bottles, jars and jugs. A jig such as in figure 7-6 should be prepared to hold the portable grinder conveniently during the scoring operation. The procedure is as fol-

Fig. 7-6. Jig for scoring cylinders.

Fig. 7-4.

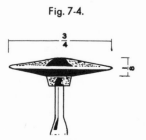

ALL STOCK ¾"

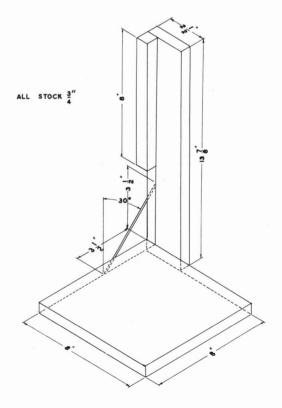

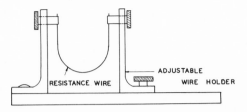

Fig. 7-9. "U" shaped loop used for heating along scored line on cylinder.

No. 22 or 24 B and S gauge with a convenient handle that enables the loop to be held tightly around the cylinder, and a variable-voltage transformer (8-amp., 117 volts) or powerstat to supply the current. (For all gallon jugs, with approximately 21" of exposed wire; the setting on the powerstat should be 17-18 volts. An amperage of approximately 6.3 amps will come through.) Turn on the current and allow the heat to remain on until the cylinder is cracked completely around. If the wire does not go completely around the cylinder, move the cylinder so the wire will cover the uncracked portion and repeat the process.

2. For cutting **tubing,** the wire may be mounted in a U-shaped loop (Fig. 7-9), between upright terminals so that the tubing may be laid in the U and turned to heat it all around. If such a loop is used, it is important to butt one end of the tube against a stop so that the wire will heat only a very narrow zone of glass as the tube is turned.

3. An **open flame** may be applied to the scored line by means of a torch or burning twine. To use the torch, place the bottle or jug that has been scored on a turntable and rotate it very slowly. (A record player turntable run at slowest speed works nicely.) Keep the flame of the torch played on the scored line as the cylinder rotates until you hear the tinkling and pinging sounds caused by cracking. When the crack travels completely around the bottle or jug, remove the torch and lift off the top of the cylinder.

4. If a torch is not available, heat may be applied as follows: Take ordinary **wrapping cord** or knitting yarn, and soak it in lighter fluid, kerosene, or denatured

Fig. 7-7. Method for scoring cylinders using a portable grinder.

lows: Insert the mandrel of the cut-off wheel in the portable grinder. Tape the portable grinder on the support column of the jig as shown in figure 7-7 so that the cut-off wheel lines up with the part of the bottle to be cut. Switch the portable grinder on, bring the bottle to the wheel, and keep it rotating to prevent cracking from heat caused by grinding. Score the circumference of the bottle six or seven times on the same line.

CUTTING CYLINDERS BY HEATING ON SCORED LINES

The heat may be applied to the scored line either by electrical means or by an open flame.

1. An **electrical means** of applying heat to the scored line is typified by the wire circle illustrated in figures 7-8. This consists of a loop of resistance wire (nichrome),

Fig. 7-8. Jig for heating along scored line on cylinder.

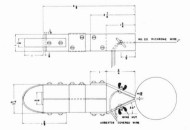

Fig. 7-10. Method for parting thin glass tubing.

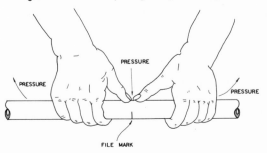

Fig. 7-11.

alcohol. Fill the bottle with cold water to the level of the score mark and put the bottle in a refrigerator. Remove the bottle from the refrigerator after 15 minutes and tie the saturated twine around the score mark. Ignite the cord and let it burn until you hear the characteristic pinging sounds of the cracking glass.

Applying Cold. If difficulty is encountered in causing a crack to develop in the scored line on the glass with the application of heat only, then it may be necessary to apply cold following the heat. This is done by speedily removing the heating device and immediately applying cold water to the scored line with either a brush or cloth wick.

CUTTING TUBING

The simplest method of cutting tubing that is applicable to sizes up to approximately 25 millimeters is to produce a scratch with a file or knife on the tubing at the point where the break is to occur. In making the scratch, use considerable pressure and one stroke, perpendicular to the axis of the tubing. Do not try to saw the tube, as this tends to widen the scratch and wears the edge of the knife excessively. Having scratched the tube, wet the scratch with water applied with the tip of the finger or small brush, then grasp the tubing firmly so that the scratch is between the hands and opposite the thumbs. Bend the tubing at the scratch, creating tension on the scratched side, and at the same time tend to pull the tube apart (Fig. 7-10). A straight, clean break should result.

Another method consists of touching the scratch on the

tubing with a very hot piece of glass rod. It may be necessary to re-heat the glass rod and apply it to the end of the crack, repeating this until the tube is cracked off completely. This method is not so satisfactory as the others, but is convenient for parting tubing that is already assembled to a piece of apparatus.

For cutting larger size tubing the procedure is to score a line completely around the tubing, apply heat, and then cold, as explained above under sections on "scoring," "heating" and "applying cold."

CUTTING BOTTLES, JARS AND JUGS

The procedure for cutting bottles, jars, jugs and similar large glass cylinders is as follows:

1. Score a line completely around the cylinder at the point desired (Fig. 7-3).
2. Apply heat to the scored line until the cylinder cracks completely around where scored (Fig. 7-11). Using a pair of pliers, lift apart the parted segment of the cylinder.
3. If the cylinder will not crack apart when the heat is applied, follow the heat with an application of cold water.

For details on the processes involved see the previous sections on "Methods of Scoring Cylinders," "Methods of Cutting Cylinders By Heating On Scored Lines," and "Applying Cold."

Fig. 7-12. Wet-cut cut-off machine. (Courtesy: Bethlehem Apparatus Co., Inc.)

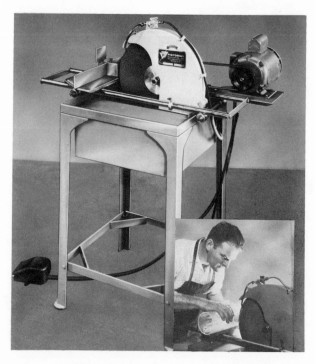

Fig. 7-13. Removing sharp edges with a silicon-carbide stone.

Fig. 7-14. The second grinding is done on silicon-carbide cloth.

Glass Cut-Off Machine

Where a great deal of precision cutting of jars, jugs, or large tubing is to be done, the wet-cut cut-off machine is most desirable (Fig. 7-12). Work need only be held by hand on this machine since the downward motion of the wheel from the applied cutting pressure holds the piece against the right-angle supports. Clamping time is entirely eliminated. A protractor gauge mounted on the rolling carriage makes cutting to any angle easy and accurate. A built-in pump plays a continuous stream of water on each face of the wheel directly at the point of cutting.

POLISHING CUT EDGES OF CYLINDERS

First Grinding (Fig. 7-13). When the cut containers or other cylinder has cooled to room temperature, fill it with water to within 1″ of the top. Dip in the water a medium-coarse silicon-carbide (Carborundum) stone. With a circular motion, use the stone to smooth irregu-lar edges at outside and inside of rim. This will prevent the glass from chipping during the second grinding.

Second Grinding (Fig. 7-14). Place a sheet of medium silicon-carbide cloth on a smooth, flat surface. Place the edges of the cylinder flat on the cloth. Exerting some pressure, rub the cylinder with a circular motion on the silicon-carbide cloth until the cut edge is smooth and a uniform gray in color.

An alternate method is to pour a small amount of carborundum powder (grit #100) into a pie tin and add water to cover the bottom to 1/8″ depth. Place the cut edge of the cylinder in the tin. Exerting some pressure, rub the cylinder with a circular motion in the paste until the cut edge is smooth.

Final Grinding Clean the edge of the cylinder, then repeat what was done during the "first grinding," only this time use a fine grit silicon-carbide stone. An alternate method is to polish the edges with fine abrasive cloth rubbed over the edges with the fingers. When a container is cut to be used as a tumbler or glass, polish the edges until well rounded and as smooth as possible.

Chapter 8
Cutting holes in glass

It is frequently necessary to make openings in glass for speaking holes and for tubing, bolts, screws and the like that may be inserted. This can easily be done provided the operation is performed with care and patience. For drilling holes up to 1/2 inch in diameter the solid type drill is used. For holes larger than 1/2 inch, the tubing type drill is used. The former type drills are used dry while the latter employs a wet abrasive process. Where a square, rectangle, oval, free-form or any odd-shaped hole is needed, it can be cut with sandblasting equipment. Sandblasting can also be used to cut holes on uneven surfaces.

DRILLING HOLES UP TO 1/2 INCH

Tools Used

There are three types of drills that are used for drilling holes in glass up to 1/2 inch in diameter. They are the "Prismatic" the "spear pointed" and the "three-cornered tool steel" types (Fig. 8-1).

The **prismatic-type glass drill** (Fig. 8-1) with the carbide tip is the workhorse of the sheet glass trade and the drill most frequently used. It can be purchased in sizes from 1/8 to 1/2 inches in graduations of 1/32

Fig. 8-1. *(Courtesy: Sommer and Maca Glass Machinery Co.)*

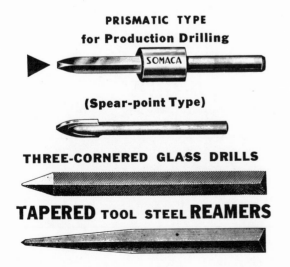

**PRISMATIC TYPE
for Production Drilling**

(Spear-point Type)

THREE-CORNERED GLASS DRILLS

TAPERED TOOL STEEL REAMERS

inch. These drills are used in a power drill (never a hand drill) at speeds of between 400 to 600 r.p.m. They hold their point for long periods of time and, when dulled, can be sharpened comparatively easily.

Fig. 8-2. Spear point glass drill. *(Courtesy: Red Devil Tools)*

The **spear-pointed glass drill** (Fig. 8-2) has a carbide tip. This drill was designed to meet the demand for a lower priced carbide drill for general purpose use. It can be used satisfactorily in a hand brace or "slow-speed" portable electric drill as well as all regular glass drilling machines. It is the best drill for the school or home work shop. It comes in sizes from 1/8 to 1/2 inch in graduations of 1/16 inch.

The **three-cornered steel glass drill** (Fig. 8-1) is the least expensive of the solid drills. It consists of a triangular piece of hard tool steel ground to a pyramid point on the end. It can be purchased in the same sizes and graduations as the spear drill mentioned above.

Note: When the above listed drills become dull, the original radius of the point and clearance should be duplicated as closely as possible.

Shop made glass drills are usually made from three-cornered files or a piece of hard tool steel of the same shape. The cross section of the file must be equal to the size of the hole desired. The tip of the file is ground to a pyramid point each side of which is at an angle of 72 degrees.

The **tapered tool steel ream** (Fig. 8-1) is not used for drilling holes but can be used to good advantage for enlarging holes in glass.

Drilling The Hole

The procedure for drilling a hole that is less than 1/2 inch in size is as follows:

Fig. 8-3. Tube drills. (*Courtesy: Sommer and Maca Glass Machinery Co.*)

1. Place the glass to be drilled on a flat board. Make sure the glass is firmly backed so that it is supported against breakage during the breakthrough of the drill. Where large pieces of glass are being bored, strips of wood can be used, making sure that one strip is directly beneath the point of the drill.

Fig. 8-4. Diamond impregnated tube drill. (*Courtesy: Sommer and Maca Glass Machinery Co.*)

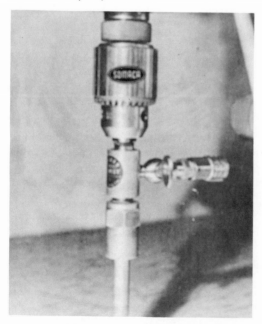

2. Locate the spot to be drilled. (To make it easy for the drill to get started, it is. recommended that the spot to be drilled be **very carefully and gently** tapped with the peen end of a ball peen hammer.)

3. Place the drill in the drilling device and bring it in contract with the glass. The speed should be approximately 25 surface feet a minute, using a steady hand feed with light pressure. The use of slow speed and the maintenance of a sharp point are important. Also, a copious flow of coolant should be used in order to dissipate the heat generated from the action of the drill. This can be applied with a paint brush. Turpentine should be used as the cooling agent.

DRILLING HOLES 1/2 INCH OR LARGER

Tools Used

Tube Drills (Fig. 8-3) are used for drilling holes 1/2 inch or larger in glass. They are made of brass and steel and can be had in sizes from 1/2 to 4 inches in graduations of 1/8 inch. Drills larger than 4 inches may be had on special order. The drill consists of the head, which fits in the drilling device, and the drill shank, which does the cutting. For high speed work there are "Diamond Impregnated Tube Drills" (Fig. 8-4). These drills are used in conjunction with a special head that has an internal pressure coolant for eliminating heat. Such a drill can bore a 1/2 inch diameter hole in plate glass, 1 inch deep in 10 seconds, using water as a coolant.

Fig. 8-5. Large tube drill that can be made in the work shop.

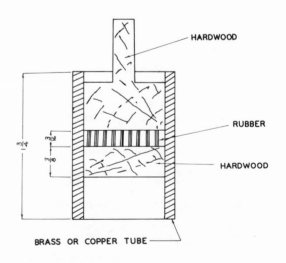

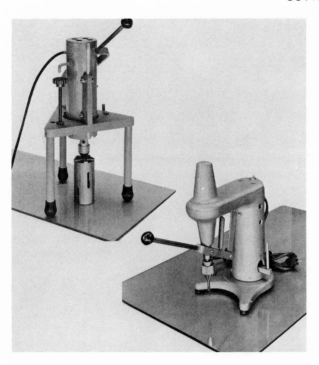

Fig. 8-7. Portable glass drilling machine. *(Courtesy: Sommer and Maca Glass Machinery Co.)*

Shop-made tube drills can be made as shown in figure 8-5. The illustration indicates a tubing of brass or copper. If these are not available, aluminum and steel will do. The cutting end of the tubing should be slit about 1/16 inch deep in two or more places, depending on the diameter. The wood is press-fitted into the tubing.

Drilling the Holes

The procedure for drilling holes with tube drills is as follows:

1. Place the tube drill in the drilling device. It should be used in a drill press or a special drilling machine as described below.
2. Place the glass on a flat piece of wood to eliminate any springing.

3. Place a "dam" around the spot to be drilled to retain the abrasive grains needed for drilling. The dam can be formed of putty or non-hardening clay. When a great deal of drilling with tube drills is done, permanent retainer rings are used in place of the hand-formed dam (Fig. 8-6). These are made of cast iron 2 inches high, with a rubber washer on the bottom.
4. Pour some 90 grit silicon-carbide grains in the dam, then add a little water and mix. The mixture

Fig. 8-8. Portable glass drilling kit. *(Courtesy: Sommer and Maca Glass Machinery Co.)*

Fig. 8-6. Permanent retainer rings that are used in place of the temporary dam.

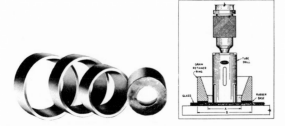

Fig. 8-9. A hole cut with sandblast gun. (Courtesy: S.S. White Co.)

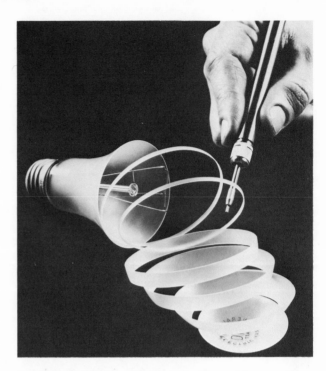

Fig. 8-10. Spiral was cut from a bulb with sandblast gun. (Courtesy: S.S. White Co.)

should be equal to a light cream.

5. Turn on the drill and bring it in contact with the glass. A speed of approximately 150 r.p.m. is best with a medium to light pressure. Extremely light pressure should be used during the breakthrough of the drill in order not to chip the edges of the hole.

Note: Holes bored by the tube drill and abrasive grit method will be slightly larger than the tube drill used due to the fact that abrasion takes place on the side of the tube as well as on the end.

There is practically no limit to the thickness of glass that can be satisfactorily bored by the tube drill and abrasive grit method.

GLASS DRILLING MACHINES

There are special electrical portable glass drilling machines on the market. Figure 8-8 shows a portable slow-speed model for use with solid drills to drill small holes. The vacuum suction cup base with lever control locks the drill onto the glass rigidly. Spring action control gives feather-light pressure on the drill to minimize chipping of glass around the hole. Figure 8-8 is a portable automatic model for use with tube drills. To use, simply place the machine over the spot where the drilling is to be done and bring the tube drill into contact with the glass. Set the adjustable stop control to the

thickness of the glass to be drilled. Turn on the motor and the weight of the tube drill alone will automatically feed itself on through the glass and stop wherever you have it set. Figure 8-9 shows a portable abrasive glass drill kit which includes everything needed to drill accurate, clean, true holes in glass.

CUTTING ODD-SHAPED HOLES IN GLASS

In glass-working, cutting holes that are other than round in shape always used to be a long, hard job even for commercial concerns. It was almost impossible for the novice. Today, with sandblasting equipment, shapes of any kind can be cut out of glass without difficulty (Fig. 8-10 and 8-11). There are two methods that can be used, depending on the type of equipment that is available.

On page 67 is given a complete description of how to sandblast glass to produce a frosted appearance. On pages 67-68 are given additional details concerning sandblasting used in connection with the engraving of glass. The equipment and materials described in the previously mentioned sections are identical with those needed for cutting odd-shaped holes in glass. The techniques are also the same except that the sandblasting is continued until it cuts completely through the glass. If the general sandblasting equipment (as described on pages 67-68) is available, the procedure is as follows:

1. Cover the glass with a masking material (see

page 68).

2. Cut out of the masking material the design you wish to cut out of the glass (see page 68). Cut around the design with an Exacto #11 knife, then remove the cut section.

3. Place the glass in the sand blast booth and play the blast on it (see page 68). Continue the sand blast until the glass is cut through. The time it will take will depend on the size of the sand blast unit being used. A modest-size unit can cut through a 1/4 inch thick glass in a matter of seconds.

If a precision sand blasting unit is available as described on page 75 and illustrated in figure 11-7, then the procedure is even simpler than that described above. With this unit you merely mark the place for the hole right on the bare glass, then proceed to cut it out with the sand blast gun, much the same way as you would with a knife. Sand blasting can be used to cut holes of all kinds in uneven glass surfaces such as found on containers, equipment, home novelties, etc.

Chapter 9
Frosting and etching glass

Glass is frosted by removing the customary smooth, slick surface, leaving it with a whitish, translucent appearance. There are, generally speaking, two reasons for frosting glass: First, for obscuring vision to insure privacy, such as on glass used for partitions in offices (Fig. 9-1), bathrooms, etc. Secondly, to enhance its beauty, such as on beads, table objects, etc. It is also used to diffuse light and reduce glare (Fig. 9-2). Frosting is accomplished by attacking the surface of the glass (etching) with an acid (chemical method), or by scratching or beating the surface with an abrasive (physical method). Both methods produce a frosted surface of approximately equal quality. The best method to use is determined by the quantity, shape, and hardness of the glass to be frosted, and the design of the frosting to be applied.

Fig. 9-1. Frosted glass gives light and privacy. (Courtesy: American-Saint Gobain Corp.)

CHEMICAL METHODS OF FROSTING

Hydrofluoric Acid Solutions

Hydrofluoric acid is the base material for preparations used to frost (etch) glass. The acid mixed with water is recommended for use only in cases where very little frosting is to be done. Where a quantity of frosting is to be done, prepared commercial solutions are recommended.

It is important to note that hydrofluoric acid sets free hydrogen fluoride fumes that are very poisonous. When working with it make sure the work is done out of doors or under an outlet hood or some place where there is very good ventilation.

A simple **etching box** for use with hydrofluoric acid is shown in figure 9-3. This consists of a box lined with wax and then filled for a depth of 1-1/2 in. with wax. Make the wax slope toward the center of the box. When a piece of glass is all prepared to be etched, the top of the wax is heated until it is soft enough so the glass can be pressed into it leaving the surface to be etched exposed. All that must be done now is to pour the acid on the glass and stand the box out on the window sill until the action is completed.

To **prepare the glass** for the etching, it is first thoroughly cleaned. To do this, submerge the glass in sul-

Fig. 9-2. Flutex frosted glass. (Courtesy: American-Saint Gobain Corp.)

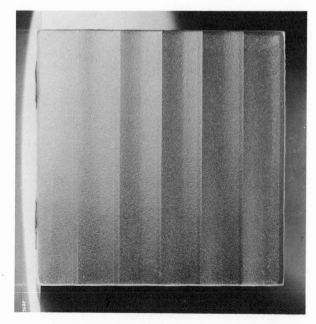

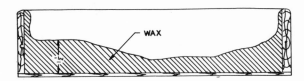

Fig. 9-3. Etching box.

phuric acid for a few minutes and then wash with clean warm water and dry. If the entire surface of the glass is to be frosted, it is now ready to be placed in the etching box. If, however, the frosting is to be applied only to a part of the glass or in the form of a design, then proceed as follows: Cover the glass with a thin layer of beeswax by heating it carefully over a small bunsen burner flame and spreading the wax slowly until a thin, even coat is over the entire glass. This beeswax coating on the glass may also be produced by painting on hot liquid wax with a brush. When the wax has hardened cut the design through to the glass. Where the wax has been **removed** the acid will act. To produce heavy lines, a jack-knife or stylus set may be used. For fine lines, a needle or engraving tool may be employed.

The actual **etching** may be accomplished either with a **vapor** or an **aqueous solution.** The vapor gives a rough surface where the glass is exposed to its action while the aqueous solution makes smooth, deep impressions. It is very difficult to use the vapor method with limited laboratory facilities as it is the more dangerous of the two. For the purpose of the school and home, if hydrofluoric acid is to be used at all, the aqueous method is more convenient.

The aqueous method may be used with several concentrations of acid. The prepared glass, which is placed in a wax-lined box, is placed under the hood or out on the window sill and the acid solution is poured over it so that it covers all the glass to be etched. The acid will eat away the glass slowly and when the proper depth has been reached, the acid should be drained off and the glass **thoroughly washed** to remove all the acid. A stream of hot water will remove the acid solutions.

The following hydrofluoric acid solutions produce an equal depth of etching on the same type of glass in various lengths of time:

Concentrated HF .... 12 minutes
1 part HF to 5 parts water .... 30 minutes
1 part HF to 1 part water .... 20 minutes
5 parts HF to 1 part water .... 15 minutes

Of the above-mentioned solutions the one that produced nice sharp lines in the fastest time was the solution of 1 part HF to 1 part water. The concentrated HF, though it acted very fast, produced very irregular lines. The 1 HF to 5 parts of water also is a very fine solution.

All the four solutions mentioned will produce deep lines in the glass and the effects produced with them are commonly referred to as deep lined etches. The results obtained are similar to those produced by the sandblast or by grinding; namely, a snow-white, translucent glass.

In connection with the foregoing, there are a few **precautions** that must be observed: When mixing the solution, **always add the acid to the water.** When using hydrofluoric acid, do so only in a well-ventilated place. Do not let the acid get on your skin as it will cause a wound which is very painful and very difficult to heal. As all the solutions mentioned eat away glass, they

1. Trace design on special gummed stencil foil.

2. Cut design with stencil knife.

3. Heat stencil slightly against warm lightbulb.

Fig. 9-4.

4. Adhere to glass.

5. Apply etching cream to open area of stencil.

6. After 2 minutes, wash with water. Remove stencil.

Fig. 9-5. Suggested designs for etching glass. (Courtesy: Etchall Co.)

must be stored in wax or lead containers.

There is another solution that may be used when a mat or dulled surface is desired. This consists of: Ammonium fluoride 5; hydrofluoric acid; water 5. This mixture is best applied with a brush.

COMMERCIALLY PREPARED SOLUTIONS

There are on the market a few "commercial" prepared frosting solutions. These may be secured in the form of a liquid, paint, spray paint, and paste. The kind used is determined by the glass to be frosted.

Liquid Frosting Solutions are known by such company names as "Jack Frost", "Velvet Glass Frosting Base No. 77", "White Acid", etc. These are primarily used as dipping solutions. The general procedure is simply to dip, wash and dry as follows:

1. Be sure the glass to be frosted is reasonably clean.
2. Although it is not necessary for regular glass, it is recommended that a pre-etching dip tank of approximately 10% hydrofluoric acid be used for opal or especially troublesome glass. The time of immersion depends on the glass. Generally from 1 to 5 minutes is sufficient.
3. Immerse the glass in the compound and allow it to remain for 30 seconds or more depending on the type of glass.
4. Remove and allow surplus compound to drip back into the tank.
5. Wash off thoroughly with cold running water and dry glass.

Cold Weather Note: These compounds will not work at maximum efficiency below 60° F.

DETAILS ON PROCEDURE FOR USING
LIQUID FROSTING SOLUTION

A. Tank. In recent years plastic containers have been used as satisfactory tanks for liquid frosting solutions when their size is suitable for the glass to be frosted. When special tanks have to be built, use rubber-lined steel. For limited runs a wax-lined wooden box is sufficient. The tank should be kept covered when not in use. Use as small a tank as possible for ease in handling. Also, never keep more than a couple of inches of frosting material in depth over the ware to be frosted.

B. Exhaust. Adequate ventilation should be maintained during the frosting operation. The fumes to be dispelled are those of hydrogen fluoride. Since they are heavier than air, they tend to remain just over the frosting tank. Therefore, to be effective, the exhaust should be just above and to the rear of the tank. The average 10" or 12" exhaust fan should be more than sufficient to dispel any fumes. Technically, the permissible hydrogen fluoride content is 3 parts per million in air.

C. Washing. The washing of the glass is an important step in the frosting procedure. The wash can be accomplished using water as it comes from a faucet or through a hose. Care should be taken to use as much water pressure as possible to remove all the spent acid. Work the water over the entire piece or pieces as quickly as possible to prevent "eating back" of the piece to a semi-clear state. This is usually caused by a mixture of water and acid remaining on the glass too long during the washing. For volume work a washing machine can accomplish this step very readily. However, even with a machine it is usually advisable to remove the excess acid before the glass is put into the machine. Otherwise the machine would have to be constructed of acid resistant plastic, rubber, stainless steel, etc.

Washing is not complete if there is any white powder remaining after the ware is dried. For hard-to-wash glass a rinse in a 10% solution of sulfuric acid is good. For opalware, a 10% solution of nitric acid is sometimes desirable. The ware should also be washed again in water after either sulfuric acid of nitric acid has been used.

Where the compound is used in volume, we recommend the use of overflow tanks containing porous, granular limestone through which the waste passes and is neutralized before going down the drain. The pH can be readily determined from litmus paper of which there are several types available. The purpose of the neutralization is, of course, to prevent any possible damage to the plumbing.

D. Drying. For the small user, a small fan blowing on the ware will be sufficient to dry it. However, for production work either hot air or heat of some sort is recommended. The drying procedure is quite important because bad frosting results can easily go by unnoticed when the ware is wet. Also, the spent fluorides form a white powder that may be unobserved. This powder in many cases will turn brown if the ware is to be decorated and fired later.

Paint Frosting Solution is painted on to glass with a paint brush in the same way as ordinary paint. When dry, it leaves the surface with an even, white appearance that diffuses light. It is particularly good for frosting glass that is already in such places as a window, skylight, flower house, transoms, partitions, lamps, etc.

Preparation. Have glass clean, dry and free from oil, soap or grease. On windows apply the paint to the

Fig. 9-6.

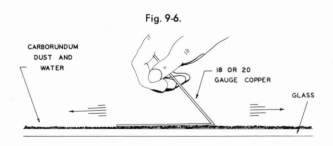

CARBORUNDUM DUST AND WATER

18 OR 20 GAUGE COPPER

GLASS

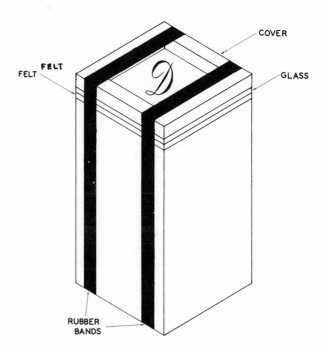

Fig. 9-7. Pellet box.

inside surface. Avoid application on glass at extreme temperatures.

Application. Stir the paint solution well and apply with a soft clean brush using long even strokes. For best results, the application should be allowed to "set" for 5 to 10 minutes and then tapped lightly with cheese-cloth—this eliminates any possibility of brush marks. The full frosted effect is gained in a few hours when the film is completely dry. The frosting may be tinted by adding oil colors to the paint.

Spray-Paint Frosting Solution is sprayed onto glass just as any ordinary paint is sprayed. When applied in light, even coats it produces a satin-like frosting. This material is particularly good for places where you want a temporary frosting as it can be removed after application.

Preparation. Surface should be clean and dry. Remove grease, wax or polish. Use paper to mask adjacent areas not to be sprayed. Shake container vigorously for a minute until rattle of agitator ball is clearly heard. Remove cap and make a test application.

Application: Spray in temperature range of 60° to 90° F. Allow adequate ventilation but avoid strong drafts or prolonged breathing of vapor. Hold nozzle about 18" from surface. Press down on valve and spray with a steady back-and-forth motion until area is covered. Apply thinly, following with a second coat in 15 minutes if denser frosting is desired. Apply to inside of windows. When used to reduce glare, spray a light "mist" coat. Dries in 15 minutes. After use invert container and direct spray onto newspaper until only gas escapes. This clears

valve for future use. Frosting may be removed at any time with turpentine and a cloth. A 12 ounce aerosol container will cover 20 square feet.

Paste Frosting Solutions are packed in squeeze tubes and are ideal to use with stencils for frosting small designs, monograms, etc. They are safe and easy to use

When etching (frosting) with a paste frosting solution, you first cut a design out of a piece of aluminum design foil (or purchase aluminum foil with letters or designs already cut in it), then glue the foil to the glass where you want the design. The paste is spread over the opening in the foil and allowed to remain about 3 minutes or more, then washed off and the stencil removed. This is a brief summary of the entire process. The details are as follows (Fig. 9-4):

When executing an **original design,** a piece of tracing paper, slightly larger than the design, should be used to make a **tracing.** (Figure 9-5 shows some typical designs.)

The tracing should be centered over a sheet of **Aluminum Design Foil.** The outline of the design should be **retraced** onto the foil, using a blunt-pointed pencil and exerting an even pressure to properly impress the design onto the foil. Then use a pointed knife, or scissors where practicable, to **cut** the **design out** of the foil.

Aluminum Design Foil usually has a specially prepared backing. When subjected to slight heat, the backing becomes activated and will adhere to any surface. Place dull side of foil close to source of heat, such as a lighted match, open flame, or place close to a lighted light bulb, until backing is tacky. (The foil should be heated only to make it tacky and not to a point of stickiness. The latter might result in the loss of a bit of the backing.)

Apply stencil to object. Be certain stencil is placed exactly where desired and proceed to secure it by rolling over the foil with any round object such as a round lead pencil or an orange stick, to be certain that all edges are adhering firmly. If etching paste finds its way under a loose edge, it will produce uneven results.

Be sure the object is thoroughly **clean.** If soap and

Fig. 9-8. Design frosted on glass by sand blasting.

Fig. 9-9. *(Courtesy: American-Saint Gobain Corp.)*

water are used, rinse thoroughly or use denatured alcohol, or a solution of vinegar and water. Be sure to dry surface after cleaning.

Etching paste is then **applied** generously to the open portion of the stencil (be certain to spread immediately and thoroughly with a soft brush, match stick, cotton swab, or pat on with finger to remove any air bubbles), and allow to remain thereon for at least three minutes (no harm if longer). Using a sponge or cloth, **wash off** cream and stencil in hot water (temperature of water should be hot enough to burn your skin). The hot water will cause the stencil to come off readily and be suitable for many additional uses.

For a more uniform etch, wash off first coat with cold water. Dry thoroughly (but do not move stencil). Then apply second coat of etching paste and complete operation as mentioned above.

After you have cut a design from Aluminum Design Foil, save the cut-out for further use. You may take another piece of Design Foil and create a shield so that you can make use of both positive and negative design.

Notice: If you do not have nor do you wish to purchase precoated aluminum design foil you may use any thin aluminum or lead foil and glue it to the glass with a thin solution of turpentine asphaltum or wax. Just make sure that none of the asphaltum or wax is on the glass where you want the design.

PHYSICAL METHODS OF FROSTING

Frosting By Rubbing With Abrasive

The surface of glass can be frosted by rubbing it with an abrasive. Where a limited amount of frosting is to be done, this is the easiest and cheapest way. Figure 9-6 shows how this is done.

Start by cleaning the glass, then **wet** the entire **surface** to be frosted with water. Sprinkle an even, thin coat of **carborundum dust** (or any good abrasive) over the glass. Take a piece of sheet 18 or 20 gauge **copper** approximately 3 inches wide by 5 inches long and bend up 2 inches for a handle **as** shown in figure 9-6. Holding the copper as shown, **rub** it back and forth over the carborundum dust and water mixture. Continue rubbing the glass until the desired frost is reached. The more rubbing that is done the deeper will be the frost. From time to time it may be necessary to add a few drops of water to the surface to enable the copper to work across it freely.

Frosting By Pounding Surface

Pellet Box frosting is a simple process by which monograms and designs can be frosted on glass. The frosting is accomplished by pounding the surface of the glass with abrasives and lead shot. The apparatus needed is a pellet box that is approximately 6 inches wide by 6

Fig. 9-10. Sand blasting booth that can be made in the work shop.

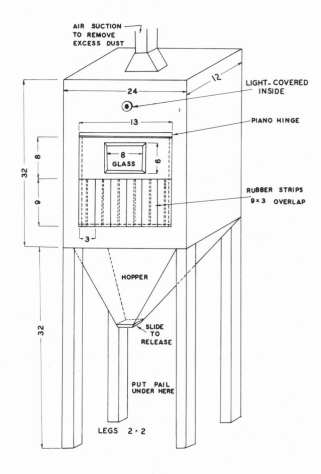

inches high and 10 inches long as shown in Figure 9-7 and has a tight fitting lid. On the open end of the box is cemented a felt stripping. A hole is cut in the removable lid about 5 inches square, leaving a 1/2 inch margin all around.

To prepare for the frosting, first **clean** the glass with any good glass or kitchen cleaner and rub it **dry. Cover** the glass with heavy paper **masking tape** or a special rubber masking tape (both can be secured in widths up to 36 inches). Draw or **trace** the desired design on the masking tape. **Cut** around the design with a sharp knife, then remove the portion you wish to frost.

Put in the **box** a pound of coarse **abrasive** and a pound of **shot**. Place the prepared glass on the box with the taped side toward the inside of the box. **Fasten the glass** and lid to the box with two heavy rubber bands (pieces cut from a tire inner tube will do) or tie it securely to the box with string. **Shake the box** up and down so that the abrasive and shot will strike the surface of the glass with as much force as possible. This should be continued for about 15 minutes.

When the box is opened it will be found that the particles of abrasive have been embedded in the lead shot and each of the latter has become a cutting tool. These can be used over and over again.

SANDBLASTING

Sandblasting permits the frosting of glass in infinite variations. It lends itself to quantity frosting for commercial purposes, as well as individual renditions of artistic beauty.

Fig. 9-11. Commercial sand blasting booth.

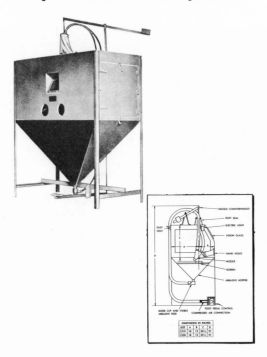

Fig. 9-12. A small sand blasting outfit that is available. *(Courtesy: Sommer and Maca Glass Machinery Co.)*

The frosting is produced in sandblasting by beating the surface of the glass with a spray of sand or some other hard abrasive under pressure. If the spray is run across the entire glass, much as with a paint spray, the entire surface will be frosted. This, however, is seldom done. Sandblasting is almost always used to produce a design or picture on the glass (Fig. 9-8 & 9-9).

Sandblasting Booth. If only a small amount of Sandblasting is to be done, the apparatus can be set up in a ventilated corner of a room with a canvas shield blocking it out. In such a set-up the operator should wear a respirator. If, on the other hand, sandblasting is to be done quite frequently, then it should be performed in a booth made for this purpose. All sandblasting booths are similarly constructed—a box with an opening in the front covered with rubber shields through which the operator puts his hands; a window so the operator can see what he is doing inside; an exhaust fan at the top; and provision to draw off the abrasive that has fallen to the bottom of the booth. Figure 9-10 shows a booth that may be constructed in the school or home shop. Commercial booths may be purchased that are as small as an apple crate or as large as a full-size room (Fig. 9-11).

Blasting Equipment. To spray the abrasive the following is needed:

(A) An air compressor of 1-1/2 H.P. or larger.

(B) An air line fitted with a pressure regulating valve that would permit reducing pressure from 100 lbs. down to 30 lbs. as required by the nature of the work being blasted.

Fig. 9-13. An abrasive gun for small accurate work. *(Courtesy: Paasche Airbrush Co.)*

in the booth. Set the air-line pressure between 60 to 80 lbs. The blasting gun is held approximately a foot from the glass and run back and forth across, keeping the gun an even distance from the glass at all times. The speed at which the gun is moved depends on the depth of cut desired. The slower the gun is moved the deeper and darker will be the frosting.

Shading. The frosted surface can be beautifully shaded with spray-on colors or oil colors. The frosted surface holds the color excellently and imparts to the surface an attractive sheen.

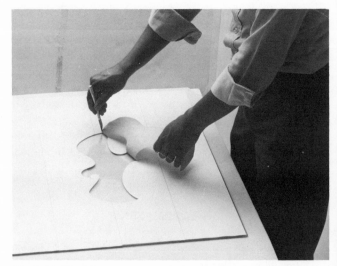

Fig. 9-14. The design to be sandblasted is cut from the masking material that was placed over the glass. *(Courtesy: C.R. Laurence Co., Inc.)*

(C) A Sandblasting gun (Fig. 9-12 & 13) complete with accessories.

Abrasive. Although the process is called, "Sandblasting", sand is seldom used as the abrasive. The reason is that sand forms a dust that can develop silicon poisoning when inhaled in large quantities. Today, Aluminous Oxide abrasive size #80 is most frequently used. This forms very little if any abrasive dust and can be reused almost indefinitely. If the sandblasting is going to be done only once in a while, then economical 70 mesh blasting sand or crushed garnet can be used.

Masking. To produce a frosted design you first cover the glass with masking material such as heavy paper masking tape or a rubber masking tape (these may be secured in widths up to 36 inches). The glass is covered with the masking material, then the desired design is drawn or traced on it. The outline of the design is cut with a very sharp knife (usually an Exacto #11). The masking material is removed from the portion of the glass to be frosted. The glass is now ready to be subjected to the blast of abrasive.

A design could be cut from cardboard and held in front of the glass for a mask in place of the glue-on type of masking. For large production runs, designs are frequently cut from hard board and repeatedly used as a masking.

Blasting Procedure. Place the masked glass in position

Fig. 9-15. The exposed glass is then subjected to blasts from the abrasive gun. *(Courtesy: C.R. Laurence Co., Inc.)*

Chapter 10
Chipping glass surfaces

Chipped surface glass is translucent, that is, it admits passage of light but diffuses it so that objects beyond cannot be clearly distinguished (Figure 10-1). In addition, the surface has a varied flaky appearance that gives it a distinctive attractiveness all its own.

The actual chipping of the glass is produced by a coating of animal glue applied to the surface—time and nature do the rest.

MATERIALS NEEDED

1. Glass. Select a sheet of glass the size desired. This may be regular clear glass; however, frosted glass is superior (see frosting page 61). When clear glass is used, rub the surface with coarse emery paper to "break" the high gloss.

2. Animal Glue. Almost any good grade of animal glue will do; however, noddle glue has been found to be the best. Animal glue is prepared by first soaking the dry flakes in water until thoroughly softened, then enough water should be added to cover the resulting gelatinous mass. This should be heated in a glue pot or

Fig. 10-1. Chipped glass gives a room beauty as well as light and privacy. *(Courtesy: American-Saint Gobain Corp.)*

double boiler to a temperature of 140° F. and kept at that temperature while it is being used. The consistency of the glue is considered good when it is creamy and when long strings form as it runs off the brush.

3. Wooden Supports. Blocks of wood should be used

Fig. 10-2.

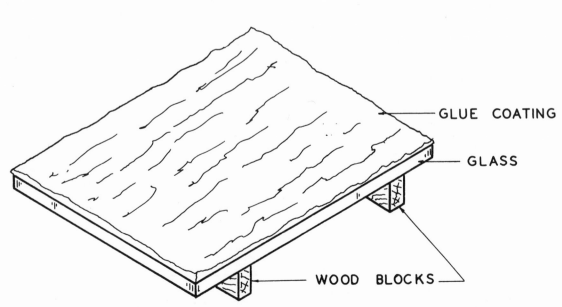

GLUE COATING

GLASS

WOOD BLOCKS

to support the glass above the table so it will be well ventilated on all sides.

Procedure:

A. Cleanse the glass with a strong alkaline cleanser to free it from grease and oil. Then rinse off with warm water and allow to dry.

B. To prevent chilling the glue while being applied, the glass should be slightly warmed.

C. Pour on a coat of the warm glue and see that all parts of the glass to be treated are covered (Fig. 10-2). Nothing is gained by using too much glue.

D. When the glue has set hard enough so that it will no longer run off the glass, set the work in a place where there will be circulation of warm, dry air around it. Support the sheet on blocks (Fig. 10-2).

E. Allow the glass to remain undisturbed for about 24 hours. By this time, if the conditions of temperature and humidity have been in any way favorable, the glue will peel off in flakes of various sizes. If the coating of glue has been too thin, it may be necessary to apply a second coat over the top of the first and repeat the treatment. If the design produced has a number of wide open spaces on it and you desire a closer chipping to strengthen the effect, repeat the treatment.

Safety Precautions

A. Always wear goggles when going near the glass when it is chipping as the chips fly far and wide.

B. Do not run your hand over the glass until it is cleaned after the chipping because of the loose sharp edges that may be on it.

Ornamental Designs

It is possible to produce both clear and frosted margins, lines, borders, or letters on the chipped glass. Where a clear design is desired, a thin coating of paraffin wax is placed on the glass in the shape desired before it is frosted, and left on through the chipping process. When a frosted design is needed, the glass is first frosted, then the wax design is put on, followed by the glue for chipping.

Wax will prevent both the frosting acid and the chipping glue from affecting the glass surface.

Chapter 11
Engraving glass

Engraving consists of cutting away the surface of glass in various shapes and to various depths to construct designs, pictures or monograms. There are many extremely attractive items which can be made with engraving: plate glass fire screens, bird or floral designs on glass doors, mirrors, serving trays, glass book ends, signs and name plates, monograms on drinking glasses, pitchers, bowls, bottles, etc.

Four methods of engraving are used: (1) Hand Engraving, (2) Portable Hand Grinder Engraving, (3) Sand Blasting, and (4) Copper Wheel Engraving. Each engraving method produces its own charming and characteristic effects.

The actual engraving process is very similar to drawing on paper with a pencil, with one very important difference: when engraving on glass you use shading techniques opposite to those used when drawing on paper. When drawing on paper, you fill in the areas to be dark and leave untouched the areas to be light. When engraving, the areas that are to appear *light are engraved* (cut with the tool), while those that are to appear *dark are not engraved* (left uncut). The novice finds this reversal of technique a bit confusing when first starting to engrave, but with a little experience overcomes any difficulty.

Hand Engraving

Hand engraving is accomplished by forcing the hard tip of an engraving tool into the surface of glass. The tool is forced into the glass in the form of "lines" or "stipples" or a combination of both. The lines and stipples cuts are not deep; they are shallow, delicate markings.

Engraving Tools: The hand engraving tools are shaped very much like an ordinary pencil with a tip that is made of either a commercial diamond or tungsten

Fig. 11-1. Engraved glasses with horseback riding design. (*Courtesy: Parker- Garrick, Inc.*)

Fig. 11-2. Floral design engraved glasses (*Courtesy: Parker-Garrick, Inc.*)

carbide steel. Some engravers prefer the diamond-tipped tool while others like the steel tip. It is a matter of personal taste and depends on the type of line or stipple preferred. Many engravers will have both types available for their use.

Diamond Tip Tools: These come in three basic forms. The least expensive type is called a writing diamond and consists of a tiny piece of diamond set in the tip of a wooden, plastic, or aluminum holder. A second and more expensive type has for a tip a piece of diamond that is approximately one-eighth inch long. The third type of diamond tip tool, and the most expensive, has a conical shaped black diamond point that can be resharpened when it becomes dull. Diamond tip tools will usually cost anywhere from seven-to-ten-times as much as steel tip tools.

Tungsten Carbide Steel Tip Tools may be had with two types of tips. The steel tip most frequently used for marking fine lines on glass has an extremely fine point on the tip. A second type, with a slightly blunted tip, is more frequently used for stippling on glass.

Glass To Use: Lead Crystal Glass is the best to use for hand engraving. This glass actually contains lead in varying amounts. It is a soft, brilliant glass. The more lead in the glass the softer it is. It is doubtful, however, that the novice will have an opportunity to work on lead crystal glass. It is only used for expensive glasware that has already been engraved when it appears in stores.

Soda Glass is a much less expensive type used to make the orindary bowls, dishes, and goblets that are usually found in the stores. They are rarely engraved. This is the glass that will most likely be available to the novice. The hardness characteristics of soda glass varies greatly. The novice will have to experiment with small quantities of different pieces of soda glass until he/she finds those that respond favorably to his/her ''touch''.

Float Glass is a flat glass that is suitable for engraving. This glass is made in a special process that makes it very clear without polishing. It can be purchased in many local glass shops. As with soda glass, the novice should experiment with small quantities of float glass to make sure it has the right texture before buying large quantities.

Design Preparation is the first thing to do when preparing an engraving. To see how the finished engraving will look, make your drawing with white chalk on black paper. The novice should keep the beginning drawing as elementary as possible. Work out the shading on the drawing as completely as possible on the black paper before starting to engrave.

Most engravers transfer the design from the drawing to the glass on which they will work. This is done by placing the drawing under the glass and copying it on to the glass. The design can be traced with any of a number of different glass marking pencils. The important

thing to remember is that the point of the pencil must be kept very sharp at all times.

Some engravers place their working drawing tight against the inside of the glass and proceed to engrave directly on the glass being guided by the drawing under the glass. This procedure is more difficult and should not be tried except by the experienced engraver.

Once the design is in position on the glass, you are ready to start your actual engraving. Whether you use lines or stipples is a matter of personal choice.

Hand Engraving Procedures: The engraving tool is held in a vertical position with the tip straight down on the glass. Lines are produced by drawing or pushing the tip of the tool against the glass. Light pressure will produce a light line while heavier pressure will produce a deeper (heavier) line. The shading effect on the glass is produced by the weight of the lines, how closely they are placed, and whether they are continuous or broken lines.

To produce a stippled effect on the glass, gently tap the glass with the tip of the engraving tool. Shading variations are produced on the glass by placing the stipple marks closer together or farther apart, and by making the impressions lighter or heavier. It is important to remember that the tip of the diamond point tools are very fragile and therefore should not be pounded very hard at any time.

When using a new engraving tool for the first time you will have to test it in several positions to find the one that will give you the best cut. Make test lines as you rotate the engraving tool into different positions. When you find the position that gives you the best cut, mark it for future reference.

GLASS ENGRAVING WITH PORTABLE HAND GRINDER

Any portable hand grinder, similar to that shown in figure 11-3 or flexible shaft having a chuck arrangement can be used to engrave glass. The only additional thing needed is a half dozen or more differently shaped silicon carbide grinding wheels that can be clamped in the portable grinder. The mounted grinding wheels can be purchased individually in dozens of different shapes and sizes or they can be bought in assorted sets similar to that shown in Figure 11-3. Shown with the set is a dressing stone that can be used to reshape the mounted wheel.

Any design that can be drawn with a pencil can be engraved on glass (Fig. 11-4). The design to be engraved may be prepared on paper or directly on the glass. If the glass to be engraved is transparent, the design drawn on paper may be placed under the glass. With non-transparent glass (such as mirrors) the design may be drawn directly on the glass with process white to which a little gum arabic has been added, or India or lithographers' ink may be preferred, and the drawing may be protected from rubbing off by a thin coat of clear quick-drying varnish or clear spray.

The actual engraving is similar to drawing with a pencil. To familiarize yourself with this work, take any simple **line drawing** appearing in a magazine or newspaper and place it **under** a piece of clear **glass.** The glass can be plate or ordinary window glass; in fact you can use a drinking glass for the purpose. Then, using a mounted **abrasive wheel** with a sharp edge, follow the **outline** of the drawing.

You will ,be amazed at the ease with which you can make a direct copy indelibly on the glass. By increasing the amount of pressure on the wheel as you grind you can cut a deeper line. By changing the arc at which you hold the tool in relation to the work you can make a hair line or broaden it to the width of the mounted wheel you are using. As you become familiar with this technique you will find that the tool responds in much the same fashion as brush and oil on canvas in the hands of an artist. Tonal shades in drawings are achieved through the amount of pressure you place upon the mounted wheel and the width of the line you make.

For stability and accuracy in cutting it is best that you rest your elbows on the table and **hold the grinder as shown** in figure 11-5. If you are going to remain in this position for any length of time it is wise to place a cushion under your elbows.

When engraving on small glass objects it is sometimes best to clamp the grinder in a rigid position (Fig. 11-6) and press the glass against the rotating mounted wheel.

Fig. 11-3. A portable grinding set used to engrave on glass. (Courtesy: Chicago Wheel & Mfg. Co.)

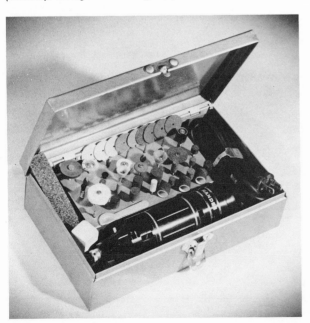

GLASS ENGRAVING WITH SANDBLAST GUN

In the section on "Frosting" (pages 61-67) are included comprehensive details on equipment, materials and procedures for sandblasting of glass. Everything mentioned in the section on frosting applies to engraving with these few additional points: The beauty in sandblast engraving derives from the various depths

Fig. 11-4.

that are engraved into the glass (Figure 11-7). The variations in depth are accomplished by exposing certain parts of the glass to longer or shorter periods of

Fig. 11-5.

blasting with the abrasive. To do this, the glass is first **masked** (see Masking **page 68**) with paper or rubber masking tape. The design to be engraved is drawn or **traced on the masking material.** The decision is then made as to which part of the design is to be the deepest, which is to be less deeply engraved, etc.

The part of the design to be engraved most deeply is **cut from the masking material** with a sharp knife and removed, exposing the glass. The **exposed glass** is given

Fig. 11-6. A handy portable grinder holder. *(Courtesy: Chicago Wheel & Mfg. Co.)*

a blast with the **sandblast gun.** The part of the design to be **next** in depth is **now removed** from the masking material, leaving the additional glass exposed. All the **exposed glass** is given a **second blast** from the sandblasting gun. Now the next part of the design is removed and the entire glass is given another blasting. In this way, the first-exposed glass will be engraved three times as deeply, the second-exposed will be twice as deep, and the last exposed only once as deeply engraved. By this process any number of variations in depth of engraving can be had, depending on the thickness of the glass.

For extremely accurate engraving in glass there are special precision sandblasting guns on the market (Figure 11-8). With guns like these, glass can be engraved to various depths without the necessity of masking as described above. Because the blast of abrasive can be controlled to cut within thousands of an inch, it can be used in much the same way as a paint brush. The design is placed under or on the glass as described above under "Engraving with Portable Hand Grinder"

(page 72). The design is then engraved on the glass by concentrating the blast of abrasives on different parts of the design for longer or shorter periods of time (Fig. 11-9 & 10). Many superb orginial art renderings are prepared in this way.

COPPER WHEEL ENGRAVING

Copper wheel engraving is a process by which the most delicate and beautiful designs are engraved on glass. It is a very exacting handicraft requiring a great deal of self-discipline and long practice.

The Lathe. The equipment needed is not elaborate. It consists of a very simple kind of lathe—an upright stand or head-stock carrying a revolving spindle (Fig. 11-11). The bearings of the lathe are of soft white metal which are easily adjustable and easily replaced when they become worn. The lathe spindle has a tapered chuck in which the spindles carrying the engraving wheel are inserted. The electric motor usually used to operate the lathe is 1/4 h.p. with a variable speed control so it can operate from 1,500 r.p.m. down to just "ticking" over—the fast speeds for truing and shaping wheels, the relatively slow speeds for engraving. (As it is the rim speed which counts, it follows that each size of wheel has an optimum speed of its own, varying slightly with the nature of the operation.) On a rod extending from the top of the lathe is a strip of sheet metal on the bottom end of which is riveted a piece of leather.

The Copper Wheel. The device that does the engraving is the copper wheel that is mounted in the lathe. The copper wheel is mounted on the end of a mild steel shaft that is jacketed in lead. The shaft is fitted in the chuck of the lathe which has grooves to prevent turn-

Fig. 11-7. A sandblasted engraving on glass 30" by 100" in size. On a building in Tarrytown, N.Y. *(Courtesy: Glo Glass Corp.)*

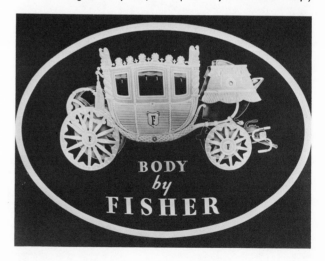

Fig. 11-10. *(Courtesy: Pangborn Corp.)*

They vary in size from 3 to 4 inches in diameter to little things like the head of a pin, and in thickness they vary from 1/4 inch to 1/64 inch or so. Some are often in use, and some only occasionally for special operations.

The preparation and mounting of wheels are important. They must be dead true, must "sleep" on the spindle. When looked at in profile while running not a quiver should be visible. Careful preparation of a true circular disc of copper, perfectly flat, and with the hole dead centre, saves much time in truing, and much wear and tear on the lathe bearings. Final truing of the wheel, provided that the spindle is running true, should not take long. A sharp steel blade is firmly held at an angle to throw the little slivers of copper aside, and it lightly and steadily nicks the wheel into shape. Safety goggles should be worn when truing the wheel. The final shaping of the wheel is done with a file, emery cloth, and lastly pumice stone to give it a good clean finish.

Fig. 11-11. *(Courtesy: Steuben Glass)*

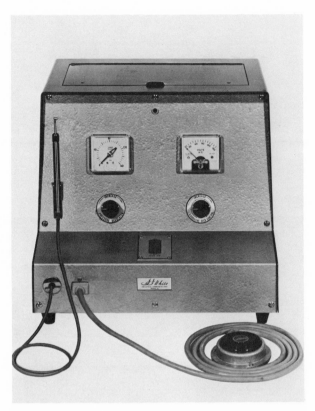

Fig. 11-8. Machine for precision sand blasting. *(Courtesy: S.S. White Co.)*

ing. The lead has a kindly grip and the mild metal of the spindle is easily trued with a compound tool which also serves as a lever to eject the spindle when changing over from one wheel to another. Spindles can be obtained ready-made from makers of lathes. Many craftsmen make their own, which is not a difficult job.

A considerable number of different wheels are needed, from sixty or so to one hundred and fifty or more. (Notice rack of wheels next to craftsman in Figure 11-11).

Fig. 11-9. Engraved designs made by sand blasting. *(Courtesy: Pangborn Corp.)*

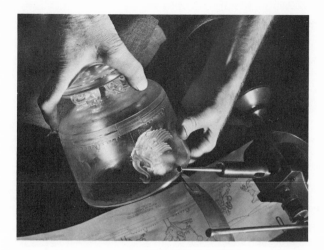

Fig. 11-12. *(Courtesy: Steuben Glass)*

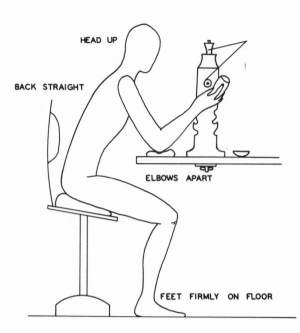

Fig. 11-13. Proper position when working at copper wheel engraving.

The varieties of profiles used on the copper wheels are endless, but they fall into only a few general groups. Some wheels are flat at the rim, some are more or less rounded or U shaped, some are V shaped. If you look closely at engraving you will be able to differentiate the distinctive traces of the different shapes.

Polishing Wheels. In addition to the copper wheels there are also polishing wheels of cork, wood and lead. The cork produces a very soft effect, the wood a brighter, and the lead a brilliant finish. Polishing media are fine pumice powder, jewellers' rouge, or other polishing powders obtainable from suppliers of optical materials. Rubber wheels impregnated with diamond dust are now also available for polishing.

Preparing Glass. A very accurate, full-size drawing is prepared of the design to be engraved (Fig. 11-12). The design is then drawn or traced on the glass. The design may be drawn on the glass with process white to which a little gum arabic has been added, or India or lithographers' ink may be preferred, and the drawing may be protected from rubbing off by a thin coat of clear quick-drying varnish or clear spray. Alternatively it may be very lightly "scrolled" in with a very small engraving wheel.

Engraving. The engraver places conveniently before him a cup of the abrasive to be used on the wheel. This consists of linseed oil and emery powder or linseed oil and silicon-carbide (carborumdum). It is this abrasive, when placed on the copper wheel, that does the actual work.

The abrasive mixture is applied to the wheel by the engraver dipping his finger in the mixture, then dabbing it on the wheel with the tip of the finger. This is done for almost every cut. The strip of leather touching the wheel distributes the abrasive mixture evenly around the edge of the wheel.

The paper with the exact design on it is hung before the engraver for convenient and constant reference.

The engraver sits comfortably at his work (Fig. 11-13) —feet flat on floor, back straight, chair of proper height so that his line of sight is directly past and just below the spindle at the point of attachment of the wheel. His elbows are placed well apart on the bench (on cushions) so that when he brings the glass into contact with the wheel he is in an attitude of ease. The light should be of sufficient intensity and diffused so it completely illuminates the work in all positions.

The glass being engraved is not held in a rigid grip. The work should rest in the tips of the fingers, the fingers should be bent (Fig. 11-12). In general, the grip should be firm yet have a mobility to it so the glass can be moved freely in any given direction. The pressure should be light, yet firm and even.

The general attitude of the engraver must be one of effortless concentration. There must not be any "fighting-with-the-wheel" as there is no need for it. The action of the copper wheel is almost automatically precise and needs only the gentlest guidance to progress in the direction desired.

All engraving in glass is a shallow intaglio. It is an optical illusion that makes it appear to be in bas-relief. The deepest cuts appear to be the most prominent. The engraver is skilled when he can blend these cuts of different depths into an eye-appealing combination of delicacy and beautiful exactness.

Chapter 12
Decorating glass

Today the volume of glass decorating by all methods has assumed enormous proportions. The modern trend has been to color: glass containers, tableware, and cosmetic jars have all yielded to the artist's inspiration. Inexpensive ware decorated with bright colors has caught the eye of millions, and food packers have not been slow to realize the sales appeal of a useful package studded here and there with bright-colored dots and stars.

Other articles decorated today include drinking glasses of all kinds, tableware, ash trays, vases, lamp bases, glass paneling, advertising signs, and numerous novelties. It is the opinion of many that the day is not far distant when permanent ceramic labels and decorations will be universally used in ways of which today we do not dream.

There are three types of materials used to decorate glass. They may be classified as Semi-Permanent, Permanent, and Vitrified decorations.

The **Semi-Permanent** includes those decorating materials that will chip off when subjected to heavy use, intense cleaning, or eventually with the passage of time.

Permanent decorations are those that are baked onto the glass but do not become a part of the parent glass.

These decorations can withstand a great deal of abuse and cleaning with mild detergents but could be scraped off.

The **Vitrified** decorations are the type that are "fired" and in the process become an integral part of the parent glass. These decorations can withstand almost any kind of abuse that the original glass could take.

PREPARING THE DESIGN

There are sufficient possibilities for variations in glass decorating to entice and fascinate beginners, professional artists, and designers alike. Many different possibilities present themselves for creating unusual and attractive results. For example, the designs and color scheme in your home can be repeated on decorated articles for surprising and unusual matching effects (Fig. 12-1). Try monograms and lines from poetry, too, for lettering plays an important role in the field of design (Fig. 12-2 & 3). It may be used in a practical way for labeling kitchen canisters, objects in the nursery, and many other home furnishing articles.

When decorating glass, the simplest motifs will bring the best results (Fig. 12-4). For the beginner, very primitive designs lead to entirely different effects and help

Fig. 12-1. Textile designs suggest interesting designs for glass decorations. *(Courtesy: The American Crayon Co.)*

Fig. 12-2. Lettering designs on glass. *(Courtesy: The American Crayon Co.)*

Fig. 12-3. *(Courtesy: The American Crayon Co.)*

Fig. 12-5. A typical "rhythm-and-repeat" design. *(Courtesy: The American Crayon Co.)*

the novice to become acquainted with this delightful medium. The same is true when working on a home craft project. It is not necessary to work out a very elaborate design which has been painstakingly planned. It is better to take a brush and start putting lines on the background in a very free way, trying to get a clear "rhythm-and-repeat" out of the motion of the brush (Fig. 12-5). Changing colors will add to the effectiveness of a simple motif of this kind. In each case, fit the design to the form and function of the object that is being painted. By handling designs in this manner it is possible to achieve effectively a logical reason for using the same design in harmonious combinations and arrangements on companion articles, both as to size, shape and purpose.

Fig. 12-4. Simple motifs are the best designs. *(Courtesy: The American Crayon Co.)*

Figures 12-6 and 12-7 include some suggested simple designs and letters that will lend themselves to designing on glass. The designs can be used by themselves or in combination (Fig. 12-8). Use one basic color for the majority of the strokes, with a very limited number of strokes in a second and possibly a third color for emphasis.

An example of shape emphasis can be shown by creating vertical lines of flowers, curves, loops, crosses, etc., for the decoration of a glass. Or the height can be minimized by the use of horizontal designs—which still conform to the shape of the glass. This horizontal effect can be carried out with flowers, stripes, stars or similar motifs.

To Transfer the Design to the Glass requires different techniques, depending on the type of glass and location of the design. Where the glass is transparent and smooth, such as a tumbler, the procedure could be as follows:

1. Make a paper pattern to fit the object to be decorated (Fig. 12-9). Make sure the pattern exactly fits the area to be decorated.
2. Draw the design on the pattern. Copy the design on a transparent acetate film with India ink (Fig. 12-10). Cut the acetate film to the same size as pattern. (A very thin paper may be used in place of acetate film.)
3. Clean the glass surface to be decorated very thoroughly (See Cleaning, page 83) (Fig. 12-11).
4. Place the acetate film inside the tumbler and press so it will fit tight against the glass. Hold it in place with tape. Lining it with a piece of paper makes it easier to see the design (Fig. 12-12).

When the above method cannot be used, the design may be drawn directly on the glass with a china mark-

Fig. 12-6. *(Courtesy: The American Crayon Co.)*

ing pencil, red or yellow preferred. Or the design may be traced on the surface—a fresh piece of carbon paper is taped on the glass; the design is taped in place over the carbon paper and traced with a hard pencil.

FOR PAINTING ON DECORATING DESIGNS use soft camelhair brushes. A couple of small brushes ranging from #1 to #4, which hold a sharp, firm point are recommended for designs such as shown in figure 12-13. Reasonably large areas can be readily and smoothly covered with such large brushes, well loaded with color. Always start painting large areas from the center and work to the outside.

For opaque and glossy effects load your brushes with color. Apply with a slow, deliberate brush stroke. When this application has partially set, apply an additional thin coat over the first. Quick application of the one-stroke technique (free brush techniques) results in a more transparent and less glossy appearance. In most designs it is advisable to outline the color with a small brush, allowing it to "set" for a few minutes and then fill in the other color with a brush. This outline acts as a retaining wall for the larger area to be colored.

For items that will be subjected to more than usual

wear and more or less continuous washings, it is important to "flow the colors" on so that they form a smooth, even, continuous coat. In general, it is easier to obtain color permanence and wearability if larger areas of color are applied than with designs of too fine detail.

During the painting the brushes can be cleaned with thinner, but to avoid stiffening of the brushes they should be given a final cleaning with turpentine and then washed with soap and water before putting the brushes aside for future use.

SCRATCH-DESIGNING is an attractive method for decorating glass. It is especially effective where it can be observed through the glass, such as on candy dishes, ash trays, serving trays, picture frames, etc. It is done by first thoroughly cleaning the surface to be decorated (See Cleaning, page 83), then, using a brush or spray, paint the entire surface with a quick-drying oil, enamel, or lacquer base paint. When the paint is completely dry, scrape out designs with a sharp instrument, re-exposing the clean glass beneath. A contrasting color or a gold or silver can be painted over the scratched area so that it will be seen in the openings. It is best to cover the painted surface with felt or cardboard to protect it from injury.

MARBLEIZING is another design method that goes well with glass. It imparts to the surface of the glass long,

Fig. 12-7. *(Courtesy: The American Crayon Co.)*

sweeping lines of color similar to those found in marble. This can be done with any oil base paint that will not mix with water. Dek-All colors work particularly well. The procedure: Take a shallow pan of water large enough to contain the glass you wish to decorate. Sprinkle on the water some of the different colors you wish to use. The paint will float on the surface. Using a match stick or similar pointed object, spread the paints around until they form a marble design that appeals to you. Clean the surface of the glass (See page 83). Carefully place the surface of the glass to be colored in contact with the surface of the water. The paint will leave the water and adhere to the surface of the glass. Set the glass aside to dry.

Fig. 12-8. *(Courtesy: The American Crayon Co.)*

THE RUBBER STAMP METHOD of decorating glass is good to use where a single design or labeling is going to be repeated several times and the operator does not care to get involved in a great deal of work or expense. Being a monocolor process, it provides an easy, simple method of applying any design which can be made into a rubber die or stamp. The rubber stamp process has its limitations, chief of which is lack of color intensity, due primarily to the very thin layer of color applied.

The operation is essentially as follows: A small quantity (1/2 teaspoonful) of color is placed on a piece of

Fig. 12-9. Start with a paper pattern. *(Courtesy: The American Crayon Co.)*

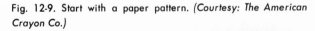
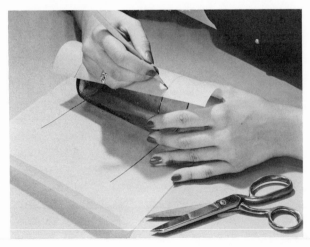

Fig. 12-10. Copy the design on a transparent acetate film with India ink. *(Courtesy: The American Crayon Co.)*

glass and by means of a brayer (rubber roller) rolled out into a very thin film. The rubber stamp is then used to transfer a print to the glass to be decorated. If the color seems to have piled up on the edges of the design, too much has been used. This condition can be corrected by working the coloring medium out still thinner.

To apply vitrifiable decorating colors with a rubber stamp you would use printing oil instead of a color and proceed as described above. After the oil impression is applied to the glass, the selected vitrifiable color is applied to a tuft of cotton which is used to powder and rub it into the wet, tacky impression left by the stamp. It is then fired according to instruction on page 85.

For printing round objects such as rods, tubing or table glasses, it is often more convenient to insert the rubber stamp, design upwards, between two narrow wooden tracks which are about four inches apart. In this case, the roller is used to coat the rubber stamp, after which the object to be decorated is rolled over it.

DIPPING is a process used for applying decorating material on glass where an over-all color is desired. When this method is used, provisions have to be devised for holding the glass item so it will be properly dipped and turned up and down so the coating will flow evenly over the item. Dipping is only used on items that have smooth, unobstructed surfaces, such as light bulb envelopes.

All decorating materials must be thinned to a light cream consistency before they can be used for dipping. An alcohol suspension of vitreous enamels is used for dipping. The enamels used are both of the lead and leadless types. Lead base enamels are far more weather resistant, although the leadless enamel presents no problem of reduction and darkening when fired.

SPRAYING is the second most important decorating operation from the standpoint of volume. The spray gun provides a comparatively simple means of creating all-over color effects which could not be achieved by incorporating the color directly into the glass itself.

The simplest spray set-up consists of a spray booth, spray gun, air pressure regulator, supply of color, and a revolving turntable to support the glass. An elaborate set-up is not necessary but it is important to install such means of ventilation as will remove completely all spray fog so that none will be inhaled by the sprayer.

The spraying of glass requires no special techniques. Any good publication on spraying will explain all that is needed.

SILK SCREEN PROCESS for the decoration of glassware has grown rapidly during the past few years. Not only has its field of application been substantially broadened, but the various methods themselves have been greatly improved. At the present time, with the proper equipment, it is hardly more difficult to print on rounded or curved surfaces than on flat areas. The silk screen has already become a very important tool for decorating ceramic ware and, judging from past development, the future holds much in store for the process.

Silk screening (sometimes called serigraph) is a method of printing that is accomplished by forcing a coloring material through the open meshes of a screen or fabric to which a stencil has been attached. The stencil is applied to the screen so as to permit passage of color through open areas and obstruct passage in colored areas.

The heart of the screen printing method is the screen plate itself. Whether it is to be used on a machine for decorating round ware or on a table for flat articles, the construction details are essentially the same. Today, excellently prepared-to-order screen plates can be purchased for a modest price from silk screen supply houses located in all larger communities. An original art sketch

Fig. 12-11. Clean the glass surface to be decorated. *(Courtesy: The American Crayon Co.)*

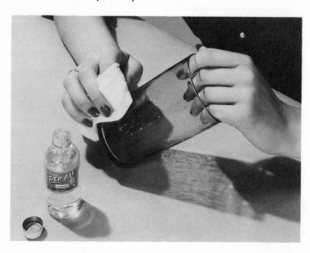

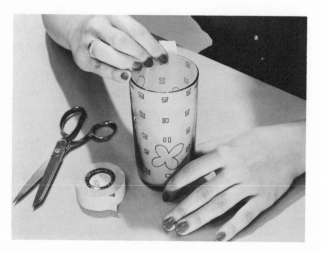

Fig. 12-12. Place the acetate film inside the tumbler. *(Courtesy: The American Crayon Co.)*

or label design can be sent to these concerns and they will reproduce it exactly on a screen mounted on a frame all ready for use.

Those who wish to prepare their own screen plates and silk screening facilities can find books on the subject in almost every library. The subject has been covered so completely in other publications that it seemed unnecessary to repeat it here. The following are incidental remarks that concern the silk screening of glass:

A. The silk screening of glass is no different than the screening of any other non-porous material.

B. Frequent changes are being made in the silk screening paints used for glass. When you are ready to use some, get in touch with a supply house. They will send you complete information and up-to-date recommendations on the paints to use.

C. The color to be used in printing is ordinarily supplied by the manufacturer in powder forms, though the ready-mixed paste may be secured if requested. To prepare a printing paste, the powder should be thoroughly mixed with a suitable oil commonly known as squeegee oil. If only a small amount of color paste is needed at one time, the mixing may be done by means of a spatula or a printer's knife on a glass tile. However, in all production work a paint mill of the burr type should be used.

D. There are no definite standards as to paint consistency. Some operators prefer a rather stiff mixture; others use it as thin as ordinary mixed cold paint.

E. Ceramic colors that are to be used for squeegee work must be kept **bone dry.** This point cannot be over-emphasized. If the color powder absorbs moisture it will, after being mixed with the oil, lose its fluidity and become stiff and unprintable.

F. Without taking the time to go into detail regarding methods of preparation, it is apropos at this point at

Fig. 12-13. Soft camel hair brushes sizes #1 to #4 can be used to paint designs such as these. (Courtesy: The American Crayon Co.)

least to make mention of the various types of screen stencils being used: Long runs on bottles are printed with **metal mesh stencils,** having a metal frame covered with 165 mesh **stainless steel** wire cloth on which a photosensitive film of polyvinyl alcohol has been applied. **Gelatin-coated screens** are still used to a limited extent, but these are less durable, prone to pin-hole, and are greatly affected by changes in humidity. Deterioration of gelatin stencils in storage is far more rapid than those coated with polyvinyl alcohol. **Phosphor bronze wire** and even **copper** may be used, but these metals lack the resiliency of stainless steel. On the other hand, the gelatin-film-coated silk screen stencil is used for the printing of practically all other types of ware except for long runs of bottles.

DECALCOMANIAS. These must be mentioned in a discussion of glass decorating. Printed in several colors, precisely registered with line and shading work not possible by any other decorating means, the decal has much to offer. There are **two** general types of decals—those printed by the well known **lithograph** press on Duplex Paper and the **screen printed** decals which may be either on Duplex Paper or on the Simplex slide-off type of decal paper. The best decals for glass are the screen printed, slide-off type because of their heavier deposit of color. Best results are secured if the transfer is made to opal glass which offers a good, strengthening background for the print.

CLEANING THE GLASS

Before **any type** of decoration is applied to glass, the surface to be decorated must be cleaned carefully and completely. None of the decorating materials will stick to a glass surface that is dirty or greasy. **Merely washing with soap and water is not sufficient.** "Energine" types of cleaners are very good, or denatured alcohol can be used. Dek-All Cleaner No. 1292 is excellent.

After the cleaner has been applied, be sure to wipe off the excess with a clean cloth. Take care that the surface is not touched again by the hands (the skin leaves a film of oil or moisture), and that no other film or moisture is allowed to collect on it before decorating. Use a glove or clean cloth to handle the glass. Large users keep articles to be decorated at slightly higher than room temperature to avoid moisture film.

Note: Do not use turpentine as a surface cleaner.

SEMI-PERMANENT DECORATING MATERIALS

Semi-permanent decorating materials for glass include non-firing oil paints, enamels, and lacquers. All will adhere to clean glass and remain in good shape for years if protected from abuse. These decorations will not withstand strong detergents or cleaners and certainly cannot be cleaned in a dishwasher.

The paints used on glass must be of good grade. Cheap colors will be hard to apply, will chip off, and the colors will not be brilliant or attractive. In a short time, inferior paints will craze and peel off.

There are two good ways of increasing the permanence of these colors: One is to rub them into a design that was previously sandblasted or etched into the glass. The second is to cover the completed design with

Fig. 12-14. Attractive ornaments made with Dek-All colors. (Courtesy: The American Crayon Co.)

a clear spray-on material that will not disturb the coloring material. Test the spray-on material before you use it on your finished work. Spray-ons have different vehicles (lacquer, varnish, phenol resins, etc.) which may not in all cases be compatible with the paint used, so pre-testing is important.

PERMANENT DECORATING MATERIALS

Permanent decorating materials are not ordinary paints, lacquers or enamels, but media containing special thermo-setting ingredients that makes them flow easily onto glass surfaces and adhere firmly to these surfaces when they dry and are heated in an oven, kiln, or lehr. These are not "Fused" ceramic glaze, but thermo- (heat) setting materials which, when properly applied and carefully set, are highly resistant to sunlight, washing, alcohol, water, vinegar, and many acids and caustics. They should not be subjected to extremely hot water and strong detergents such as are found in mechanical dishwashers. Decorating the inside of ash trays, cocktail glasses, plates, etc., is not recommended.

The most popular material in this category that is used in handicrafts is Dek-All, a product of the American Crayon Company which manufactures Prang Products. These colors can be used on all glass surfaces—dishes, mirrors, windows, trays, lamps, glass bowls, etc.

Fig. 12-15. (Courtesy: The American Crayon Co.)

Fig. 12-16. For heavy-duty use the colors should be fired. (Courtesy: The American Crayon Co.)

Very attractive Christmas tree and other holiday ornaments can also be made (Fig. 12-14 and 12-15). The unique feature of Dek-All is that it can be fired at a temperature that can be obtained in a kitchen oven. After the design has been applied to the glass, proceed as follows:

Allow the Dek-All decorated article to **dry 12 to 24 hours,** then place it in a **cold oven** and allow the temperature gradually to reach **300 degrees F.** (Fig. 12-16). Hold this temperature for, **15 minutes,** then turn off the heat, allowing the painted article to remain in the oven with the door closed till cool. Heavier color applications call for a longer heating period. An extreme degree of heat will of course cause a change in the hue of certain colors, but in the case of items to be subjected to "heavy duty use", more heat is advisable rather than not enough —right up to, but just below, the point where discoloring begins.

Other Permanent Decorating Materials that are commonly used are Liquid Bright Gold, Roman Gold, Bright Gold Squeegee Paste, Lusters, Amber Stains, and Copper Ruby Stains. These require a kiln or lehr to be fired, inasmuch as they must be fired at temperatures far above those previously mentioned. They are not, however, of the vitrified type and therefore are subject to wear and vulnerable to abuse.

Roman Gold or burnished gold is pure gold powder fluxed to fire at glass decorating temperatures. Subsequent to firing, the decoration is burnished with round grain sand and water or with a glass fiber rubbing brush in order to smooth down the surface and develop the true character of the metal.

Bright Gold Squeegee Paste is also made in a heavy oil consistency for silk screen printing. Silk screen stencils made of #25 silk (approximately 200 mesh) are used.

Lusters. A number of metallic resin compounds known

Fig. 12-17. View of a direct fired decorating lehr for glass containers. (Courtesy: Surface Combustion Division Midland-Ross Corp.)

as lusters are used to produce various transparent color effects such as ruby, mother of pearl, orange, pink, amethyst, green, etc. The lusters are restricted primarily to purely decorative applications since they lack the durability of vitreous enamel. Luster colors may be applied with a flat lustering brush or by spraying. Spraying is not often resorted to, however, owing to the rather costly nature of the lusters since whatever is lost in the over-spray cannot be reclaimed. Lusters are extremely sensitive to any contamination from foreign matter such as dust, lint, etc., and for this reason the atmosphere where they are applied should be dust free. It goes without saying, of course, that the glass surface should be clean. Ruby lusters develops a beautiful depth of color if the ware is given two applications and two firings.

Amber Stains. The only true glass stain for direct use is an infusible preparation containing silver chloride which may be dipped or painted. During the firing-in period, a penetration of the glass by silver ions produces a permanent transparent amber color. The average firing range for amber stains is in the neighborhood of 1000°F. Also it is interesting to note that firing an amber stain application for 24 hours, say at 800°F., will often produce a color possibly as intense or even more so than could be obtained if the same were fired at 1000°F. for 30 minutes.

Copper Ruby Stain. Another glass stain indirectly achieved is the copper ruby. Here again, the stain owes its color and permanence to an ion exchange. Subsequent to contact with the copper compounds, the glass is submitted to reduction at temperatures between 1000° and 1100°F. Surface contact of the glass with the copper compound is usually brought about by 1) exposing the glass at high temperatures to copper halide vapors or, 2) dipping in a bath of a molten copper salt or 3) by screen printing, painting or spraying of a copper salt and subsequent heating to approximately 1100°F.

VITRIFIED DECORATING MATERIALS

Vitrified decorating materials or glass enamels are essentially low fusing, lead borosilicate glasses, to which an oxide or mineral pigment has been added. After passing through the various processing steps, the finished product is produced as an extremely fine powder. Whether mixed with oils by the color manufacturer or by the ultimate user, it is this finished powder that is used as a starting point in practically all phases of decorating work.

Although every manufacturer can supply the various enamels ready mixed, it is safe to estimate that better than 90 percent of all color used is shipped as dry powder. Minimum stocks may thus be kept on hand by the user; for from the same powder, with the addition of suitable oils, paste may be made for printing, banding, or lining. Using turpentine, or alcohol and water, spraying and dipping mixtures may be prepared.

All manufacturers furnish firing temperatures with their vitrifiable colors. The manufacturers' recommendations should be followed in all cases. Generally, these colors are fired at 1050° to 1200°F (cone 022 to 020).

Vitrified colors, when properly applied and fired, become a part of the parent glass and cannot be removed by washing of any kind, or even through rather heavy abuse.

The colors that are generally available are black, white, blues, browns, gray, greens, carmine, reds, purples, vermillon, and yellows. Colors made by the same manufacturer may be mixed to produce other colors. A palette should be made of the colors to have available for reference when planning future designs.

Other Vitrified Decorating Materials may include any material that will soften at approximately the same temperature as the glass to which it is applied. This could include glass crushings, chips, and glass jewels.

Glass Crushings and Chips, when fused to a glass surface, give it a pleasing textured appearance. The crushings and chips may be colored or uncolored, depending on effect desired. They may be used as an overall design, a border, or a segment of a design.

The crushings and chips must be held to the surface of the glass with an adherent. The adherent could be a very thin coating of "baby oil"; or distilled water if the object is flat and is handled carefully; or, if vitrified enamel is used, this would be sufficient to hold the crushings and chips in place. No matter which adherent is used, it must be completely dried before the glass is fired.

The crushings may be sifted onto the glass or sprinkled from between the fingers. Small chips can also be sprinkled on, but most times each piece is individually placed with a tweezer.

The fusing of crushings and chips offers the craftsman an excellent opportunity for experimentation. It is hard to predict just how they will respond under firing—some will lose their coloring and others will not; some will produce a high texture while others will flow almost flat. Samples should be prepared and kept for ready reference.

To make crushings and chips (See page 108), place the glass in a strong cloth bag or in a flat pan that is covered with a few thicknesses of plastic. Pound the glass with a hammer until the pieces are the sizes you desire. If you desire uniformity, the crushings can be sifted through different sizes of wire mesh screening.

Glass Shards and Jewels may be fired to the surface of glass to give the effect of gems. The glass **shards**—being raw pieces of glass—can be safely fired to glass items that have a higher melting point than the shards. In this case, the sharp edges on the shard will round over in the process. Where the decoration piece and the glass to which it is being applied have approximately the same melting point it is best to use **jewels**—which are shards that have previously been fired to round over their edges (See Glass Jewels, page 108, and Firing Jewels, page 103).

The shards and jewels are applied to the glass item in the same way as the chips mentioned above—on cleaned glass apply adherent where shard or jewel is to be. Carefully place jewel with tweezers. When adherent is dry it is ready for firing.

Decorations for Laminated Glass may include all previously mentioned vitrified decoration materials recommended for use on the surface of glass, plus additional materials. In other words, any decorating materials that can be fired to the surface of glass can be fired between the two fused layers of laminated glass. On the other hand, there are certain materials, such as the following, that can be used for decorating between fused layers of glass but cannot be used on the surface.

Copper Enameling materials can be used to decorate glass provided they are confined between two fused (laminated pieces of glass. They cannot be used on the surface as they will eventually craze (chip) off. Transparent enamels are ideal for laminating. Some will craze, giving a beautiful effect between the two layers of glass, while other colors will be clear with no trace of crazing. The only way to know how a given enamel will respond is to work a test and record the results for future reference. Use the enamel more generously than in copper enameling. Small quantities (particularly at high sagging temperatures) will tend to burn out and leave a weak color. **Lumps and threads** are simple to use and particularly effective in combination with glass colors or transparent enamels. Smaller lumps are the easiest to use because of the difficulty in "sandwiching" large pieces.

Gold and Silver Lustre are used just as in metal enameling except that the firing cycle must be slowed for glass. The kiln MUST BE VENTED until 1000° or corrosive gases from the lustres will frost or discolor the glass. Flat laminating to be fired to 1350°-1400° F. can be fired with the decorating mediums between the layers and lustre on the top simultaneously (with the kiln vented). Since lustres will burn out at temperatures in excess of 1400° F., sagged glass can only be decorated with lustres in a second firing.

Gold and Silver Foil are the simplest and most versatile decorating mediums for glass fire. They may be used with any other decoration. Color may be applied above and beneath foil for various effects.

Thin Silver and Copper Wire can be used between layers of glass. Mosaic effect can be achieved by shaping wires into free forms, fish, etc., and applying color to these areas. When forming designs with wire do not close off any area completely with wire. (Opening must be allowed for gases to escape during firing process.)

Mica in the form of flakes produces a special texture when fused between layers of glass. Each flake of mica causes a bubble in the glass during firing. The size and location of the bubbles can be controlled by the size and placement of the mica.

Chapter 13
Finishing glass edges

The cut or "raw" edges of glass are always sharp and therefore capable of inflicting a severe injury to anyone who might rub against them. In addition, the "raw" edge on glass is usually unattractive. By grinding the edge until it is smooth and uniform both deficiencies can be eliminated.

Glass used for unframed table tops, mirrors, desk tops, etc. must have its edges polished or beveled. There are several popular shapes that are used to edge glass as indicated in figure 13-1. The large bevel types number 13 through 20 are usually reserved for finishing the edge of mirrors.

The edges may have different degrees of finish on them depending on taste and preference. Different degrees of finish may be applied to the same edge, producing a very interesting effect. The degrees of finish that are commonly used are as follows:

1. **Natural Cut**—Untouched by any abrasive.
2. **Ground**—Coarse abrasive used on natural cut, producing a white appearance.
3. **Smooth**—Ground, finished edge worked with finer abrasive, producing an opaque glass appearance.
4. **Polished**—Smooth finished edge worked with fine rouge abrasive, leaving it with a clear glass appearance.

Note: Always wear goggles or a face mask and gloves when grinding or sanding glass as protection against possible injury or breakage.

EDGING GLASS IN SCHOOL AND SMALL SHOPS

Where a limited amount of glass work is performed, the edge can be given a very satisfactory finish with simple materials and equipment that are found in most shops. The abrasive Silicon-Carbide can be secured mounted on a cloth backing, as a hand sharpening stone (Carborundum Stone), as a grinding wheel, or on the cloth belt that is used on a sanding machine.

Portable Hand Sander. The most popular device used for edge finishing of sheet glass in school shops, as well as in most neighborhood glass shops, is the portable belt sander (Fig. 13-2). This machine is particularly good for this work inasmuch as it requires a minimum of

glass movement during the edging process. The procedure used is as follows:

1. The glass is placed on the table with the edge to be ground hanging over approximately 1/4 inch. "Bumping sticks" are placed around the glass if there is a chance it might shift during the edging process.

2. A number 60 (coarse) grit silicon-carbide belt is placed on the portable belt sander and the edge is ground to shape. At this point the edge will have a "ground finish".

   Note: The sander should be raised from the glass intermittently to avoid overheating. The sander should be moved slowly back and forth so the entire surface of the belt is used (Fig. 13-3). Concentrating on one small section of the sanding belt will form a groove and ruin it.

3. To produce a "smooth" finish on the edge, place a number 100 to 220 (medium) grit silicon-carbide belt on the sander and go over all edges. This belt will eliminate the coarse scratches left by the previous belt.

4. For a "polished" finish use a number 320 to 400 (fine) grit silicon-carbide belt to go over the edges. To produce a "high gloss polish" the edge can be gone over with a rouge belt.

5. Wash all chips and abrasive material from the glass.

Fig. 13-1. 1. (See p. 88) Flat Polished 2. Semi-Round Polished 3. Round Polished 4. Bullnose Polished 5. Ground Swiped 6. Seamed 7. Swipped 8. Ground Seamed 9. Polished Miter 10. Ground Miter 11. 1/8" Bevel-Ground and Seamed 12. 1/4" Bevel-Ground and Seamed 13. 3/8" Bevel-Ground and Seamed. 14. 1/2" Bevel-Ground and Seamed 15. 1/2" Bevel-Semi-Round Edge 16. 3/4" Bevel 17. 3/4" Bevel-Ground and Seamed 18. 1" Bevel. 19. 1-1/4" Bevel 20. 1-1/2" Bevel
G — Ground   N — Natural Cut   P — Polished   S — Smooth

EDGES

BEVELS

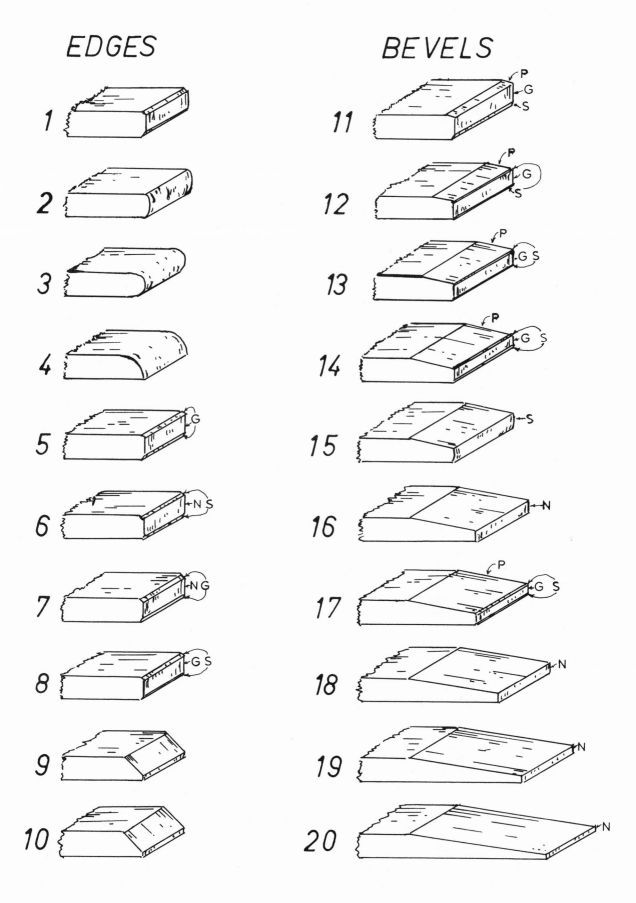

Fig. 13-2. This type of portable sander is good for beveling the edge of glass.

Upright Belt Sander. The upright belt sander can be used to produce the same results as the portable belt sander described above. The only difference is that with the portable belt sander the glass remains relatively stable and the sander is moved, while with the upright sander the glass is moved past the machine. The latter can be a real handicap when working on a very large piece of glass. The procedure recommended for the portable hand sander—starting with the coarse abrasive and following with fine—is the same for the upright sander.

Power Grind Stone. The power grind stone can be used to finish the edge of glass. A silicone-carbide grinding wheel, from 100 to 220 grit is best for grinding

glass. To prevent the wheel from glazing, a jet of water should be directed on it as close to its center as possible—outside the flange, of course. The water will spread over the wheel side and edge by centrifugal force. Any arrangement having a suitable splash guard, and sump or drain, can be used. A simple arrangement could have a can hanging above the grindstone with a small pipe leading from it to the wheel. In this pipe should be a petcock to control the flow of water. A five gallon can cut the long way and placed below the stone could be the sump to catch the water.

The edge as well as the side of the grinding wheel can be used. Using the side will wear the wheel thinner and make it more susceptible to breakage. A wheel of one inch thickness is best. Keep it dressed and replace it when necessary. Avoid touching the glass to the wheel-retaining flange and nut or the glass will chip or crack. Contact with the wheel should be intermittent to avoid heating the glass.

Sanding Disk. The sanding disk should be prepared as follows to finish the edge of glass: The disk itself (usually 6″ to 10″) can be metal or plywood with a flange arrangement on one surface to mount it on a motor arbor. The side opposite the flange should be covered with 1/4 inch sponge rubber or 1/8 inch felt. The silicon-carbide cloth disk is mounted over the rub-

Fig. 13-4. Grinding laminated safety glass. *(Courtesy: Libby-Owens-Ford Glass Co.)*

Fig. 13-3. Beveling the edge of glass with a portable sander.

ber or felt. For a ground finish the silicon-carbide disk should have a 60 grit; for the smooth finish a 100 to 220 grit; for a polished finish use 400 grit.

The glass is sanded dry on the sanding disk. It is important to remember that the glass will heat quickly, therefore short, intermittent contacts are necessary to prevent the glass from cracking.

Note: Always wear face protection.

Hand Edging. A very satisfactory job of finishing the edge of glass can be done without the aid of any machines. Silicon-carbide in a few different forms can be used for this purpose. The two most frequently used are the very popular combination grit sharpening stone (#49 Carborundum) and the cloth backed.

Proceed as follows when using the sharpening stone: Set the glass so the edge is convenient for working. Near by place a container of water that has a little kerosene mixed with it. Dip the stone in the water and rub it across the edge with a circular motion. Dip the stone in the water-kerosene mixture frequently. Avoid working a groove into the stone. Work from the coarse to the fine stone until the desired finish is readied.

The silicon-carbide cloth can be tacked or cemented on a piece of wood and rubbed across the glass like the sharpening stone, or the cloth can be attached to a flat surface and the glass rubbed back and forth over it.

For a final high polish the edge can be rubbed with rouge cloth.

PREPARING THE EDGE ON SAFETY GLASS

Preparation Of Safety Glass For Grinding. Be sure that all excess plastic is cut off flush with the glass. Otherwise, the extended plastic will cause bumping and possible chipping of the glass while on the grinding wheel.

"Be a good housekeeper" applies particularly to your glass-grinding equipment. Keep your wheels in first-class condition, properly dressed and clean. Recommendations on types of grinding wheels for various operations in your shop, how to use them, and their care, are given below.

Roughing. The horizontal roughing grinding wheel (Fig. 13-4) is not recommended for grinding safety glass with plastic, because it is likely to cause excessive chipping along the edges. However, if you do have this type of equipment, most of the chipping can be eliminated if the edge is first seamed (chamfered) and then only moderate pressure is used in grinding the edge of the glass.

Belt Machine. The wet belt grinding machine (Fig. 13-5), that is made by different glass machinery manufacturers, is found to be efficient. It turns out a first-class

Fig. 13-5. The wet belt grinding machine being used to finish the edge on glass. *(Courtesy: Libby-Owens-Ford Glass Co.)*

Fig. 13-6. Smoothing the edge of the glass after it has been ground. *(Courtesy: Libby-Owens-Ford Glass Co.)*

edge job in less time than other types of grinding equipment.

Diamond Edging Machine. Extensive progress has been made in grinding equipment, especially in equipment for grinding and edging automotive safety glass. Diamond wheels for edging may be purchased from several leading manufacturers of grinding equipment. They are being used advantageously in both large and small shops throughout the country.

Upright Smoothing Wheel. The upright smoothing wheel, (Fig. 13-6), while slower, does a very good edge job, particularly where there is little or no excess stock to be removed from the safety glass after it is cut from the block size.

Using a small amount of kerosene in the water of your grinding wheel will speed up the grinding, and act as a solvent. It prevents the plastic from filling up the pores of the grinding wheel. **Do NOT use any other type of oil as a substitute for kerosene.**

Polishing. It is recommended that, for the more finished job, the edges be polished by making several quick passes over a cork or composition polishing wheel, using pumice as the polishing agent.

All edges to be inserted in channels should be seamed, or chamfered, to avoid breakage. The rounded edge is much less vulnerable than the sharp edge, because pressure is distributed over a greater area. Then, too, sharp edges may cut through the various types of packing used to set the glass in the channels, destroying their effectiveness and often making an immediate replacement of this material necessary.

In grinding the glass, do not hurry the job. You avoid excessive chipping of edges and save yourself extra grinding. Remember that chips developed during the grinding operation often start minor runs, causing breakage while the glass is being set in channels, or after it is installed in place.

Do not seal the edges of Laminated Safety Glass. This operation is not required. Thus you save all the trouble, time and expense of undercutting and sealing the glass.

Figure 13-7 shows a special jig that is available for pressing channels on the edge of safety glass without injuring the glass.

BEVELING PROCEDURE USED IN GLASS FABRICATION SHOPS

The beveling of plate glass as performed in commercial shops is a treatment by a series of experts in each particular process, from the abrasive grinding away of the bevel section to the restored burnished surface of polished glass.

Five divisions of skilled workmen are engaged in the beveling operation, namely: roughers, emeryers, smoothers, white-wheelers and buffers (polishers), using different abrasive or polishing materials, such as sand or carborundum, emery, sandstone, pumice and rouge.

Roughing. The roughing mill or wheel is a horizontal circular cast-iron disc having a fine cut corrugated surface about 30 inches in diameter. An abrasive is conveyed to the mill from above through a hopper with a fine stream of water. When the edge of the plate is brought into contact with the swiftly revolving roughing wheel, the abrasive, crushed between the iron and the glass, cuts the bevel to the desired depth, while the water minimizes the friction heat. Curved pattern plates, miters, etc., require an expert practiced eye and special skill on the part of an operator.

Emerying. In the first roughing process the beveled surface is cut and scored so deeply by the coarse sand that it is necessary to follow with a finer abrasive on another mill to bring the bevel to a smoother finish.

Smoothing. The rough ground edge is smoothed on the stone mill, or smoother, which consists of a horizontal fine grain sandstone wheel revolving with water flowing upon it to reduce friction.

Polishing. Preliminary polishing of the bevel is accomplished by pressing the beveled edge of the glass against the edge of an upright wood wheel which brings

Fig. 13-7. A jig used for pressing channels on the edge of laminated safety glass. (*Courtesy: Libby-Owens-Ford Glass Co.*)

the bevel to a dull, milky polish by the use of powdered pumice in solution automatically splashed upon the wheel.

Finishing. The final high gloss polish is imparted to the bevel surface by contact with the rouge-impregnated felt-covered edge of a rotation upright polishing wheel.

The preceding outline describes the beveling process generally used by glass fabrication shops.

Precision beveling machines for volume production of beveled plates for mirrors or plaques in rectangles, circles or shapes are used by most of the larger manufacturers.

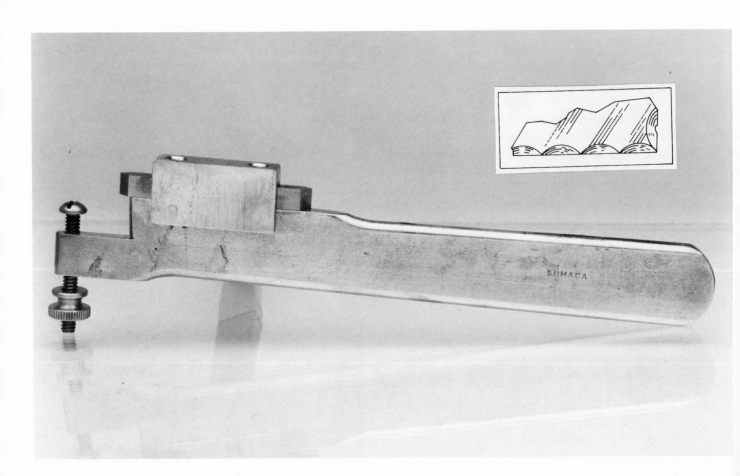

Fig. 13-8. Glass chipping tool. This unique tool makes "scallop" clips in glass edges as illustrated. (*Courtesy: C.R. Laurence Co., Inc.*)

Chapter 14
Removing scratches from glass

Glass that has been scratched can be unsightly, unattractive, and even dangerous when it might obstruct necessary vision—as is the case with a scratched windshield on a car. Shallow scratches can be removed without difficulty. Deep ones should not be removed as they leave the glass with a wavy surface. Such a wavy surface may be more objectionable than the original scratch.

LIMITS OF POLISHING

Light scratches and rubs can be removed satisfactorily. Deep scratches and rubs can be lessened only, not completely removed. A simple way to test the depth of a scratch is to move your fingernail slowly across the groove of the scratch (Fig. 14-1). If your fingernail does not catch in the groove, the scratch is usually shallow enough to attempt removal by polishing. However, if your fingernail does catch in the groove, it is too deep for removal.

REMOVING SLIGHT SCRATCHES BY HAND

Slight or even moderately deep scratches can be removed with hand rubbing. For this purpose **black** rouge is used as the abrasive. (White and red rouge are not coarse enough for this purpose.) Mix the black rouge with water to make a medium thick paste. (A shallow dish is good to hold the mixture.) Pick up the paste rouge in a stiff felt pad and rub the scratch with a firm pressure. Fresh supplies of paste rouge should be picked up on the felt pad as the polishing progresses. When the scratch is removed wash the glass with a detergent or strong soap.

Fig. 14-1. (Courtesy: Sommer and Maca Glass Machinery Co.)

LIMITS OF POLISHING

REMOVING SCRATCHES WITH A PORTABLE POLISHER

There are on the market portable scratch polishers that are capable of removing light surface scratches, rubs, fine lines, stains, and discolorations on automobile windshields, mirrors, or glass of any kind. The polisher (Fig. 14-2) consists of an all wool felt wheel 3 inches in diameter by 1-1/2 inches thick that is cemented to an aluminum adapter that has a shank on it for chucking in a power tool. A speed of 1300 r.p.m. is best for this operation.

The abrasive used on the felt for polishing is optical quality cerium oxide polishing powder No. 12. The polishing powder should be kept tightly closed in a dust proof container so it will not become contaminated with dust or moisture.

Mix the polishing powder with water until it is the consistency of light cream. Clean the glass with water and a soft cloth. Start the felt wheel rotating and gently apply it to the scratch (Fig. 14-3). Move the felt wheel so it not only polishes the scratch itself, but also covers an area a few inches on both sides of the scratch. Apply a gentle pressure on the felt wheel. Every few minutes

Fig. 14-2. Portable scratch polisher. (Courtesy: Sommer and Maca Glass Machinery Co.)

Fig. 14-3. The polisher should be applied flat against the glass and a minimum of pressure applied.

wipe the polishing compound from the scratch to see if it is gone. Continue the above procedure until the scratch is removed. Then wash the glass clean with soap or detergent and water.

Tips On Polishing Scratches

1. Do NOT attempt the polishing procedure until you have practiced a bit on scrap glass. Get the knack first. Don't practice on the actual job.

2. Work the polisher in a scrubbing motion **around** the area of the scratch, never just back and forth.

3. At first, some pressure may be necessary, but gradually reduce this pressure until just the weight of the polisher is used. Avoid using too much pressure (practice will tell you how much to use) since this will increase the possibility of breakage of the glass. Too much pressure also will cause polishing in one spot, resulting in distorted vision.

4. Keep the polisher moving at all times, that is, don't grind only in one local spot. Remember that the polishing operation should be performed in a circle, gradually tapering upward and outward.

5. DO NOT LET THE FELT DRY OUT. Keep it wet and supplied with sufficient polishing slurry.

6. DON'T KEEP THE FELT SPINNING ON THE GLASS LONGER THAN 45 SECONDS AT A TIME. If the glass becomes hot, LET IT COOL BEFORE PROCEEDING FURTHER.

7. Hold the pad flat against the glass. If only one edge touches, deep polishing will result.

8. If the face of the felt pad becomes bumpy, true it by touching the face to a piece of sandpaper tacked down to a flat surface.

9. SPECIAL CAUTION SHOULD BE TAKEN TO MAKE SURE DIRT AND DUST DO NOT GET ON THE FELT OR IN THE POLISHING AGENT.

10. As with many other tasks, polishing glass is a simple procedure once you get to know it. However, it requires a "touch" obtained only by experience and intelligent following of directions. Good polishing jobs are not done haphazardly; they are the result of knowing what can be done, using good materials, and most important of all, calling upon a bit of craftsmanship and pride in doing it right.

Chapter 15
Firing glass

The word "firing", as related to glass work, refers to the heating of glass in any manner. Glass may be "fired" by being placed directly in a torch or burner flame, as it is in glass blowing, or it may be placed in a kiln or lehr, where it is not subjected to a direct flame but is evenly surrounded by the heat resulting from a fuel source such as electricity, gas, fuel oil, etc.

All forms of heat produce the same general results in glass—render it molten when subjected to sufficient amounts of heat. The answer as to which form of heating should be used is determined by the size and kind of glass to be worked on. If a small piece of glass, or the end of a rod of glass is to be heated, it can be done speedily in a direct flame or lamp. However, if a large piece of glass is to be heated it must be done in a kiln or lehr, where the entire item can be given a controlled, even heating and cooling. This is necessary with a large sheet in order to avoid strains and stresses that could cause the glass to break.

EFFECT OF HEAT ON GLASS

When glass is subjected to heat of even the slightest amount, a change takes place in its structure. The alterations that occur with slight rises in temperature are not sufficient to be of concern for ordinary purposes. However, when a great deal of heat is applied to glass the structural changes are significant. These changes take place at definite viscosity reference points in the firing process as follows:

Fig. 15-1. A universal blast burner being used to assemble glass laboratory equipment. *(Courtesy: Corning Glass Works)*

Fig. 15-2. A liquified petroleum burner. *(Courtesy: Central Scientific Co.)*

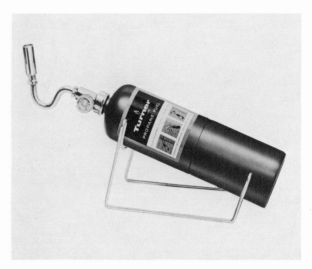

Working Point is the temperature at which glass has a viscosity of 10^4 poises . At this temperature glass is soft enough for most lamp working or sealing operations.

Softening Point is the temperature at which glass has a viscosity of $10^{7.6}$ poises. In this temperature range glass will deform noticeably (sagg) under its own weight.

Annealing Point is the temperature at which glass has a viscosity of $10^{13.4}$ poises. At the annealing point the internal stress caused by rapid cooling from lamp working or forming temperatures may be substantially removed in 15 minutes.

Strain Point is the temperature at which glass has a viscosity of $10^{14.6}$ poises. This is the temperature at which the internal stresses are reduced to low values in four hours. This is not a directly measured property but for most glasses is approximately 30 Centigrade degrees below the annealing point.

Expansion Coefficient is the average increase in length per unit length per °C. change in temperature over the range of 0 to 300° C. Since the expansion coefficient is affected slightly by annealing the values given are for annealed glass.

The above viscosity reference points are reached at different temperatures in different types and kinds of glasses. On page 116 are given descriptions of general types of glass. From these descriptions it is obvious that the glasses you most frequently will be using in the preparation of different items will be Lime Glasses, Lead Glasses, and Borosilicate Glasses. The following chart indicates the approximate temperatures at which these glasses reach their viscosity reference points with their corresponding alteration in structure. It is important to know these temperatures so that you can visualize what is happening to the glass that is enclosed in a kiln or lehr.

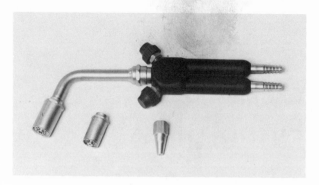

Fig. 15-3. The hand torch. *(Courtesy: Central Scientific Co.)*

SEPARATORS FOR GLASS

During the firing process glass becomes soft to a greater or lesser degree, depending on the amount of heat to which it has been subjected. A characteristic of softened glass is that it will adhere to practically any surface it touches. To avoid having the glass adhere to an undesired surface, we use materials called "separators". Separator material is applied to kiln shelves, lehr trays, molds, planchon, and other similar devices used to support glass during the firing process.

There are several different materials that can be used as separators, such as **whiting** (calcium carbonate), **plaster of Paris, dry clay, alumina hydrate,** and special commercial preparations. Whiting is by far the most frequently used material.

Whiting is a fine, white powder. It may be used dry.

Fig. 15-4. Meker burner. *(Courtesy: Central Scientific Co.)*

Type of Glass	Working Point		Softening Point		Annealing Point		Strain Point		Expansion Coefficient (per° c)x10
	°C	°F	°C	°F	°C	°F	°C	°F	
Lime	1000	1832	696	1285	510	950	477	890	92
Glasses			710	1310	525	976	493	920	87
Lead	971	1780	625	1157	427	800	399	750	91
Glasses	975	1787	630	1166	432	810	404	760	89
Borosilicate	1220	2228	821	1510	554	1030	516	960	32.5
Glasses	1210	2210	780	1436	516	960	477	890	34

The ultimate determination as to when glass has been properly fired is a visual one. In other words, when the temperature approaches those listed above, the craftsman should closely observe the glass and make the ultimate determination as to when the heat should be turned off by what he sees happening to the glass.

In this form it is sifted onto the surface to be separated from the glass. The whiting is sifted on until the color of the surface being protected cannot be seen. When whiting is applied in this manner it leaves the surface with an uneven, textured appearance. Glass fired until soft on such a surface will have the same uneven, textured appearance. This is considered very attractive by many craftsmen. However, should you wish to avoid the textured surface, the whiting may be flattened with a roller, or the back of a spoon.

Brushing and spraying are other ways of applying whiting. The whiting has to be mixed with water for these methods. Both of these methods of application result in a smoother surface than can be achieved with sifting. However, if the brushing and spraying strokes are frequently overlapped, a texture will develop in the surface.

Whiting can be used for one firing only, after which it must be discarded.

When plaster of Paris, dry clay, and alumina hydrate are used as separators, they are sifted onto the surface to be protected in the same manner as whiting. When commercially prepared separators are used, they should be applied as indicated on the package.

FIRING DEVICES

Glass firing devices may be classified into two general types: direct firing and indirect firing. Direct firing devices include the different burners that can be used to apply a concentrated heat on a small area of glass. Such devices are used in glass blowing and laboratory glass forming (Fig. 15-1). The indirect firing devices in-

Fig. 15-5. The universal blast burner. (Courtesy: Central Scientific Co.)

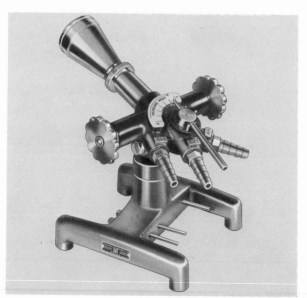

Fig. 15-6. Gas lighter. (Courtesy: Central Scientific Co.)

clude the kilns and lehrs that are used to fire uniformly large sheets of glass. These are used in glass sagging, laminating, decorating, etc.

Direct Firing Devices that can be used to work glass include the very elementary **Liquified Petroleum Burner** (Fig. 15-2) which is a lightweight, portable burner utilizing liquified fuel supplied in an I.C.C.-approved heavy-gauge, disposable tank. It usually includes a wire support cradle for holding the burner during heating operations.

The Hand Torch (Fig. 15-3) can use artificial, mixed natural or bottled gases, and oxygen or compressed air. Needle valves on both air and gas lines control flame size and intensity. Three interchangeable tips provide a range of flame shapes from needle flame to wide brush shape.

The Bunsen and Meker Burners (Fig. 15-4) are used on artificial or mixed gases only. They have a needle control valve for gas control and a sleeve vent for air regulation. The meker burner is capable of developing

Fig. 15-7. Floor model kiln. (Courtesy: American Art Clay Co., Inc.)

a higher temperature than is possible with the ordinary bunsen burner.

The Universal Blast Burner (Fig. 15-5) can be used with compressed air or oxygen and gasses having B.T.U. values of 500 to 1000 and above. A convenient mixing valve with lever control regulates the mixture of gas and compressed air or oxygen. Extremely hot flames can be obtained, ranging in shape from a sharp, short needle type to a very wide brush. Interchangeable nozzles permit use with any combination of air or oxygen and any type of commercial use.

Of all the burners listed above only the Universal Blast Burner is capable of heating lime and borosilicate glasses sufficiently for working. All the other burners can heat the very soft lead glasses sufficiently for working but are inadequate for the harder ones. The smaller torches can also be used for preheating the glass to be worked on the universal blast burner, and also for the annealing process. By using the smaller torches for these purposes, the larger blast burner can be used a longer time for the actual working of the glass.

For lighting gas burners use a **File Gas Lighter** (Fig. 15-6) that produces sparks by rubbing a spark-metal tip across a steel file. To fire-up the liquified petroleum and bunsen type burners, first turn on the gas slightly and light, then adjust the gas for a clear blue flame about two inches long. To fire-up a burner when using an oxygen-gas flame you first ignite the gas, then adjust the flame by adding oxygen. When oxygen-air-gas is used, the proper procedure for lighting is to ignite the gas first, then turn on the air, and finally introduce oxygen. It is well to remember that the hottest part of a flame is at the tip of the inner cone.

Indirect Firing Devices that are used to fire glass include kilns and lehrs. Both are more or less glorified ovens with special controlling devices.

Kilns are all constructed in about the same manner. They consist of a steel exterior, an interior firing chamber of a high fire refactory, heating elements and control devices. Within this framework, however, there can be dozens of variations, such as the thickness of the steel jacket, the type and thickness of the refactory, the kind and amount of heating surface, the number and kind of control devices, whether it is a floor model (Fig. 15-7) or a table model (Fig. 15-8), whether it has a door (Fig. 15-8) or lid (Fig. 15-9), etc. These differences can affect the quality and price of the kiln and therefore are of concern to the person buying the kiln. As far as the firing of glass is concerned, only a few differences need be considered.

Kilns for school and home workshop use are of two general types; the ceramic kilns and the metal enameling kilns. Both are good for firing glass.

The ceramic kilns (Fig. 15-7, 8 & 9) are particularly good for firing glass because they generally increase their temperature at a slow, even rate. This is true of both the standard and the "high fire" models. The "high fire" kiln differs from the standard in that it is constructed to safely develop a higher temperature.

The Metal Enameling Kiln (Fig. 15-10) differs from the ceramic kiln in that it develops its heat at a much faster **rate.** Glass heated at this fast rate will fracture. To retard the heating rate of this type of kiln, the door is left ajar about a half inch until it reaches a temperature of approximately 1000°F. This is not particularly objectionable since the door on the ceramic kiln must also be left ajar when firing glass, but for a different reason to be explained later. Incidentally, the Hot-Plate Type Metal

Fig. 15-8. Table model kiln with door. (Courtesy: American Art Clay Co., Inc.)

Fig. 15-9. Lid type kiln. (Courtesy: American Art Clay Co., Inc.)

Fig. 15-10. Metal enameling kiln. *(Courtesy: American Art Clay Co., Inc.)*

Enameling Kiln (Fig. 15-11) can be used for preparing glass "jewels," but is not acceptable for the general firing of glass because of the fast rate of heating and cooling.

Kilns can be purchased in knock-down kit form rather inexpensively and constructed in the workshop (Fig. 15-12) shows such a kiln sold by the American Art Clay Company.

The Heating Elements In Kilns are fueled by electricity, gas, kerosene, or fuel oil. Electricity is the most frequently used and, all things being equal, it is the best for two reasons: First, it is the easiest to control. Secondly, it is the cleanest. Glass must be fired within a rather close tolerance. A few seconds of overfiring or underfiring can make the difference between failure and success. Electrically powered kilns respond to heat-

ing and cooling in a manner more in keeping with the needs of glass. This is not to say that the others will not do. It merely means that they would have to be watched more closely during the firing process to make sure their temperature stays within acceptable limits.

The cleanliness of electricity is a factor in firing glass because any dirt or contamination in a kiln will cause a blemish in the glass. In some cases gas, kerosene, and fuel oil will give off fumes and residues that will cause discoloration or fogging of the glass being fired. Most of the time such fumes and residues can be eliminated by venting the kiln during the heating and cooling cycles but this is not always true.

Control Devices For Kilns may be secured in a few different forms. Some devices are used to indicate the firing chamber temperature, while others time the starting and stopping of the firing cycles. All the control devices listed below can be used with any cabinet type kiln. The devices you select will be determined by the convenience you desire and the amount of money you have available to spend.

Pyrometric Cones (Fig. 15-13) deform progressively under advancing heat at fairly regular temperature levels; they therefore serve as accurate indicators of firing chamber temperatures. Usually three cones are selected and set in a clay pot at an 82° angle. One cone indicates the recommended temperature to be fired; one is a higher temperature and one lower for a warning signal that the desired temperature approaches or has been passed. The three cones are placed in the firing chamber of the kiln in such a position that they may be observed through the peephole during a firing. When the cone bends, it indicates that the desired temperature has been reached. The chart below indicates the temperature equivalents of cones.

Fig. 15-11. Hot plate type metal enameling kiln. *(Courtesy: American Art Clay Co., Inc.)*

Fig. 15-12. Kiln made from a knock-down kit. *(Courtesy: American Art Clay Co., Inc.)*

Fig. 15-13. Pyrometric cones used for indicating temperature levels. *(Courtesy: American Art Clay Co., Inc.)*

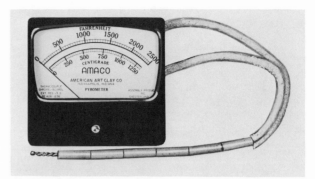

Fig. 15-15. Pyrometer. *(Courtesy: American Art Clay Co., Inc.)*

Pyrometer (Fig. 15-15) installed on kiln indicates fire chamber temperatures on a dial which can be read at a glance. Kilns can be purchased with the pyrometer already installed (see Figures 15-7, 8, 9 & 10) or the pyrometer may be purchased separately. Kiln-Gard (Fig. 15-16) is inserted through the peep-hole of the kiln. A ring of high-temperature-resisting metal at the end of a pivoted rod encircles and rests upon a junior size pyrometric cone. When the cone bends, the ring lowers and actuates an electric switch outside the kiln. **Kiln Starters and Cut-Offs** (Fig. 15-17) can be had in many different price ranges. These devices can be set to start and stop your kiln at a predetermined time and temperature. Figure 15-18 shows a combination pyrometer and automatic cut-off.

Fig. 15-14. *(Courtesy: The Electric Hot Pack Co.)*

Important Notes Concerning Kilns. Each kiln has its own individual characteristics of heating and cooling. This is true of supposedly identical kilns made by the same manufacturer. This being the case, you should prepare a very careful log for your kiln. Some of the information the log should include is the kind and number of pieces of glass that were fired, rate of heating, amount of venting required during heating, temperature at which kiln was turned off, rate of cooling, amount of venting during cooling, etc. After just a few firings you can then predict with accuracy how long it will take to fire properly each type of glass. The log should be periodically checked because kiln performance changes with age.

To vent a kiln during the firing of glass you must prop the door or lid open a half inch or more for various lengths of time. This venting is required when the firing is begun, and frequently for a short time right after the kiln has been turned off. Just how long this venting should be done will vary from kiln to kiln.

Most kilns will be sufficiently vented at the beginning of the firing cycle if the lid or door is opened one half inch until the temperature reaches 1000° F or when an 022 cone bends in half. The amount of venting required

Fig. 15-16. Kiln-Gard. *(Courtesy: American Art Clay Co., Inc.)*

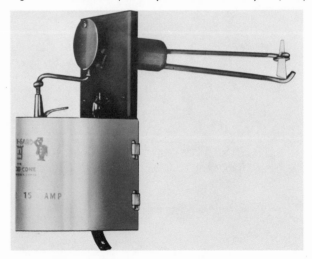

Temperature Equivalents of Cones

Cone Number	When fired slowly 20° C per hour ° Cent.	° Fahr.	When fired rapidly 150° C per hour ° Cent.	° Fahr.
022	585	1085	305	1121
021	595	1103	615	1139
020	625	1157	650	1202
019	630	1166	660	1220
018	670	1238	720	1328
017	720	1328	770	1418
016	735	1355	795	1463
015	770	1418	805	1481
014	795	1463	830	1526
013	825	1517	860	1580
012	840	1544	875	1607
011	875	1607	905	1661
010	890	1634	895	1643
09	930	1706	930	1706
08	945	1733	950	1742
07	975	1787	990	1814
06	1005	1841	1015	1859
05	1030	1886	1040	1904
04	1050	1922	1060	1940
03	1080	1976	1115	2039
02	1095	2003	1125	2057
01	1110	2030	1145	2098
1	1125	2057	1160	2120
2	1135	2075	1165	2129
3	1145	2093	1170	2138
4	1165	2129	1190	2174
5	1180	2156	1205	2201
6	1190	2174	1230	2246
7	1210	2210	1250	2282
8	1225	2237	1260	2300
9	1250	2282	1285	2345
10	1260	2300	1305	2381
11	1285	2345	1325	2417
12	1310	2390	1335	2435
13	1350	2462	1350	2462
14	1390	2534	1400	2552
15	1410	2570	1435	2615
16	1450	2642	1465	2669
17	1465	2669	1475	2687

Fig. 15-17. Kiln starter and timer. (Courtesy: American Art Clay Co., Inc.)

when the kiln is turned off varies greatly from one kiln to another. Some kilns require no venting at all at this time. Others may require the door to be opened a half inch or more from 2 to 25 minutes. The important thing to remember is that the temperature in the fire chamber can be controlled by opening and closing the door or lid of the kiln. By experimenting and recording the results, your kiln can be found to do an excellent job.

Many kilns are heated on three sides, leaving the spot just inside the door cooler than the rest of the fire chamber. Large pieces of glass frequently do not sag or laminate evenly because of this condition. To overcome it, the glass must be rotated. When the temperature reaches 1450° F, or by observation the glass to the back of the kiln appears a bright orange, the kiln door

Fig. 15-18. Combination pyrometer and automatic cut-off. (Courtesy: American Art Clay Co., Inc.)

can be opened and the glass quickly turned. The kiln shelf can be pulled out onto a large flat piece of metal, turned, and slid back in. If the glass placed on a iron or steel turn table (Fig. 15-19) in the kiln it will make this job much easier.

Note: Do not use an aluminum turn table as it will melt!

Care must be exercised when stacking glass items in the kiln. To assure proper circulation so that objectionable gasses and fumes can be safely dissipated there should be a space of at least 2-1/2 inches between the top of the glass and the shelf above. Make sure that pieces of glass that are not to be joined do not touch.

Lehrs are specially designed ovens used in the decorating and annealing of glass. The modern lehr used in industry is designed around an automatically controlled conveyer belt that carries the glass items through preheating, firing, and annealing cycles in one continuous operation. Decorating lehrs vary from 25 to 125 feet in length, and from 2 to 8 feet wide. Figures 15-20 through 15-22 show typical lehrs used in industry.

Fig. 15-19. Turn table. (Courtesy: American Art Clay Co., Inc.)

PROCEDURE FOR FIRING GLASS IN A KILN

The following procedure is the same for all decorating, and sagging of glass. The only difference is in the amount of heat to which the glass is subjected. The heat will vary with the different kinds of glass, the size and thickness, the amount of formation you desire, etc. In the following descriptions suggested temperatures will be listed. These temperatures may not be correct for your kiln. Only experience will dictate the one which is correct for your particular kiln.

It is important to remember that glass must be put into a cold kiln and removed only after the kiln is cold again, because "thermal shock"—quick heating or cooling—will crack the glass or cause it to crack at a later time.

Review of Preparations for the kiln:

After thoroughly cleaning glass with soap and water, wipe it with alcohol and **do not touch the surface again** with fingers.

Apply decorating mediums following recommended procedures (See Decorating pages 77-86).

If laminating two or more pieces, assemble them together for placing on the mold, planchon, turntable, or kiln shelf.

Apply separator to prevent the glass from sticking to surface on which it is resting (See Separators page 96).

When separator and surface of mold or shelf are dry, place the glass carefully in position on mold, planchon, etc.

Carefully slide the glass into the kiln, making sure it is not upset or that it does not touch some glass it is not supposed to. You are now ready to fire the glass.

Firing Procedure

1. Open the door or lid approximately 1/2" and keep it open until the kiln reaches a temperature of 1000° F. (or when an 022 cone bends half over) to provide vent for corrosive gases from lustres, adherents, etc. These gases can frost and discolor your glass.

2. Heat kiln slowly, taking at least one hour to reach 1000° F. Start kiln on low and determine for yourself when to switch to medium and high since different kilns vary.

3. Close kiln door or lid when temperature reaches 1000° F.

4. When temperature reaches approximately 1450° F the door may be opened for quick observation of glass to determine when it is completely fired. To compensate for the "cold side" in kilns heated on three sides, the glass can be rotated at this time. This can be done by quickly and carefully removing the mold or shelf, turning and replacing in the kiln.

> **Notes: Underfired** glass will have squared-off edges which have not fused completely. **Overfired** glass may become distorted in shape—tiny needlepoints form around the edges of the fired shape—the color may burn away from the edges of the sagged glass, and the edges may even recede or flow from the edge of the mold.

In the average enameling kiln glass will sag, edges will fuse and round off at approximately 1500° F. or when an 014 cone bends half over. Laminating will occur about 50° sooner.

5. When the glass appears to have been completely fired, turn kiln off. **Remember** that when you turn the kiln off the glass will still be subjected to high temperature for a short time.

6. Do not open door once the kiln is shut off until it is absolutely cold. The long cooling period is needed to anneal the glass.

> **Notes:** Some kilns retain intense heat after the kiln is shut off, causing overheating of the glass. To overcome this, such kilns should be vented for a short period immediately after the kiln is turned off. To vent, open the door approximately 1/2 inch for a length of time that is appropriate for your kiln (it could be 2 to 25 minutes). Experimentation will indicate the appropriate time. When venting is completed, close the door and do not open it again until the kiln is absolutely cold.

Firing Glass Decorations

The firing of decorated glassware is, within certain limits, a function of time and temperature. In other words, a fast firing at a given temperature often produces the same results as a slower firing at a lower temperature. The weight and thickness of the glass and the type of vehicle used in applying the color are determining factors. Some oils require a slow preheating treatment. On the other hand, heavy glassware must be heated and cooled slowly for reasons of annealing. Light, thin blown tumblers can be run through a good lehr (Fig. 12-17) or kiln in 30 minutes, while heavy glass may require as much as three hours time.

Fig. 15-20. Annealing lehr for processing glass containers. *(Courtesy: Surface Combustion Division, Midland-Ross Corp.)*

Fig. 15-21. Reannealing lehr for processing plate glass. *(Courtesy: Surface Combustion Division, Midland-Ross Corp.)*

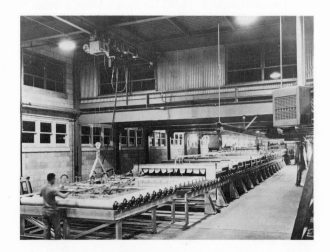

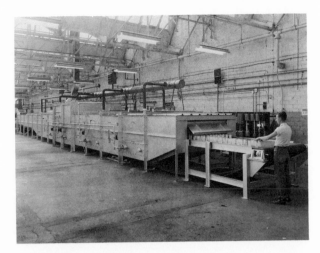

Fig. 15-22. Direct fired decorating lehr for glass containers. (Courtesy: Surface Combustion Division, Midland-Ross Corp.)

It is best to fire glassware as soon as the decoration has been applied. Some oils oxidize on standing and become difficult to remove later in the firing operation.

Allow colors and adherents to dry thoroughly before firing, or bubbling will result.

Complete the firing operation, carefully observing the procedures recommended above on pages 102-103.

Firing Jewels differs from the firing of large pieces of glass. Jewels, being very small pieces of glass, are not subjected to the same internal strain and stress large pieces are subjected to when being heated and cooled. The tiny pieces of glass can be heated in a hurry and removed from the heat when still hot. The Hot-Plate Type Metal Enameling kiln (Fig. 15-11) and the Hand Torch (Fig. 15-3) are good for this purpose as they enable the pieces to be easily watched during firing. See **Glass Jewels,** page 108 for detailed description of these attractive decorating materials.

Chapter 16
Glass sagging (bending)

The sagging of glass—or bending as it is sometimes called—is the result of firing sheet glass placed over a mold until the rising temperature causes the glass to soften and sag down into the mold (Fig. 16-3). The sagged glass will not only take the shape of the mold, but also the texture of the surface.

Not every shape can be sagged in glass. Best results can be obtained in a mold with sides that **slope gradually** and are not too deep (Fig. 16-4). The novice should choose a design that has a fairly large horizontal rim at the top (Fig. 16-5). Such a rim prevents the melted glass from "crawling" out of shape.

The mold for sagging can be made of metal, grog mix clay, soft fire brick, and asbestos.

Plaster of Paris and regular clay are not satisfactory for glass sagging molds as they will not withstand the repeated firing to which they would be subjected.

METAL MOLDS

Metal molds are used commercially in the manufacture of large quantities of a bent glass item. The cost is prohibitive to make up a cast or machined metal mold for a limited number of saggings.

Almost any of the hard metals (iron, steel, copper,

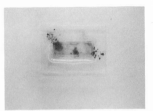

Fig. 16-2. Samples of sagged glass items made by students. (Courtesy: Mary L. Scarpino)

brass, etc.) can be used for a glass sagging mold. Although it is quite expensive to make an elaborate metal mold for a bowl or similar object, there are metal designs, which can be found already made-up or which can be easily fabricated, that can be used. For example:

Fig. 16-3. (Courtesy: Mary L. Scarpino)

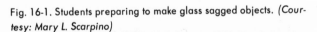

Fig. 16-1. Students preparing to make glass sagged objects. (Courtesy: Mary L. Scarpino)

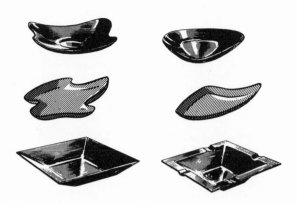

Fig. 16-4.

(a) Old cast iron pans and trays.
(b) Metal letters and numbers that can be purchased at local hardware stores.
(c) Fairly thick (12 gauge or more) wire can be bent into initials or designs. Such designs should be kept simple and have sufficient space between the parts for the sagging glass. Making up the design of a few short pieces of wire rather than one long piece eliminates warping of the wire during firing.
(d) Miscellaneous items such as door keys, hair pins small nails and screws, bolts, etc., can be used.

Any hard metal item that does not include undercuts—which means that it can be withdrawn from the sagged glass—can be used as a mold. It must be remembered that items such as those listed in b, c, and d above must be placed on a kiln shelf or block of fire brick in preparation for firing.

GROG MIX MOLDS

Grog clay is firing clay that has fine particles of previously fired clay mixed in it. When this clay is fired it becomes porous, enabling it to withstand the shock of repeated heating and cooling.

Fig. 16-5.

The grog clay may be formed into an original shape in much the same way as any other clay (Fig. 16-6). There are some things, however, that must be kept in mind when forming a sagging mold:

(A) Unevenness in the thickness of the clay may cause cracking or separation.

(B) The mold must be even in height or the glass will shift.

(C) The bottom of the mold must be flat or the final sagged glass will rock or tip.

(D) Make small vent holes in the mold to allow air to escape when the glass sags into the mold. Remember, you are making a mold and not a finished piece of ceramic ware.

An excellent idea is to pattern the inside of the mold where the glass will sag. The pattern could consist of short dashes, crosses, semi-circles, etc. pressed into the soft clay with a pointed stick. This imparts an additional dimension to the sagged glass that is extremely beautiful.

Sagging Molds

If you wish to make a sagging mold by copying an existing bowl or similar object, proceed as follows:

1. The bowl selected to form the mold cavity is covered carefully and smoothly with tissue paper and inverted on a plaster bat larger than the bowl (Fig. 16-7).

2. A slab of grog mix clay, rolled to a uniform thickness is draped over the inverted bowl. The slab is large enough to extend several inches onto the bat (Fig. 16-8).

3. The mold is turned over when the clay can be handled safely and will support its own weight without deforming. The rim is marked for trimming (Fig. 16-9).

Fig. 16-6. Free formed sagging molds. (Courtesy: Mary L. Scarpino)

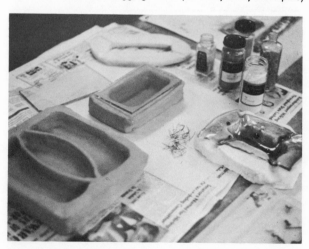

4. Excess clay is trimmed off with the paring knife or potter's fettling knife. The rim may be even in width or a free form shape, as desired (Fig. 16-10).

5. If a design is desired in the surface of the mold it should be sculpted in at this time.

6. Place the mold upside down on a plaster bat. Dry slowly. A light weight placed on top will prevent warping.

7. Fire the mold to keep the rim even. The mold should be fired upside down. Molds last longer if they are fired 2 to 4 cones lower than the recommended maturing temperature of the clay.

FIRE BRICK MOLDS

When one wishes to create his own sagged glass design it can be done by carving the desired design in soft insulating fire brick. The soft fire brick can be easily carved and shaped to the desired form. The bricks are usually 9" long by 4-1/2" wide by 2-1/2" thick. If a design too large for one brick is desired, two or more bricks can be taped together, carved and then glued together, using kiln cement.

The fire brick can be left with the carved surface which would impart a rather interesting rough texture to the sagged glass, or the brick can be given a smooth face hardening by brushing with a water diluted solution of kiln cement.

ASBESTOS MOLDS

The asbestos molds are made from powdered asbestos that has been mixed with warm water to form a dough-like mixture. The mixed asbestos can be molded and formed by lightly pressing and tapping it into shape (Fig. 16-11). Inasmuch as this material does not have very much strength, it should be formed into rather large

Fig. 16-7.

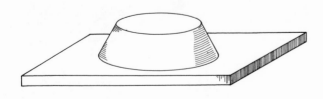

Fig. 16-8. The slab must be large enough to extend several inches on to the bat.

segments and thicknesses. The desired shape should be prepared on a shelf of the kiln and not moved unless absolutely necessary and then carefully.

To duplicate a bowl with the asbestos mixture proceed as follows: Invert the bowl on a clay bat; cover with tissue paper; apply hands full of asbestos mixture and lightly pat to shape; when dry enough to handle, carefully remove the mold. When the asbestos mold is thoroughly dry, paint or spray it all over with kiln wash.

GLASS SAGGING OPERATION

Any ceramics or enameling kiln is acceptable for glass sagging except for the small hot plate type. Cut the glass disk to be sagged equal to the largest diameter of the mold. For example: If the mold is 5" across the top and 1-1/2" deep, cut the disk to 5" in diameter. (It appears that sagging is a "stretching" process.)

1. After thorough cleaning with soap and water, wipe the glass with alcohol and do not touch the surface with fingers.

2. Apply any surface decorating mediums at this point, following recommended procedures (Fig. 16-11) (See Decorating, pages 77-86).

3. If laminating two or more pieces of glass for sag-

Fig. 16-9.

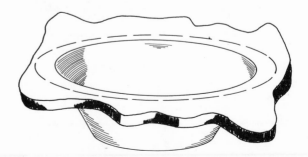

Fig. 16-10.

ging, (which is a very popular procedure) assemble them together with desired decorating medium between them for placement on the mold (See laminating, page 86).

4. To prevent the glass from sticking to the mold, coat it with an even layer of whiting or other separating material (See Separator, page 96).

5. Fire the glass carefully as recommended on pages 101-102.

MISCELLANEOUS GLASS OBJECTS THAT CAN BE SAGGED

All glasses, when subjected to sufficient heat, will become soft and flow. This is true of sheet glass as well as glasses used in bottles, dishes, lenses, jugs, jars, and other items. This characteristic makes it possible for all of them to be sagged, opening up a wonderful opportunity for experimentation and design.

Objects such as bottles, tumblers, etc. may be flattened by sagging them on a kiln shelf, or given a varied shape by sagging them into a mold. They may be sagged individually, or placed together in combination so they

will fuse together when they are sagged and produce interesting combined design variations.

Some glass containers have thick bottoms and tops that tend to excessively distort the object when sagged. If you find this objectionable, you may cut off the bottom or top or both before firing (See Cutting Jars and Bottles, pages 49-53.

The sagged glass container can be decorated with vitrifiable materials while it is being fired just as can any other glass object. It must be remembered that decorations applied to the outside of the sagged piece must be restricted to those decorative materials that can be used on the surface of glass (see Decorating Glass, pages 77-86) and those used on the inside of the sagged bottle, tumbler, etc. can include the same decorative material as those recommended for use between laminated glass (see Laminating, page 86)

The procedure for firing the bottle, jar or other glass object to be sagged is the same as that recommended for regular sagging on page 107, with these additions: Clean the object inside as well as outside; when starting to fire, raise the temperature **very slowly;** keep the kiln door ajar at least an inch until the temperature reaches 1000° F. A bottle will sagg somewhere above 1500° F.

GLASS JEWELS may be prepared from any small pieces of glass. The pieces may be collected from broken glass objects of all kinds—containers, sheets, stained glass, etc. or they may be cut to exact sizes from the different glasses or made from glass marbles. These glass shards from broken objects usually have extremely sharp edges and therefore should be handled with extreme care.

The jewels are created by firing the glass shards until the sharp edges have rounded over. The jewels can be made flat by firing them on a flat surface, or they can

Fig. 16-11. Applying decorative materials to the surface of the pieces of glass to be sagged. *(Courtesy: Mary L. Scarpino)*

Fig. 16-12. Bending lehr used to bend automobile windshields. *(Courtesy: Surface Combustion Division, Midland-Ross Corp.)*

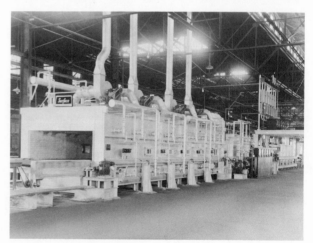

be made to sagg into a mold, giving them a formed shape. The molds may be made of the same materials and by the same methods described for regular sagging molds (page 105).

Glass shards taken from the same original piece of glass will fire with the same characteristic. However, those taken from other sources may have entirely different responses when fired. At certain temperatures some

Fig. 16-13. through 17. Hand Blown Glass Sculptural and Functional Glassware by Paul Rendzunas (*Photos by Tony Tantillo*)

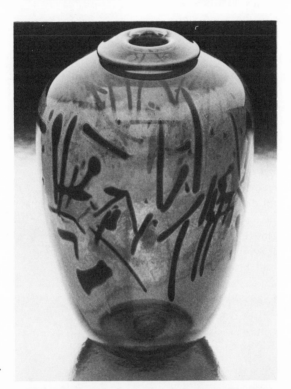

Fig. 16-13.

Fig. 16-14.

may have their edges rounded over, while some may flow out flat, while still others will almost retain their original shape. At the same temperature some pieces may retain their original colors while others may lose all traces of color. When the pieces are fired for the first time, their characteristics should be observed and recorded on the receptacle in which they are stored.

Glass shards, unlike larger pieces of glass, can be heated quite fast and can be removed from the kiln while still hot. This being the case, several different kinds of glass pieces can be fired at the same time. They can be constantly watched and each piece re-

Fig. 16-16.

moved from the kiln with tongs or a spatula as they sag to the point desired. The hot plate type of enameling kiln is best for this purpose as it affords an excellent opportunity to observe the progress of the pieces being fired. Remember that a separator must be applied to the kiln shelf and mold when the jewels are fired or these small pieces will stick as would large pieces.

Multicolored jewels can be created by laminating a few pieces of different colored glass. Another method is to fill a small glass bottle with layers of different colored enamels and threads! When sagged in a well-vented kiln, the bottle and contents become a multicolored single lump of glass. It can then be broken into different sized pieces for jewels with a sharp cold chisel and hammer, or with tile nippers, and the smaller pieces refired.

The jewels formed as described above may be mounted as individual gems, or may be refired onto another piece of glass as a decoration. (For more precise information on firing, see pages 95-103.)

Fig. 16-17.

Fig. 16-15.

Chapter 17
Laboratory glass blowing

Glass blowing is the practice of forming and shaping glass, when heated to a plastic state, by inflating it with air through a tube (Fig. 17-1). It is an ancient art that still flourishes today. It is a craft that is surrounded by misconceptions and misunderstandings that tend to frighten and discourage the novice and, in most cases, the technician alike. With a modest degree of application, the amateur can do a great deal of glass blowing with surprising success.

It is true that blowing a large globe on the end of a four-foot pipe is beyond the ability of the amateur. But there are many useful and exciting objects and fun-filled projects that require no more than a day or two or practice and a very minimum of equipment. The novice must remember these two basic, elementary requirements of glass blowing if he is to realize any success at all: First, the glass must be **evenly heated** to a plastic state over an area slightly larger than the part to be worked and the proper heat must be maintained until the operation is completed. Secondly, the formed and heated object must be cooled (annealed) **very slowly** to avoid strains that could shatter the piece.

The processes and procedures covered below are those that can be accomplished in a small shop or lab. They are basic procedures used in forming tubing that can be applied to many different needs and objects limited only by the imagination of the operator.

MATERIAL

There are several thousand different kinds and grades of glass. Only a very few of these are used in glass blowing. There is one used far more often than all the others put together; that is Pyrex Brand Glass No. 7740. It is this glass with which the following material is primarily concerned.

EQUIPMENT

The equipment for lamp (burner) working with Pyrex Brand Glassware need not be elaborate. The tools required depend to a great extent upon the type and amount of work to be done.

Most laboratory supply houses list the equipment and supplies required for glass blowing. However, many of the simpler tools may be made by the operator if he has access to a shop. The equipment described below will be found sufficient for most of the glass blowing operations. Figures 17-2 & 3 show a glass blower's table and an assortment of commonly used tools.

Work Bench. A table, or bench, on which to work is essential. The top should be approximately 54 inches by 30 inches, and covered with a fire-resistant material such as transite or asbestos. The heights of the table may vary considerably, depending on the preference of the glass blower to sit or stand while working. Thirty-one inches is the customary height. It is advisable to mount the burner on a wooden or metal slide instead of directly on the bench top so that it can be pulled out over the edge of the table, thus allowing space for free rotation of larger pieces of apparatus. Air and gas con-

Fig. 17-1. A laboratory glass blower at work. *(Courtesy: Corning Glass Works)*

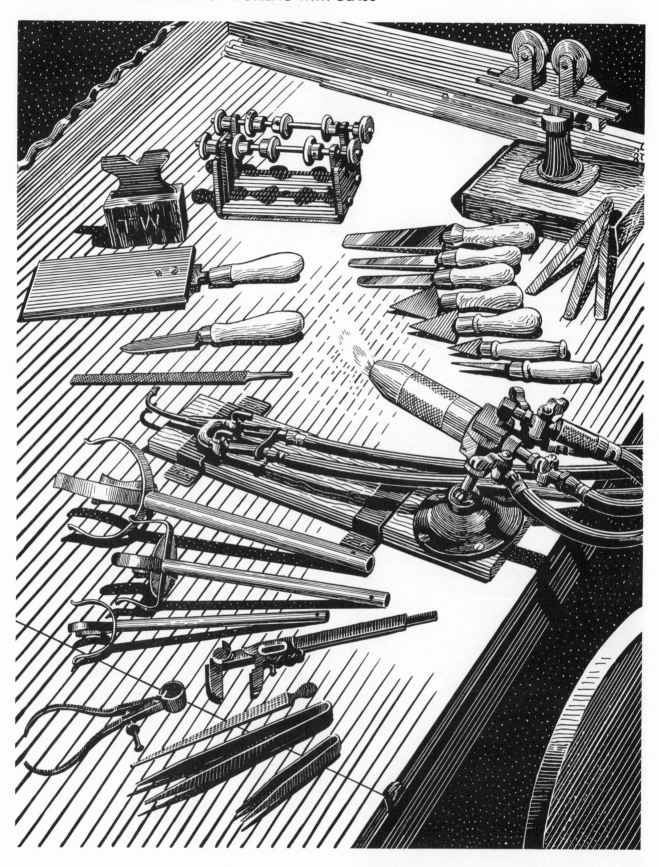

A glass blower's table and an assortment of commonly used tools

Fig. 17-2. (Courtesy: Corning Glass Works)

Fig. 17-3. The tools used by the glass blower are few and comparatively inexpensive. *(Courtesy: Corning Glass Works)*

nections should be located on the front of the table, or at least close to one side. Oxygen is usually used directly from a tank, and there should be sufficient space at the side of the table for this container. Good light is essential, the ideal location of the source is from the sides or back, not in front of the glass blower.

Burner. The burner is a very important part of the equipment, and should be selected according to the type of gases available. It should be so constructed that all variations from a sharp, intense flame to a soft, bushy flame may be readily obtained through adjustment of its valves, and so placed that the glass blower can make

adjustments easily and quickly. Some type of pilot light arrangement affords a further convenience.

For working Pyrex Brand Glasses, the temperature of the flame must be higher than that developed by an air-gas flame. Hence, the burner must operate on either air and oxygen or pure oxygen. Most commercial burners are provided with connections for air, oxygen and gas, and have valves for regulating the flow of each (Fig. 15-5). If the only burner available is the type with two connections, air and gas, it may be used with oxygen by bleeding this gas into the air line by means of a "T" connection. The opening of the oxygen inlet into the air line should be smaller in diameter than the opening in the air line.

Hand Torch. Another desirable type of burner is the hand torch (Fig. 15-3). This is useful to the technician in mounting apparatus on stands, and assembling it in place. Usually the oxygen-gas types are more desirable; however, the oxygen-air-gas type will be found fairly satisfactory. The torch need not be large, since it will ordinarily be used in sealing small tubes.

Pressure Regulator. A pressure regulator for the oxygen tank is essential if the glass blower intends to regulate his burner by the burner valves. A pressure of from 8-12 pounds is very satisfactory. If a regulator is not available, the oxygen flow can be regulated with the tank valves. In this case the burner valve is open at all times.

Rollers & Stands. Glass blowing rollers are not necessary pieces of equipment, but they aid and simplify the rotation of tubing on the flame. The simplest form of roller is made of two fibre or metal rollers about two inches in diameter so mounted that the distance between the rollers may be varied from one quarter to three inches. They are mounted on a base which permits adjustment for height (Fig. 17-4).

Another type of roller consists of two shafts four to six inches in length, on which are mounted six rollers approximately one inch in diameter. The frame is usually four inches wide and four to six inches high. These rollers are also constructed so that space between them may

Fig. 17-4. Glass blowing rollers.

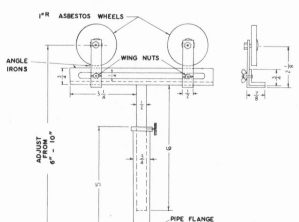

Fig. 17-5. Asbestos glass stand.

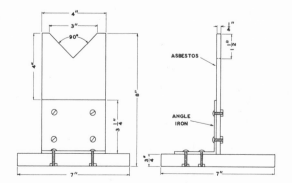

be varied (see Figures 17-2 & 3).

They may be driven mechanically by a belt arrangement, in which case they cause the glass to rotate, or the roller may be used to support the glass which is rotated by hand.

A glass support made of asbestos is a handy tool to have (Fig. 17-5). It is good for lining up glass tubing in a level plane while it is being worked or while cooling.

Glass Storage Bin. A cabinet is recommended to protect the glass from breakage and keep it clean. A horizontal or sloping cabinet with a door is best.

Glass Knife. The glass knife should be designed to produce a sharp scratch, and since it must be forced through the surface of the glass, it should be quite hard. An ordinary three-cornered file may be used, but it soon wears away at the corners, and the scratch produced is rather wide. Grinding the edges of an ordinary machinist's mill file forms corners which make satisfactory cutting edges that may be easily sharpened by re-grinding. Smooth steel knives are also available. Knives with tungsten carbide cutting edges are excellent and produce a very fine scratch. They are durable, and because of the hardness of the tungsten carbide do not wear or dull easily.

Reaming Tools. Reaming tools are usually in the form of flat triangular pieces of brass fitted with handles. The brass varies from one-sixteenth to one-eighth inch in thickness, depending upon the length of the tool, which ranges from one to four inches. Sharply pointed tools, as well as blunt ones, will be found useful. Metal tools produce a better finish and are less likely to stick to hot glass when a lubricant such as beeswax is used.

Rods & Wires. In addition to the brass reamers, pointed carbon rods, ranging in size from one-eighth to one-half inch in diameter, will be useful. A flat piece of carbon, either with or without a handle, may be used for squaring ends of tubing, or blowing flat bottoms in apparatus. Pointed tungsten wire, mounted in a glass rod or handle, is useful for putting small holes in light weight ware.

Tweezers. Glass blowing tweezers are used for grasping and pulling off hot pieces of glass, and also for flattening or squeezing heated tubes together. A small pair with tips or jaws about 1/8" wide, and a larger pair with 1/2" jaws, will be sufficient for most work.

Flask Holder. Since Pyrex Brand Flasks are often used as blanks in apparatus construction, some simple holder is necessary. This should hold the flask squarely, and facilitate easy rotation and manipulation of the ware. A simple form is made from three or four prongs of spring wire or spring metal strips mounted on a round handle of approximately 12" in length. The prongs are bent to fit the flask, with sufficient tension to hold it rigidly. The springs allow the flask to be easily removed. More elaborate holders with adjustable grips are available from laboratory supply houses. These are more flexible and will accommodate several sizes of flasks.

Goggles. Special glass blower's goggles should be worn to protect the eyes from the intense sodium glare of the heated glass. These are filtering goggles similar

Fig. 17-6. The spandle is used to hold glass tubing in alignment. (Courtesy: Bethelehem Apparatus Co., Inc.)

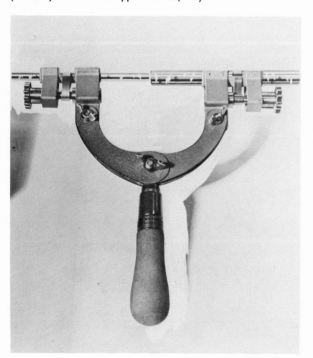

Fig. 17-7. Glass blowing kit. (Courtesy: Bethlehem Apparatus Co., Inc.)

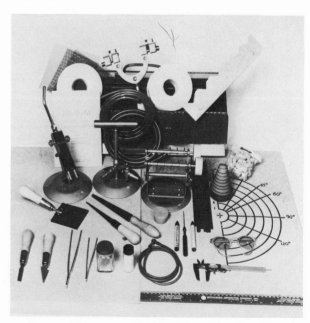

to those used in arc welding.

Asbestos Glove. An asbestos glove is handy to have around especially for the novice.

Spandle. A spandle (Fig. 17-6) reduces the skill requirement of conventional glass working by holding in alignment tubes being sealed.

Glass-Blowing Kit. There is on the market a glass blowing kit (Fig. 17-7) that is intended primarily for those required to perform glass-working operations intermittently. It contains all the essential tools for bench glass working which, with the accompanying instructions in the Manual, enable a relatively unskilled person to make the full range of sample apparatus as well as to construct, modify and repair glass systems for chemical or vacuum use. The kit is made by the Bethlehem Apparatus Co. of Hellertown, Pa.

SUPPLIES

The supplies needed for laboratory glass blowing are very limited and inexpensive. Most of them can be purchased in local drug stores and supply houses.

Glass Tubing. A small supply of Pyrex Brand Glass No. 7740 in sizes from six to ten millimeters will be sufficient for the amateur's need. For those who wish to vary their experience, a little soft glass (lime-soda glass) may be added. The tubes are usually four feet in length and are sold by the pound.

The novice should not be misled to believe that soft glass is easier to work than Pyrex glass. Both require about the same degree of skill to fashion; however, Pyrex glass has a lower coefficient of expansion and greater resistance to thermal shock, which greatly reduces its tendency to crack.

Stoppers. Two kinds of stoppers are used by the glass blower. First, regular corks are needed to close the ends of tubing. Secondly, rubber stoppers with glass tubing inserted through them are needed to insert in tubing so a blow tube can be used. The different stoppers needed will depend on the number of different size tubings worked on.

Asbestos Tape and Paper. For holding and storing heated glass it is wise to have handy a roll of asbestos tape at least one inch wide. It is also nice to have a few sheets of asbestos paper handy.

Rubber Tubing. At least one three foot length of eight millimeter soft rubber tubing should be available at all times.

PROCEDURE

General Instructions

It is essential that the glass blower wear filtering goggles, similar to those used in arc welding, for eye protection. The article to be worked must be cleaned thoroughly in soap and water. All foreign matter, even fingerprints, should be removed. The flame must be well controlled and the glass worked at the lowest possible temperature. It is important that the flame does not have too much force, nor should the glass be overworked, as distortions appear easily and are hard to remove.

PREPARATION OF GLASS

Many difficulties encountered in glass working are the result of using **dirty glass.** Any dust or foreign material may cause the glass to devitrify rapidly when heated. Merely running water through tubing to clean the inside is not satisfactory, as all particles of dust may not be removed.-

The best method is to force snugly fitting **wads of cotton,** wet paper, or cloth through the tubing by means of a **wooden stick. Do not** use a metal or glass rod, as this may scratch the tubing. The plug can also be forced through the tube by a stream of water. The outside

Fig. 17-8.

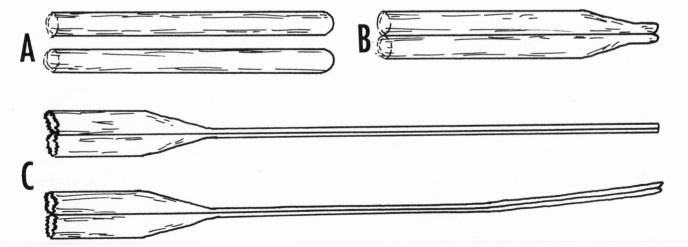

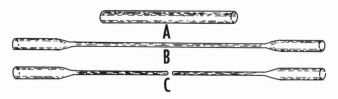

Fig. 17-9.

should also be wiped, then the entire tube **rinsed thoroughly,** and **dried.** When repairing apparatus, it may not be so simple to clean the parts; however, a combination. of **solvents, soap** and **water,** and **cleaning solutions** will usually be satisfactory, providing the ware is thoroughly rinsed and dried.

Caution: Do not heat the glass in the flame until **all traces** and vapors of any solvent **have been removed;** otherwise, there is danger of explosion and fire. A glass blower should also keep his hands clean while working, as finger marks readily burn into glass, and once there, they are not easily removed.

IDENTIFICATION OF GLASSES

In repairing a piece of apparatus, the question often arises as to the type of glass from which it is made. Likewise, pieces of tubing become mixed, and it is then necessary to make identification tests.

Visual inspection is not reliable. The best method is to make a comparison with a known piece of glass. This may be done in several ways: by testing for relative softness (Tests 1 and 2); by comparing their coefficients of expansion (Test 3); and by comparing the index of refraction (Test 4).

Test 1. Hold the glass in the flame, and notice the color and the comparative rate of softening.

Test 2. Heat the two glasses in the flame and seal them together. Heat evenly, and then, by pushing and pulling, one can tell if there is much difference in the softening points. The softer glass will, of course, pull out more easily.

Test 3. A comparison of the coefficient of expansion may be made by heating the ends of the known and the unknown glasses in the form of rods 3 to 5mm in diameter, so that the ends are rounded and softened for a few millimeters (See Fig. 17-8A). In case a piece of glass not in the form of tubing is to be tested, the glass may be worked in the flame so as to form a ball and then drawn into a rod. While hot, place the two pieces side by side in such a manner that the heated ends coincide. Re-heat, and then press together with flat tweezers to assure good contact between the two (See Fig. 17-8B). Hold the two pieces of glass in the left hand, heat the joined ends evenly, and then grasp the heated ends with tweezers, being sure to catch both glasses. Draw out the heated part to a small fibre about 1/4 to 1/2 mm in diameter, let cool while holding taut, and cut in two in the middle. If the drawn-out part remains straight, the glasses have approximately equal coefficients of expansion. However, if it curves, the glass on the side toward which it curved (inside) has the higher coefficient of expansion. (See Fig. 17-8C).

Test 4. A quicker method of identifying Pyrex Brand Glass No. 7740 is based on the refractive index of the Glass. Pyrex Brand Glass No. 7740, having a refractive index of 1.474, will become practically invisible when immersed in liquid with a matching refractive index. However, should there be another glass with the same index of refraction it, too, will disappear when placed in this liquid.

Method: Mix 16 parts of Methyl Alcohol and 84 parts Benzene in a closure-top jar. Into the jar insert the sticks of tubing in question so one end of each stick is

Fig. 17-10.

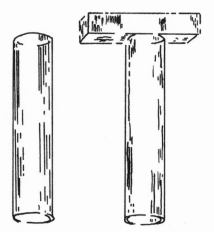

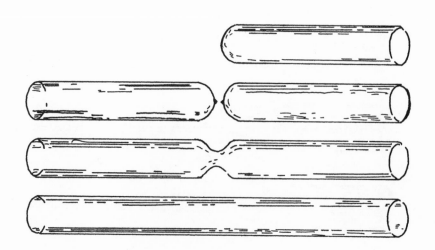

submerged in the liquid. If any stick of tubing is Glass No. 7740 its submerged portion will be practically invisible (a faint, vari-colored outline may be detected if viewed against a light.) The submerged portion of other lengths of tubing, not Glass 7740, remain entirely visible.

This chart shows how a variety of types of Corning Glass react in the solution:

Glass	Type	Principal Use	Refractive Index	Reaction
0080	Soda Lime	Lamp Bulbs	1,512	Visible
7720	Borosilicate	Electrical	1,487	Visible
7052	Borosilicate	Kovar Sealing	1,484	Visible
7740	Borosilicate	Chemical	1,474	Invisible
7900	96% Silica	High Temp.	1,458	Visible

Note 1. When not in use keep the jar tightly covered to prevent evaporation.

Note 2. Make sure tubing is dry when immersing in solution. Water turns the solution cloudy, making it useless for this test.

CUTTING TUBING—GENERAL INSTRUCTIONS

The cutting of tubing is explained in detail on pages 45-53. In general, the method used for cutting is determined by the size of the tubing and its accessibility. For cutting tubing up to approximately 25 millimeters the simplest method is to produce a scratch with a file or knife at the point where the break is to occur, then bend the tubing at the scratch, creating tension on the scratched side.

For cutting large sizes of tubing, the electrically heated wire is very satisfactory (see page 51). Tubing that is already in place can be cut by the hot glass rod method (see page 52).

METHODS OF MARKING TUBING

Burning-On. Etching paste for markings to be burned on at regular annealing temperatures can be made by grinding 100 grams of No. 2330 red enamel* with approximately 30 cc of glycerin in an ink mill. To insure paste smoothness, the mixture should be milled at least three times. The amount of glycerin may vary from time to time, depending upon the temperature and humidity conditions at the time of mixing and application.

Spread a small amount of the paste in a thin layer on a smooth surface such as a glass plate. With a good, cushioned-type rubber stamp apply the paste to the surface of the glass to be marked. Anneal in oven at regular annealing temperatures (See Page 125).

Etching Ink. If a marking material that does not require annealing of the glass is preferred, one of the commercially available etching inks should prove satisfactory. The ink is applied to the surface of the glass

Fig. 17-11.

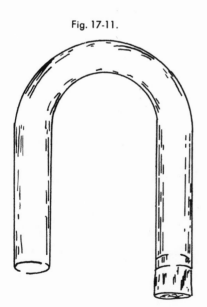

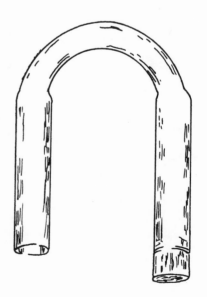

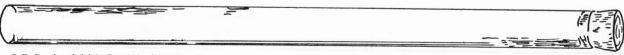

* B.F. Drakenfeld & Co., Washington, Pa.

with a rubber stamp or, with a steel pen. Before applying the ink, it is desirable to slightly warm the surface of the glass in a flame.

Zirconium Rod. Another marking method consists of writing on the glass, freehand, with a length of 1/16" diameter Zirconium Rod, held and revolved in a "Hobbyist" type electric drill.

Recent tests indicate this method produces well defined, white lettering of a high degree of permanency.

PROPERTIES OF SOFT GLASS

There are various types of soft glass but in general lead glass and lime-soda glass are the most common. However, since lime-soda is more commonly used in laboratories, its manipulation is described.

The range of temperature within which it is plastic enough to be worked is narrower than Pyrex Brand Glass No. 7740. Because of its high coefficient of expansion, it breaks more readily than borosilicate glasses, unless it is very carefully annealed (See "Annealing" page 125). Burners of the air-gas type are in general suitable for working soft glass. An ordinary Bunsen burner may be used for bending small tubing, and if bends of long radius are made, a "wing-top" is useful. For larger tubing a Meker burner is satisfactory. When still higher temperatures are required, a blast lamp must be used.

PROPERTIES OF PYREX BRAND GLASS NO. 7740

Laboratory glassware and tubing are made from Pyrex Brand Chemical Glass No. 7740. The general characteristics of this glass which, are of importance to the glass blower are listed below:

softening point	820° C.
strain point	510° C.
annealing point	555° C.
linear coefficient of expansion	32.5×10^{-7}

Burner Adjustments, Pyrex Brand Chemical Glass No. 7740

The glass requires an oxygen-air-gas or oxygen-gas flame. The proper procedure for lighting a burner is to ignite the gas first, then turn on the air, and finally introduce oxygen. The oxygen-gas burner is adjusted by adding oxygen, alone. The temperature of the flame should be varied according to the size of the piece being worked and the ability of the operator. With increasing skill, higher temperatures can be used, and thus the working speed will be increased. Care should be taken not to use an extremely hot flame which might burn or cause excessive vaporization of the glass. Likewise, an inadequate amount of heat may cause devitrification of the glass, but reheating in a hotter flame will usually correct this condition. A Bunsen burner will be found useful for pre-heating larger pieces. Pre-heated ware may be placed directly in the hot burner flame, thus eliminating the necessity of readjusting the burner. It also permits speedier work, as one piece can be warmed while another is being worked. The technician can vary the effective temperature by working in different parts of the flame, as well as by flame adjustment.

Note: It is well to remember that the hottest part of a flame is at the tip of the inner cone.

Fig. 17-12.

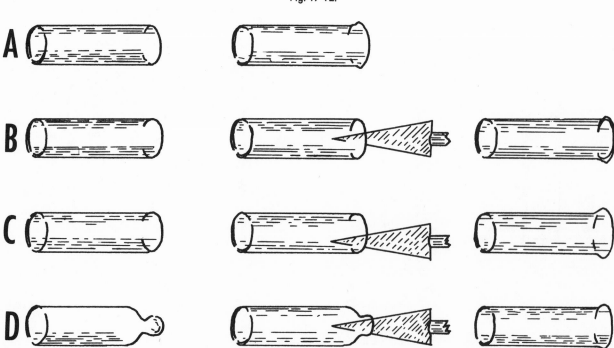

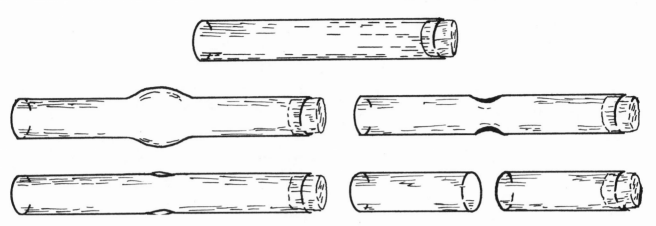

Fig. 17-13.

PROCEDURE FOR WORKING TUBING

Uniform rotation of the glass during the heating and blowing process is essential for obtaining even wall distribution and symmetrical shapes. The tubing is generally **held** by the last three fingers of the **left hand,** which act as a bearing. The thumb and forefinger are used to rotate the glass. The **right hand** supports the **other end** of the tubing.

Here again, the thumb and forefinger are used to rotate the tube while the other fingers are used mainly for support. The palm of the **left hand** is **downward,** while that of the **right hand** is **upward.** These positions permit the glass blower to **blow into** the **right end** of the tubing, which should be the shorter end. Rotation may be in either direction, or back and forth; however, the better method is to **rotate** so that the **top** of the tubing moves **away** from the glass blower. The important factors in rotating and holding tubing are:

1. Synchronize right and left hand movements so as to prevent twisting of the tubing.
2. Hold the tube in a straight line, and do not bend at the heated zone.
3. Do not push or pull the tube unless it is necessary to constrict or change the wall thickness.
4. Do not try to hold a long or heavy piece of tubing in the left hand alone, but use a roller for supporting the end.

Points. "Points" as designated by glass blowers are elongations on the ends of tubing, formed by pulling the tube to a small diameter. They form convenient handles for holding short pieces of tubing, as well as providing a means for closing the tube.

The correct procedure in pulling a "point" consists of **rotating** the tube in the **flame** so as to heat a length of about 1/2". When the glass has become pliable, **remove** it from the flame, while **still rotating, pull slowly** to a length of eight or nine inches (see Fig. 17-9B). It is important that the resulting "points" have the same axis as the original tube. The drawn out portion may be melted apart at the center, thus closing both "points" (see Fig. 17-9C). If the "points" are not in line, it will be necessary to heat at the junction with the tube to straighten them.

In some cases it may be necessary to handle or pull off a piece of large tubing which is too short to hold conveniently with one's fingers. The glass blower will find **tweezers** very useful for this, although uniform rotation is difficult to accomplish. By fusing a small diameter tube or rod to the large tubing, and bending it properly, one may center it so as to form an effective handle similar to a "point".

ROUND AND FLAT BOTTOM TUBES

To form a round bottom tube, proceed in the same

Fig. 17-14.

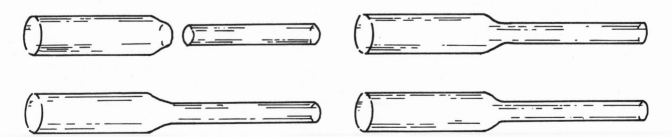

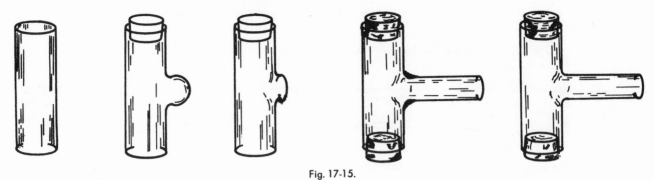

Fig. 17-15.

manner as for pulling a "point", except that the tube is **pulled apart** in the flame as it is heated, causing the hot glass to be constricted. The **closed bottom** of the tube is **heated uniformly** so as to allow even distribution at the tip where the glass was pulled apart. If a heavy drop or globule remains in the bottom it may be removed by touching a piece of glass rod or tube to the softened bottom, and pulling off the excess glass. Tweezers may be used for the same purpose. **Re-heat** the closed part until the glass **distribution** seems **uniform,** then remove from the flame, and **blow** the closed end into **spherical shape.** Several small puffs, together with rotation, will be found satisfactory.

Flat bottom tubes are formed from round bottom tubes. Heat the round bottom tube, press and blow the **bottom against a carbon block,** or rotate while holding the carbon against the bottom (see Fig. 17-10).

BENDS, U-L

The diameter of the tubing, and the distance between the tubes after the bend, will determine the length of tubing which must be heated.

Close one end of the tube with a cork, and hold the tube so that the **stoppered end** is in the left hand. **Rotate** in the flame, heating a sufficient length necessary for the bend. When the tube has become **pliable, remove** from the flame, and quickly **bend the ends upward,** allowing the heated portion to move downward. As soon as the bend is completed, and the tubes are parallel, **blow** into the **open end** with sufficient pressure to eliminate any irregularities in the bend, and also to expand it to the full size of the original tube (see Fig. 17-11).

Fig. 17-16.

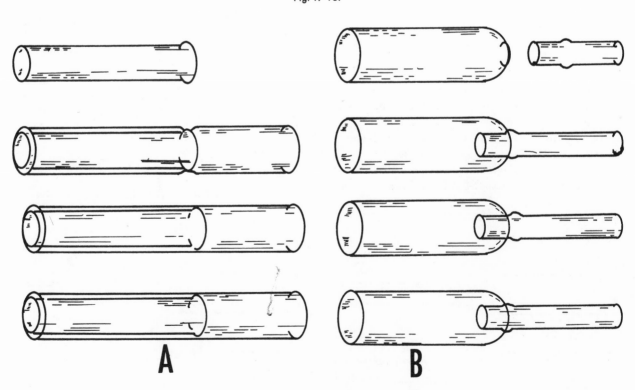

A   B

Don't blow into the tube **before** the bend is completed, as this may cause it to bulge. Do not push together or stretch the tube while it is heating; however, on larger sizes it may be desirable to pull slightly on removing from the flame. This tends to remove any irregularities.

Right angle or "L" bends are formed in similar manner. Bends should not be made in a horizontal plane, except where a template is required to form a definite arc.

END FINISH AND FLARING

A common end finish is obtained by simply heating the end of a tube in the flame until it is soft enough for the glass to smooth out by surface tension (see Fig. 17-12A). **Beaded** edges are formed by heating the end of the tube for a longer period of time, allowing the glass to thicken, that is, run together. The inside diameter at the opening is reamed out with a brass reaming tool (see Fig. 17-12B).

A **flare** is made by heating the end of the tube for a short distance until the edge begins to sag. While still rotating the tube, **insert** the glass blower's arrow-head **reamer,** so that the edge comes into contact with the glass. The edge of the tube will be forced outward and expanded, forming the flare. The uniformity of heating and rotation determines the uniformity of the edge of the flare (see Fig. 17-12C)

All of the above finishes are used to smooth and strengthen the end, the beaded edge being the strongest. The flared end makes an excellent pouring lip, but its main use is in making seals which are described later.

In some cases, a "point" is used in sealing a tube to a piece of apparatus. The seal having been completed, it becomes necessary to open the tube. This is accomplished by uniformly **heating** the **shoulder** between the "point" and the tube. When it is sufficiently heated, **pull off** the "point." **Re-heat** a small central area, and **blow** a small light **wall bulb** (see Fig. 17-12D). **Break** this with a reaming tool, and chip off the thin glass. **Heat** the **opening** in the flame until it is well softened, and then **flare** with the tool thus expanding the opening to equal the inside diameter of the tube.

STRAIGHT SEAL

The straight seal, or the joining together of tubes, is the most common type used in glass blowing, and mastery of it is well worth the effort to achieve it.

Stopper or **close** the end which is to be held in the **left hand.** The tube in the **right hand** must be **open** for **blowing.** Heat by rotating in the flame until the **ends** of the tubes, which are to be sealed together, are **softened.**

Fig. 17-17.

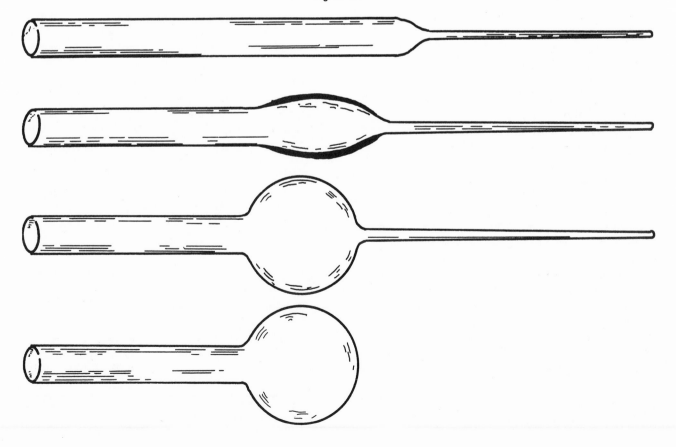

The two tubes are then **pushed together,** and as soon as the glasses touch **pull** the joint **slightly.** (All of this is done without removing from the flame.) The tubes should be brought together carefully, and on the same axis, affecting contact at all points.

Continue **rotating** in the flame until the **diameter** of the tubing is **decreased,** and the **wall thickness** is **increased** at the point of juncture. Continue rotating as the tube is removed from the flame. **Blow into** the tube until the heated portion is **expanded** to somewhat **greater size** than the original tubes. Do not pull the tube when blowing, as this will decrease the wall thickness at the seal. Re-heat the enlarged portion with a softer flame until its diameter is decreased; remove from the flame as before, again blowing to expand the tube somewhat larger than the original diameter. Before the tube has cooled appreciably, pull sufficiently to reduce the diameter of the seal to that of the tube. Figure 17-13 shows the various steps in the procedure.

The following points will serve as a guide for regulating diameters and wall thickness:

1. Pulling decreases the diameter and the wall thickness.
2. Blowing increases the diameter and decreases the wall thickness.
3. Heating decreases the diameter and increases the wall thickness.
4. Pushing an enlarged tube increases the diameter and the wall thickness.
5. Pushing a constricted tube decreases the diameter and increases the wall thickness.

The **straight seal** is the one most commonly encountered in setting up apparatus with a hand torch. In this case, the advantage of rotation is lost as one side of the seal is usually supported or held stationary, while only the other is movable.

Hold the movable tube in one hand, and **manipulate** the **torch** with the other. **Uniform heating** is necessary, and is obtained by playing the flame from the torch about the ends of the tubes. As soon as the ends are **softened, bring** them **together** and work in the same manner as the lamp.

Before making a seal of this nature, be sure that one side of the system to be joined is **closed,** and that provision is made for **blowing into** the **other.** A length of rubber tubing attached to the open side often provides a convenient means for blowing.

The hand torch method is often very useful for sealing off tubes or systems which are **evacuated.** Such a tube which is to be sealed should be **constricted** before evacuation. **Heat evenly** and, as the glass softens, the **air pressure** will force the tube together. **Pull off** the **excess** glass with tweezers, or a glass rod, using care not to overheat the evacuated portion.

A **straight seal** between tubes of different diameters is made by first **drawing** the **large** tube down to the **diameter** of the **smaller** one. The joint then is essentially the same as in the case of tubing of the same diameters, except that rotation is more difficult. In heating, most of the **flame** is **played** on the **larger** tube since more heat is required to soften it to the same extent as the small one. The application for such a seal will determine whether it should be made with a gradual taper from the large diameter tube to the smaller one, or an abrupt rounded joint. The **tapered** joint is made by **pulling** the heated **shoulder,** whereas the **rounded** type is made by **blowing alone** (see Fig. 17-14).

"T" SEALS

A "T" seal is essentially a straight seal, differing only in that the piece is not easily rotated except by use of a holder. To make a "T" seal, **close one end** of a tube with a cork. With a sharp flame, heat a **spot** on **one side,** in the middle of the tube. **Blow** on the tube so as to form a small **bulge** at the heated spot. Re-heat this bulge, and then **blow** a small **bulb** having light walls. **Break** this, and chip it off, leaving an **opening** in the side of the tube. This opening should be slightly **smaller** than the tube which is to be attached. The size of the opening is controlled in the main by the size of the area heated before blowing. If the hole is **small,** it may be **heated** and **reamed** out larger with a tool. With another cork, close the other end of the tube, or the tube which is to be sealed on, depending upon which is to be used for blowing.

Heat uniformly the **edges** of the opening, and the **end** of the **side tube** until they are quite soft. **Remove** from the flame, and **bring together** quickly, **pulling** slightly as soon as complete contact has been made. **Blow slightly** to expand and remove any irregularities. If the glass was sufficiently softened when joined, this procedure will result in a good seal; however, should it not appear uniform, small spots can be heated with a sharp flame, and then be blown to proper size. **Reheat** the **entire seal** to remove stresses and to adjust the angles between the tubes (see Fig. 17-15).

RING SEALS

Ring or inner seals can be made in various ways, as described below.

A common and useful ring seal formed by a tube within a somewhat larger one is illustrated in figure 17-16A. The **inner** tube, which should be a few millimeters smaller than the outer, is **flared** as described before, so that the diameter of the flare is of sufficient size to allow insertion into the larger tube. The **flare** should also be **uniform,** so as to fit closely along its entire circumference. **Support** the **inner** tube at the correct position with some material such as paper or asbestos, or by means of another tube held in place by a stopper. **Rotate** the **outer tube** in a rather sharp, intense flame, **heating** the area directly in **contact** with the flare. This will shrink the outer tube, and contact will be made with the flare. **Continue heating** until the **two** pieces are completely **fused** together. The **outer** tube will have been decreased in diameter about the seal, and it should be **re-heated** and **blown** to its original diameter (see Fig. 17-16B). The general heating also tends to relieve strain. A joint of this type, which is larger than 25 mm, should be cooled slowly in the flame by gradually working in the cooler part, and also by reducing the flame temperature.

Another type of inner seal is that in which a tube is sealed through the round bottom of an item such as a test tube.

The procedure for fabricating is as follows: **Flange** the tube which is to be sealed **inside.** Support it by means of a cork and another piece of tubing so that the **flange rests** lightly against the center of the **bottom.** The tube which serves as a holder should be such that air will pass freely into the larger tube. **Heat** the **bottom** in the flame sufficiently to **fuse** the **flare** to it. Remove from the flame and blow the bottom to its original round shape. If the two tubes are not completely joined, repeat the process.

The **next step** is to heat the bottom at the center of the sealed-in tube sufficiently so that a small, thin **bulb** may be blown. **Crack off** this bulb, stopper the open holding tube, and **seal** a **tube** onto the **outside** as in a straight seal. The blowing is done through this added tube. **Re-heat** the entire bottom, and **blow** to **rounded shape,** being careful to keep all tubes concentric. This type of inner seal may be used to seal a tube through the side of another tube. In this case, the same procedure is used in the final seal as in making a "T" seal.

The **third type** of seal is modified so that a very **short length** of tubing may be sealed on the inside, or in cases when it is not convenient to place or hold a flared tube on the inside. In making a seal of this kind, the **large** tube is heated with a **sharp flame** at the **point** where the seal is to be made. Blow a small, thin **bulb, crack it off,** and form an **opening** with a beaded edge slightly **larger** than the tube which is to be sealed into it. The tube to be sealed-in is heated uniformly with a sharp flame at the **point** where it is to **join** the **larger tube.** When sufficiently soft, it is **pushed together** slightly, causing a **ridge** to form. The diameter of the ridge should be two to three millimeters larger than the tube.

Now, hold the large tube in the left hand, and with the right, hold the tube with the ridge so that the **short portion** extends **into** the other tube, the **ridge** being against the edge of the **opening.** Heat the opening and the ridge with a sharp flame as in a straight seal, and as it softens, push together. When the two are fused, blow out round to the original shape, pulling slightly on the sealed-in tube, but with care so as not to constrict the tube. While the glass is still soft, adjust so that the tubes are centered (see Fig. 17-18B).

SEALING CAPILLARY TUBING

Capillary tubing is sealed in the same manner as ordinary tubing except greater care is needed to align the bores. If a capillary tube is to be sealed to an ordinary tube, the procedure is modified. The capillary tube is closed by pulling off an end. Re-heat and blow a light wall bulb on the end. Break off the thin glass. The expanded end of the capillary is now similar to ordinary tubing and can be sealed as in a straight seal.

Fig. 17-18.

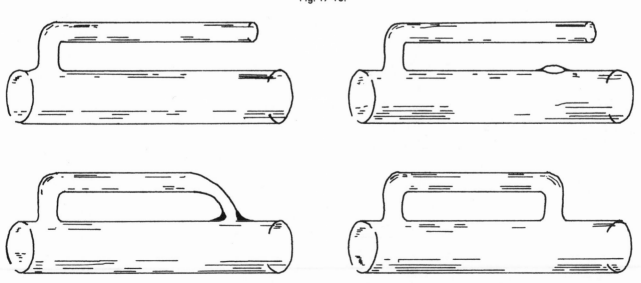

JOINING CANE

Cane is joined by heating in the same manner as tubing. Since it is not possible to blow into cane, the diameter and shape are controlled by pushing or pulling the heated area. Carbon rods, plates, or brass tools may be used to shape the heated cane.

BULBS

In the building of special apparatus, bulbs are often needed. The various shape and sizes of flasks listed in the Corning Glass Works Laboratory Catalogs will be found ideal for these applications. Small bulbs of less than 50 mm diameter can be made from tubing in the following manner.

A "point" should be pulled on one end of a length of tubing of approximately **half the diameter** of the finished bulb. The length of the tube depends upon the wall thickness required in the bulb. **Heat** the length of tubing which is to form the bulb, gradually **shrinking** it to about **one-half** its original length. By occasional puffs, the **full diameter** is maintained throughout the process. **Heat** the thickened tube until quite **soft; remove** from the flame, and **rotate** with the **left hand;** at the same time

blow into the "**point**" until a spherical bulb of the desired diameter is obtained. Use the right hand to guide the point so as to keep it in line with the left-hand tube. **Uniform rotation is essential** in order to effect uniform heating and cooling, and to prevent the hot glass from flowing unevenly, due to gravity. The wall thickness will depend upon the size of the bulb, and the amount of tubing actually worked into it.

With a sharp flame heat the "point" at the juncture with the bulb until very pliable. Pull it off and remove any excess glass from the bulb at this area. Heat the bulb and blow to the round shape as in forming a round bottom tube (see Fig. 17-17).

In blowing a small bulb, the "point" is not necessary, as even distribution in the wall can be obtained by uniform heating and rotation.

CLOSED CIRCUITS OF TUBING

In some cases it is necessary to **seal both ends** of one tube to another tube. Make the first seal in the usual manner. Then, with a sharp flame, blow open the tube at the point where the other end is to be sealed. Cut the tube, allowing sufficient length for a bend. Heat the opening and the end of the tubing until quite soft, at the same time playing the flame on the tube where it

Fig. 17-19.

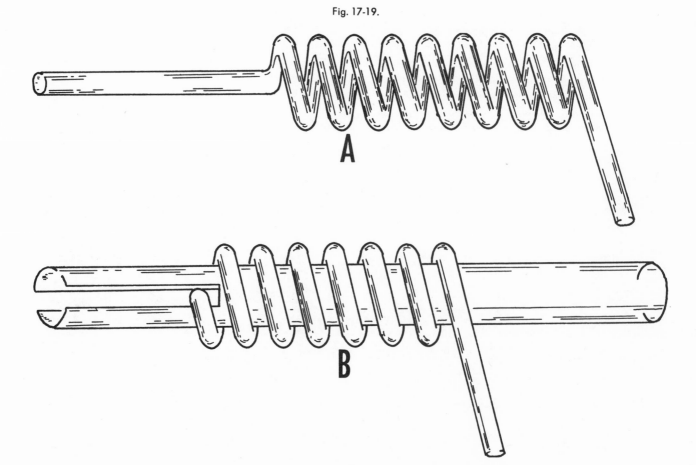

is to be bent. With a pair of tweezers, bend the tube and force the heated ends together. Re-heat with a pointed flame and blow to uniform size and shape (see Fig. 17-18).

COILS FROM TUBING

Coils may be made free-hand (see Fig. 17-19A). A short length of tubing is bent at right angles to the main length of tubing to serve as a handle. This handle is to be on the center line of the coil. Bend the first turn of the coil very carefully, as this will serve as a guide for the succeeding turns. This method is most applicable to the winding of large diameter spirals.

Smaller coils are best wound on a mandrel (see Fig. 17-19B). The mandrel may be made from an iron or brass tube or rod. A slot cut in one end serves as a convenient means for holding the glass tube. The mandrel should be warmed before winding the tubing so as not to chill the hot glass which might cause it to crack. If the mandrel tapers slightly toward the slotted end, the completed coil can be removed more easily. Hold the burner in the right hand, and play a soft, bushy flame on the tubing near the mandrel. Wind the tube as it softens, spacing it evenly. Since it is necessary to

guide the tubing, it is usually advisable to have another person turn the mandrel or guide the tubing.

In case a metal mandrel is not available, a glass tube which has been wrapped with asbestos will serve. When the coil has cooled, the entire piece is soaked in water to loosen the asbestos. As soon as the paper is removed, the coil can be slipped off the mandrel.

ANNEALING

Glass which is strained may or may not be stronger mechanically or thermally than well annealed glass, depending on how the strain is set in the article. By the use of very sensitive polariscopes, strain caused merely by supporting its own weight can be detected in a well-annealed specimen. Pyrex Brand Ware as supplied has been annealed in lehrs specifically designed to eliminate harmful strains. Strain-free, as discussed here, refers to a glass in which the excessive or harmful strain has been removed.

If a piece of Pyrex Brand Ware is heated above the strain point, harmful strains may be introduced, unless the entire piece is **re-heated** and then **cooled slowly** and uniformly. It is, therefore, desirable to anneal all ware which the glass blower makes and a discussion of two methods is given.

Fig. 17-20. Heavy duty model glass lathe. *(Courtesy: Litton Engineering Lab.)*

Fig. 17-21. Laboratory model glass lathe. *(Courtesy: Bethlehem Apparatus Co., Inc.)*

FURNACE ANNEALING

Annealing as applied to glass manufacture is a process in which a glass is heated to a temperature at which it is sufficiently softened to relieve itself of strain. In order not to introduce strains when cooling, the entire piece should cool as uniformly throughout as possible. Annealing depends upon two factors: time and temperature. As the temperature is increased, the time required to relieve strain is decreased.

The annealing temperature recommended below for Pyrex Brand Glass No. 7740 is one which will relieve strain in a comparatively short time, and which is not sufficiently high to cause the glass to soften enough to change shape under its own weight.

Transfer the article to be annealed to a furnace so that it may be heated gradually to 580-585° C., and hold at this temperature for a period of five to seven minutes. The maximum cooling rates should be as follows:

wall thickness	cooling rate to 503 C.
mm	not to exceed
1	224° C/min
2	56° C/min
3	25° C/min
4	14° C/min
5	9° C/min

Slower rates are satisfactory, and will usually be used, since most furnaces do not approach the maximum rates of cooling shown above. As soon as the ware can be handled with the hands, it will be safe to remove it from the furnace.

FLAME ANNEALING

Annealing in an oven or furnace is much more satis-factory than flame annealing, since the entire piece is heated and cooled more uniformly, and temperature control is more accurate. However, many glass blowers must depend on flame annealing. This consists of heating the worked portion in a **soft, bushy flame** until it is uniformly softened. Gradually **decrease** the working **temperature** by manipulating the **glass** in the **cooler** parts of the flame, and also by **lowering** the **flame temperature.** Continue until a **layer of soot** has been deposited from the smoky flame which is finally used. Lay aside for **cooling** on a roller or stand so that the hot glass does not come into contact with a cold table top.

This process requires good technique and judgement on the part of the worker, as it is not only necessary to heat uniformly, but undue softening must be avoided; otherwise, the apparatus may sag under its own weight.

ETCHING PYREX BRAND GLASSES

Pyrex Brand Glasses are more difficult to etch than soft glass. The resulting mark is also usually clear, and not frosted. It is difficult to obtain narrow lines, as the etching acid tends to undercut and widen them. Beeswax or a mixture of beeswax and ceresin wax is more satisfactory for the protecting agent than paraffin.

Coat the article to be etched by dipping into or brushing with **melted wax,** for protection when immersed in the acid. The **lines** to be etched are **cut** into the wax with a sharp instrument such as a **needle.** Dip the article into 60° or 70° **hydrofluoric** acid for a period of two minutes. As the acid weakens, this time must be increased. (The usual laboratory reagent has a lower H_2F_2 concentration, and if used, a longer etching period will be required.)

In many cases, a single lengthwise line is to be etched on a long tube or article which cannot be dipped conveniently. In such cases, wax an area extending about an inch on either side of the mark. **Cut through** the wax as above. **Wrap** a narrow band of **cheesecloth** around the tube directly **over** the **cut,** and **saturate** this with the acid. As soon as the necessary time has elapsed, **wash with water,** remove the cloth, and all traces of acid.

Note: Hydrofluoric acid burns are very dangerous, and the glass blower should take precautions not to breathe fumes, or permit the acid to touch any part of his body.

Wax may be removed by warming it to its melting point and then wiping with a cloth. Solvents may also be used.

GRINDING GLASS

In many instances, it is necessary that ends of tubes be ground square, or at angles, or to specific lengths.

A smooth **iron or glass plate** may be used for this purpose. **Abrasive powder** is placed on the plate, and water is added. The glass to be ground is then **rubbed** on the plate with sufficient pressure to obtain **rapid grinding.**

Various sizes of grinding materials are used, depending upon the finish desired. The common sizes of powders used are 30, 60, 120, and 600 mesh. For faster grinding, a rotating iron disc, together with grinding powder and water, may be used. The glass blower should wash all particles of grinding material and dust from the tubing before attempting to work it in a flame.

GLASS TO GLASS SEALS (GRADED SEALS)

Laboratory equipment is often of such design that it is necessary to join Pyrex Brand Glasses to soft glass. This is done by using a series of glasses with increasing coefficients of expansion and lower softening points. It is not an easy matter to make such a graded seal, and it is possible that an apparently perfect seal may develop checks soon after being made.

Corning Glass Works has developed a series of intermediate glasses for this purpose. This tubing is available to laboratory technicians who care to make their own seals. However, most laboratories will find the completed seals much more satisfactory and reliable.

GLASSWORKING LATHES

The glassworking lathe (Fig. 17-20) is a machine on which can be performed all the different operations that are used in laboratory glass blowing. The lathe with its many interchangeable chucks and burners can hold and fire almost any shaped glass in any desired manner.

Both the novice and the experienced glass blower alike can benefit from the possession of a glass lathe. The amateur need not have previous glass working experience before he operates the lathe. He can develop in a short time a glassworking proficiency on the glass lathe that he could not match on the bench except through long practice. The experienced glass blower can use the lathe to advantage for making multiple unions in one operation that would take several manipulations on the bench. For mass-produced items the glass lathe can repeat an operation without fatigue or variation. Schools and laboratories which do not have an experienced glass blower on their staff would find the glass lathe a welcome addition, for the teacher or a lab technician could master its operation with little practice.

The larger and more comprehensive glass lathes are made by the Litton Engineering Laboratories of Grass Valley, California (Fig. 17-20), who were the pioneers in the field. Recently a smaller, less expensive model was placed on the market by the Bethlehem Apparatus Company, Inc. of Hellertown, Pa. (Fig. 17-21). Both companies sell many different chucks, burners, and tools that can be used on the lathes. All the companies that sell glass lathes furnish very complete instructions for their operation.

Fig. 17-22. Artist Hans Godo Fräbel is shown next to his "Profile," selected by the Corning Museum of Glass in Corning, New York, for inclusion in its contemporary collection. Fräbel's work is found in museums and private collections all over the world, and even on the President's desk in the Oval Office.

Fig. 17-23. These striking glass chandeliers were designed by John Portman & Associates. They are 9 feet square and are located in the Peachtree Center Plaza Hotel Ballroom, Atlanta (Courtesy: John Portman)

Chapter 18
Offhand glass blowing

Offhand glass blowing is an art that has been practiced for nearly two thousand years. Today we continue the art, using methods and tools not very different from those of medieval times. Commercial items are produced by this hand method in a few different companies in the United States. Although it accounts for only a very small part of the total glass production, it includes some of the finest artistic creations to be found anywhere (Fig. 18-1).

Offhand glass-blown items are produced by very few craftsmen and artists other than those employed by glass concerns. However, a strong effort is being made at the present time to correct this situation. More and more home craftsmen and artists as well as schools and universities are investigating this wonderful art form. Workshops devoted to offhand glass blowing are being conducted in colleges and museums rather frequently. More artists who make offhand blown objects are being represented at the Museum of Contemporary Crafts in New York City (Fig. 18-2). There is a strong feeling among artists, craftsmen, and educators alike that the challenge posed by this magnificent art form has been ignored for too long. If the present efforts at mastery are continued and increased, this situation should be corrected.

The tools and materials used for offhand glass blowing are very few and inexpensive—well within the means of an individual or a school. The space needed is also small. What is required is a great deal of patience, practice, and know-how.

Following is an explanation of the offhand glass blowing method used in industry. It should be noted in this explanation that a number of people work together to produce the final glass item. The individual crafts-

Fig. 18-1. CRYSTAL ROOSTER. Ornamental rooster fashioned of solid crystal. *(Courtesy: Steuben Glass)*

Fig. 18-2. Blown glass objects made by artist, John Burton. *(Courtesy: Museum of Contemporary Crafts)*

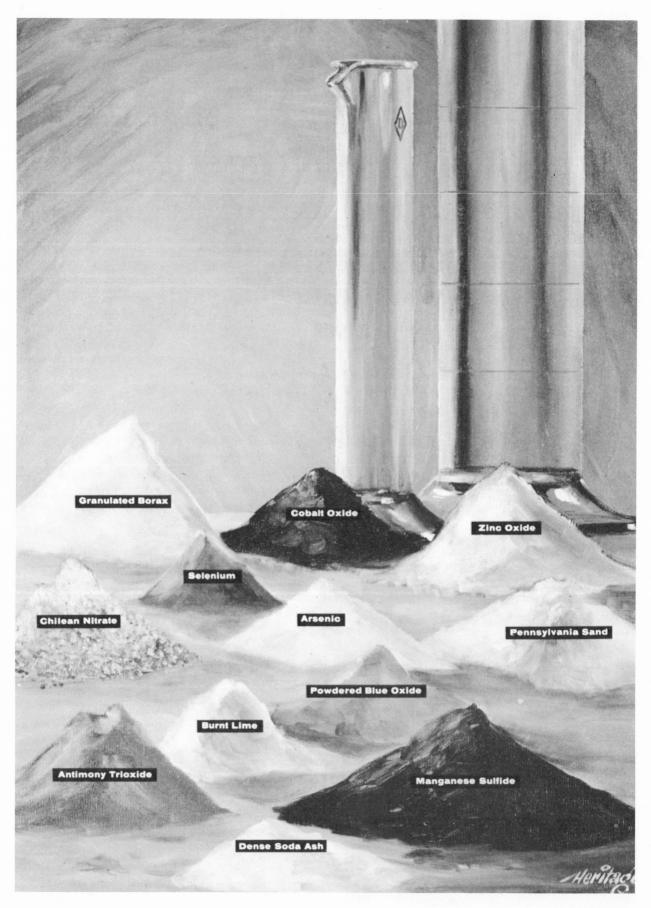

Fig. 18-3. (Courtesy: Doerr Glass Co.)

Fig. 18-4. *The raw materials in the batching bin being dumped into the batcher. (Courtesy: Doerr Glass Co.)*

Fig. 18-5. *Cullet being added to the batch. (Courtesy: Doerr Glass Co.)*

person or artist working alone must go through exactly the same procedures described and do every one of the procedures himself. There are no shortcuts. It is recommended that the individual wishing to do offhand glass blowing should seek the assistance of at least one other person.

Glass Used: The glass used in the offhand blowing process is made of materials of exceptional purity (sand, potash, lead oxide, and powdered glass) mixed in a special formula which uses no chemical cecolorizers that might dull the glass. Quantity blown glass objects, such as laboratory glass ware are made of lime glass ingredients such as shown in Figure 18-3.

Blending: The raw materials are placed in a **batching bin**. From here they are dumped into the **batcher**. In Figure 18-4 the workman is about to release

the raw materials into the batcher to be whirled about until they are thoroughly blended. After the raw materials are blended they are dumped into the glass furnace. Care is taken to be sure that no dirt or other foreign matter falls into the blended mixture.

Adding Cullet: Small pieces of glass left over from previous batches are called **cullet**. A certain amount of this cullet is necessary in new batches. The raw materials and cullet are added to the glass furnaces which are frequently oilfired and operate at a temperature of approximately 2400°F to 2700°F (Fig. 18-5).

The workmen who create the glass blown objects are organized into small **shops,** each compromising a master glass maker and his assistants (two to seven men). A shop works together as a team around its own reheating furnace or **"glory hole"** (Fig. 18-6).

Gathering: In glass-making parlance, the term **gathering** means collecting a mass of molten glass on the end of a **blowpipe**. The **gatherer** is one of the most

Fig. 18-6. THE MAKING OF STEUBEN GLASS AT CORNING GLASS CENTER, CORNING, NEW YORK. From observation galleries stretching the length of the factory, visitors have a close view of the actual working of the material, including the blowing, finishing, and engraving operations. *(Courtesy: Steuben Glass)*

Fig. 18-7. *The several ingredients of crystal glass—sand, potash, lead oxide, and powdered glass—are first melted in a special clay furnace. From this furnace, the "gatherer" takes the required amount of molton glass on the end of his blowing iron, and starts to blow the form. (Courtesy: Steuben Glass)*

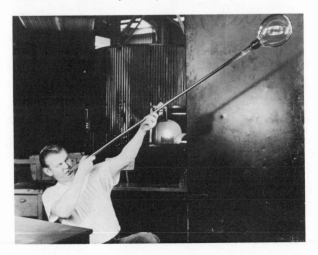

Fig. 18-8. The "servitor," then begins to shape the glass. He may elongate the mass by swinging the blowing iron—or flatten the mass by spinning—or shape it with a wooden tool while revolving the iron on the arms of his bench. *(Courtesy: Steuben Glass)*

Fig. 18-9. Assisted by the "stick-up boy", the servitor transfers the partially-formed piece from the blowing iron onto a long, solid "pontil" rod. (Courtesy: Steuben Glass)

important men on the glassmaking team. He must gather exactly the right amount of glass on the end of the blowpipe for the item being made (Fig. 18-7).

Systems Used: In the "German" system, the gatherer picks up only a small amount of molten glass on the blowpipe. The **ball boy** then proceeds to blow this small amount of glass and shape it. When it has cooled, additional glass is picked up . . . enough for the complete item . . . and this is shaped in a hollowed-out block of cherry wood to point and shape the glass so it can be blown and distributed effectively. In the "American" system the gatherer picks up on the blow pipe just the right amount of glass for the item being made. It is then taken to the **marveling table**, a smooth, cast iron surface, on which the glass is rolled into shape. This is called marveling. Generally, the German system is used in blowing large items and the American system in blowing small ones.

Glassblowing is one of the most dramatic and colorful crafts in the world. The glass on the end of the blow-pipe is cooling rapidly. The glassblower must work fast. He whirls the iron in deft circles. When it is the approximate shape desired he passes it on to the **servitor.**

Fig. 18-10. The "gaffer," who is the master craftsman, now takes over. He joins and shapes the component parts—using shears to cut off excess glass, calipers to check dimensions, and simple wooden tools to achieve the completed form. (Courtesy: Steuben Glass)

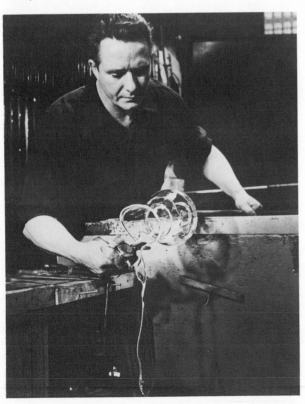

Fig. 18-11. Many designs incorporate ornamental forms contrasting with plain surfaces. The "bit-gatherer" brings extra gathers of glass from the furnace to form these decorative elements. Their application demands the gaffer's skilled and experienced craftsmanship. (Courtesy: Steuben Glass)

Fig. 18-12. In the final opening out and shaping of the glass, the gaffer manipulates the still pliable form to the designer's specifications. *(Courtesy: Steuben Glass)*

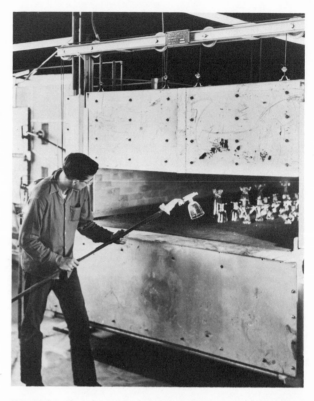

Fig. 18-14. The glass must be cooled slowly to prevent internal strains. The carry-in boy places the still-hot objects in an annealing oven, where they are carried through gradually decreasing temperatures. This cooling takes from several hours to several days, depending upon the size of the piece and the thickness of the glass. *(Courtesy: Steuben Glass)*

Fig. 18-13. After the blowing has been completed, the "carry-in boy" grasps the finished piece in an asbestos-covered prong, while the gaffer cracks the glass off the pontil rod with a sharp top. The pontil rod, on which all handmade glass is held while it is being formed, leaves a rough mark on the base of the finished piece. Customarily this mark is ground and polished smooth. On some pieces it remains discernible as a shallow depression in the crystal. *(Courtesy: Steuben Glass)*

The Servitor: The servitor then begins to shape the glass. He may elongate the mass by swinging the blowing iron . . . or flatten the mass by spinning . . . or shape it with a wooden tool while revolving the iron on the arms of his bench (Fig. 18-8). He may re-heat the glass in the glory hole from time to time to keep it soft. Assisted by the **"stick-up boy"**, the servitor transfers the partially-formed piece from the blowing iron onto a long solid pontil or "punty" rod (Fig. 18-9).

The Gaffer: The **"gaffer"**, who is the master craftsman, now takes over. He joins and shapes the component parts . . . using shears to cut off excess glass, calipers to check dimensions, and simple wooden tools to achieve the completed form (Fig. 18-10).

The Bit-Gatherer: Many blown designs incorporate ornamental forms contrasting with plain surfaces. The **"bit-gatherer"** brings extra gathers of glass from the furnace to form these decorative elements. Their application demands the gaffer's skilled and experienced craftsmanship (Fig. 18-11).

In the final opening-out and shaping of the glass, the gaffer manipulates the still pliable form to the designer's specifications. He must reheat the piece frequently to maintain a proper working temperature (Fig. 18-12).

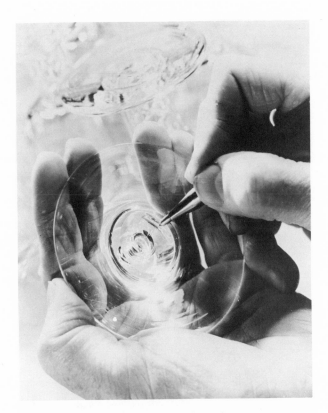

Fig. 18-15. Each completed piece is subjected to an uncompromising inspection. Any piece which fails to pass this inspection is immediately destroyed. Glass meeting the required standards is polished and then signed with a diamond point. (Courtesy: Steuben Glass)

Fig. 18-16. Forming the final shape of the glass in a mold. (Courtesy: Doerr Glass Co.)

Fig. 18-17. An early glass blowing shop. *(Courtesy: Steuben Glass)*

After the blowing has been completed, the **"carry-in boy"** grasps the finished piece in an asbestos-covered prong, while the gaffer cracks the glass off the pontil rod with a sharp tap (Fig. 18-13).

Cooling: the glass must be cooled slowly to prevent internal strains. The carry-in boy places the still hot objects in an annealing oven (lehr) where they are carried through gradually decreasing temperatures. This cooling takes from several hours to several days, depending upon the size of the pieces and the thickness of the glass (Fig. 18-14).

The pontil rod, on which all handmade glass is held while it is being formed, leaves a rough mark on the base of the finished piece. Customarily this mark is ground and polished smooth (Fig. 18-15). On some pieces it remains discernible as a shallow depression in the crystal.

Each completed piece is subjected to a careful inspec-

tion. Any piece which fails to pass this inspection is immediately destroyed. Most manufacturers sign their names on the completed piece with a diamond point instrument (Fig. 18-16).

Quantity Production Variation

When a blown object is to be reproduced in a limited quantity . . . such as laboratory glassware, fine drinking glasses etc. . . . . a variation of the above process is used. The variation is this: Instead of the glass blower passing the molten, partly formed object on to the servitor for shaping, he blows the glass until it is approximately the shape of the finished object. While the glass is still **"running"** the glass blower finally lowers it into a mold. At the same time he "pumps" it up and down and spins it constantly in one direction to prevent sticking. Finally, the **"pit boy"** opens the mold and the cylinder is removed (Fig. 18-17).

Chapter 19
Joining glass to glass and other materials

It is often necessary to join glass to other pieces of glass or to other materials such as metal, plastic, wood, etc. This is frequently done on school and home craft projects as well as commercial items of all kinds.

Glass joints may be classified into two general groups —**non-fired** and **fired** (requiring heat). There are a few different bonding materials that can be included in each group. The best to use will depend on the degree of permanence desired, the equipment available, the cost, whether the joint is to be parted and rejoined, time available for setting, color of joining materials, etc.

NON-FIRED GLASS JOINTS

Non-firing glass-joining materials require no heat in their application. In order to perform efficiently, they must be applied to surfaces that are dry, clean, and free of oil, greases, and waxes. They should be applied as directed on the container by the manufacturer and given sufficient time to be "cured" (hardened). When mixing is required to prepare the material, it should be done in a thoroughly clean, moisture-free container.

Principal Materials and How to Use Them

Dow Corning Glass & Ceramic Adhesive (Fig. 19-1) is excellent for bonding glass to almost any kind of non-porous or porous material. It is translucent and colorless when applied in a thin layer (so it can not be seen when in place), easy to use, non-shrinking, non-flammable, and will withstand temperatures up to 500° F. Unlike most of the other non-fired glass adhesives, this one is not affected when placed in the automatic dishwasher. Being heat resistant, it can be used on ashtrays, lamps, and similar objects; being water proof, it can be used on outdoor glass signs, bird baths, etc. Glass and Ceramic Adhesive contains no solvent and therefore will not damage painted or varnished surfaces.

Glass and Ceramic Adhesive requires no mixing. To use just apply a thin coating to both edges to be joined. Fit them together and support if necessary. The adhesive is dry to the touch in an hour and "sets" in four hours. Bonding action continues until optimum strength is reached, about 8 hours after application.

Duco Cement, which is an ethyl acetate product made by DuPont, is completely clear, transparent and fast setting. It works well on both porous and non-porous surfaces (except rubber), and is easy and convenient to use.

When Duco Cement is used to join glass to non-porous

Fig. 19-1. *(Courtesy: Dow Corning Corp.)*

Fig. 19-2. Laminated glass panel by the glass artist Maurice Heaton.

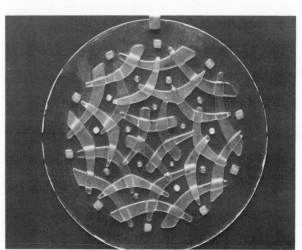

materials, such as another piece of glass, metals, etc. the cement is applied in a thin coat to each surface, held or clamped until set, then put aside to dry for at least 12 hours. When glass is being joined to a porous surface, a thin coating of the cement should be first applied to the porous surface, then permitted to dry. A second coat is then applied to the porous surface while a first coat is applied to the glass. They are held together to set, then permitted to dry as mentioned above.

Duco Cement will injure most finishes on furniture, some plastics, and many fabrics. Care should be exercised in using it. To remove it from fingers use acetone or "nail polish remover".

Epoxy Glue that is made to be used on glass is an excellent non-firing material for joining glass to glass and such other materials as pottery, porcelain, wood, fabric, marble, concrete, brick, rubber, and such metals as steel, iron, aluminum, bronze, brass, stainless steel, silver, etc. Epoxy glue consists of two components, one part epoxy resin and one part hardener, that are mixed together before using. The holding quality of epoxy adhesives is achieved by a chemical reaction. The molecules of the epoxy band with the molecules of the surfaces to be joined. Unlike other types of adhesives; there is no water or solvent to evaporate. It will make a tough, durable, waterproof bond.

To prepare a glass joint with epoxy glue you first place equal amounts of epoxy resin and hardener in a disposable clean container and mix them thoroughly. Apply a coat of the mixed glue to the surfaces to be joined. Keep the surfaces in contact, (clamping is not necessary). The epoxy glue will set in approximately 2 hours and bonds overnight.

The strength of the joint will be substantially increased if the glass surfaces are roughened by frosting (see page 61) before joining.

When using epoxy glue avoid contact with skin as it may cause irritation. To clean it from the skin, use nail polish remover or denatured alcohol, then wash with soap.

There are many brands of epoxy glue on the market that can be purchased in local hardware and novelty shops. "Elmer's" and "Weldwood" epoxy glues are two brands that produce good results. There are other good brands available. The important things to remember in selecting an epoxy glue are that it be specifically recommended for glass, and that it be of a color that is compatible with the work you plan to do. None are completely transparent—some are yellow when cured while others are a deep gray.

EC 826 is a product made by Minnesota Mining and Manufacturing Company that is good for joining glass. It is a transparent, amber-colored, high-strength, general-purpose adhesive that has unusually high adhesive strength when joining glass to metal, wood, synthetic rubber, etc. It displays excellent resistance to weathering, gasoline, oil and aromatic solvents.

It is used as follows: Apply a thin, even coat to both surfaces, then bond may be made by any of the following methods;

(a) Allow the adhesive to dry until it exhibits an aggresive tack (1 to 3 minutes), then join surfaces together, pressing firmly. On extremely porous surfaces, use two coats allowing about 10 minutes drying time between coats.

(b) Allow the surfaces to dry tack free. Apply heat (250° to 300° F.) to the adhesive surfaces. Bond as above.

(c) Allow the adhesive to dry tack free. Wipe the surface with a rag dampened in ethyl acetate. Bond as above. If desired, greater strength and oil resistance can be obtained by curing the bond at 300° F. for 20 minutes. EC-826 can be brushed, flowed, or sprayed (when diluted 50/50 with ethyl acetate). Excess adhesive may be removed from equipment and work areas with a low-boiling ketone or acetate type solvent.

Fig. 19-3. A fused glass panel by the renowned artist Maurice Heaton.

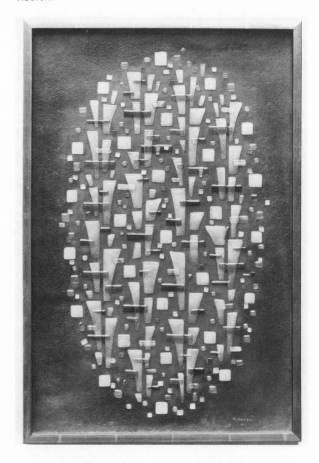

Note: When using acetate solvent it is essential that proper precautionary measures for handling such material be observed. Do not inhale or use near a flame.

EC-1103 is a colorless synthetic-resin-based sealer that is also made by the 3M company. This material exhibits good adhesive qualities when joining glass to fairly large surfaces. It is **not** recommended for edge joints that will be subjected to any amount of abuse. EC-1103 is a smooth-flowing sealer when applied with brush (or sprayed) at normal room temperatures; it will dry tack-free in 5 to 10 minutes.

Klein's #200 Cement will join glass to glass so firmly that it can be washed with hot soapy water. It is a strong, two part adhesive made by A. Ludwig Klein & Son, of Philadelphia. The adhesive consists of a cement and hardener that must be mixed. Just before use, the cement and hardener are mixed in the proportion of **two parts cement to one part hardener.** These proportions are critical. After mixing, the adhesive will stay useable for approximately twenty minutes. In cold weather the cement may thicken; to thin it out place the tightly capped jar in hot water, or on a radiator.

The **thoroughly mixed** No. 200 cement should be applied with the stiff brush that is supplied with the cement, in a very thin sheet to both surfaces (too much cement will make the joint slide apart.). The surfaces are then pressed firmly together immediately after applying the cement, and allowed to air-dry for at least 48 hours.

The adhesive quality of this cement can be greatly increased by doing the following: (1) Frost the two surfaces to be cemented. (2) Heat the glass until it is warm to the touch before applying the cement. (3) After the cemented joint has air-dried for 24 hours, force-dry in an oven or with a heat lamp (not over 150 degrees) for about three hours.

FIRED GLASS JOINTS

Glass can be joined by firing only to those materials that can withstand the high temperatures necessary to soften the glass. Generally, glass is joined in this manner to glass or to one of the hard metals such as iron, steel, silver, copper, brass, etc.

The pieces of glass involved in a fired glass joint must be thoroughly clean (see page 83) or the joining can not be successfully accomplished.

Fired Glass-To-Glass Joints

Laminated Glass is the result of joining or fusing two or more layers of glass together by firing. The word laminating, as concerns glass, always refers to the joining of the surfaces, never the edges or ends. The fundamental requirement for laminating is that the glass pieces to be joined must be compatible—they must have the same general characteristics of response when subjected to heating and cooling.

There is, theoretically, no limit to the number of layers of glass that can be fused together. In actual practice, however, it is usually limited to two layers. The two layers may consist of two full sheets; a full sheet with a partial sheet fused to it; several fragments fused to a full sheet; or several strips fused together to form a design. Figures 19-2 and 19-3 show laminated glass prepared by the famous artist, Maurice Heaton.

The laminating procedure is as follows:

1. Prepare a full-size design of the laminated item to be made. Applying color to the design will give you an idea as to how the completed object will appear.

2. Cut glass to desired sizes (see pages 25-34).

3. Clean the glass with soap and water; dry; then wipe with alcohol. Do not touch the surface with your fingers after cleaning.

4. Apply decorating mediums (See "Decoration of Laminated Glass" page 86). Color may be applied to the bottom of the top piece of glass and also the top of the bottom piece on overlapping areas for design effects. Apply an adherent such as Klyr-Fire, or a very thin layer of "Baby Oil", to the surfaces to receive coloring materials, such as dry enamels or powdered mica, which should be sifted on. Allow all materials to dry before "sandwiching" and firing.

Note: If small pieces of glass are being laminated and you are having trouble keeping them in position, they may be glued together with liquid white glue (Polyvinyl resin) in preparation for firing. Make sure the glue is thoroughly dry before firing.

5. If the laminated pieces are to be hung on a mobile or as jewelry, a piece of nichrome wire (or another metal) may be placed between the glasses to be fused in place. The looped portion of the wire must extend beyond the glass (Fig. 19-4). Use glue on the glass to keep from shifting.

6. Prepare either a kiln shelf or a stainless steel planchon to take the glass. This is done by sifting an

Fig. 19-4.

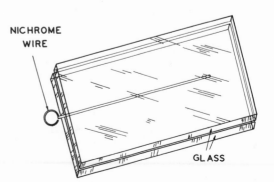

NICHROME WIRE

GLASS

even flat layer of whiting on to the shelf or planchon, covering it completely. This will prevent the glass from sticking. Sifted whiting will give the back of the glass an attractive, wrinkled appearance. For a smoother back, mix the whiting with water and brush it on evenly. A roller or knife blade may be used to smoothen the whiting.

Note: Other heat resistant powders that might be used in place of whiting are flint, plaster-of-Paris or silicon.

7. Press each set of glasses to be laminated firmly into the whiting, being careful not to touch the surface. Make sure that pieces not to be fused do not touch.

8. Slip the tray or planchon carefully into the kiln, making sure no glass is upset.

9. Fire the glass (see "Firing" page 95). Remember that you **start with a cold kiln** and you remove the laminated glass **after the kiln is again completely cold**. The temperature required to cause the laminated glass pieces to fuse together will vary with different types, thickness, and sizes of glass. Generally, it will take place somewhere between 1450° to 1500° F. Once a temperature of 1430° is reached, the glass must be closely observed to make sure the kiln is turned off when the desired results have been achieved.

JOINING ENDS OF GLASS TUBES AND RODS is frequently done in the construction and repair of laboratory equipment of all kinds. Such a joint is not a particular problem unless it involves the necessity of joining two dissimilar glasses such as Pyrex Brand Glasses to soft glass. The preparation of such joints is fully explained in the section on "Laboratory Glass Blowing," pages 111 through 128.

GLASS SOLDER is an all-important joining material used to make a permanent seal between sections of glass, which are to be opened and resealed by heating. This was originally developed by the Owens-Illinois Company to join the TV picture tube face plate to the back funnel section. It enables the working parts of the bulb to be installed or removed for repair without destroying the bulb.

Solder glasses fall into three general types. The **vitreous** type is simply a low melting glass which can be applied to higher-melting glasses at temperatures below or just above their annealing points and effects a seal in the same way that metals are soldered. It may also be used for glass-metal seals where proper matching of material properties is employed.

The **devitrifying** type might be compared to a thermosetting plastic, since the solder glass after maturing has properties which enable it to be reheated at a reasonably high temperature without showing glassy flow.

The third type is the **conductive** solder glass. This type of solder glass has been most recently developed and it, as well as the previous types, is subject to further developmental extention or improvement. Conductive

solder glasses are intended to provide both a seal and a low-resistance electrical path between component parts.

Note: It should be remembered in using any of these solder glasses that they are compounds **high in lead.** In the finely powdered state they are relatively soluble and, therefore, all precautions should be taken to handle them so that no **health hazard arises.**

Glass solder is not difficult to apply or prepare. It is important, however, that the glass solder and the glass to which it is going to be applied are compatible. It is recommended that anyone having need for this product write to the Kimble Glass Company, Toledo, Ohio, who will send all necessary information for its successful use.

FIRED GLASS-TO-METAL JOINTS

The ideal glass-to-metal seal would be one in which the metal had the same coefficient of expansion as the glass over the entire temperature range encountered in making or using the seal. The glass should adhere tightly to the metal. The metal should withstand the heating necessary to soften the glass. It has been demonstrated that a glass and metal of widely differing coefficients of expansion can be sealed if the size and the shape of the materials are such that the stress between the glass and the metal is less than the shearing strength. This is usually accomplished by using extremely thin metal.

Glass and metal may be joined by **encasing the metal in the glass** or by bonding **the metal to the outside** of the glass. Of the two, the former is the easier, as will be explained. Glass and metal are joined in different forms for either decorative or utilitarian reasons.

Decorative Uses For Joining Glass and Metal would include such things as wire loops for hanging glass ornaments and jewelry; wires, screens, and small chips of metal used for ornamentation; and metal pieces used for stands or legs to support the glass items.

There are only a few metals that can be successfully fused with glass—and even these only when used in thin and narrow pieces. For decorative purposes the most important are: sterling silver, stainless steel, brass, copper, nichrome, and iron. All of these metals turn black due to oxidation when fired with glass except the sterling silver. This must be kept in mind when preparing your design.

Laminating (see page 139) is the only practical method available to the novice for permanently joining glass and metal. The metal wire (or whatever form you choose) is placed between two pieces of glass and fired until the glass completely encases the metal (see page 139). A part of the metal may be left unlaminated to serve as a hanger or other design feature.

The metal should be shaped and formed before it is

Fig. 19-5. "Mustangs" by artist Hans Godo Frabel is an excellent example of what can be produced by joining glass segments.

laminated in glass. This includes those parts of the metal that are not encased in the glass. Although it is possible to bend exposed metal after a portion has been encased in the glass, there is always a danger that such a procedure might chip or crack the glass. When it is unavoidable and you must bend an unlaminated part of metal proceed as follows: Grip the exposed metal with a long nose **pliers just beyond** the glass, then proceed to **bend** the protruding metal with a **second pair** of long nose pliers. In this way you prevent the shocks of the bending from being transmitted to the glass.

The procedure for laminating metal in glass is as follows:

1. Prepare the pieces of glass to be used, then clean them thoroughly (see page 83).
2. Lay the bottom piece of glass over your design and position your formed metal on the glass according to the design. (Strategically place a few drops of glue (see page 139) to keep the metal from shifting.)
3. Put the top glass in place and weight it until the glue dries.
4. Fire according to instructions on page 101

There is no practical way for the hobbyiest to join metal to the outside of glass using heat. There are commercial methods which will be discussed later, but for school and home it is best that one of the non-firing adhesives be used.

Practical Uses For Joining Glass and Metal include such items as those used for laboratory equipment, electronic and electrical equipment, etc. There are hundreds of commercial items on the market that include segments that are unions of glass and metal. These things must be made with a degree of accuracy that is far superior to that acceptable for decorative purposes. The prime consideration in a decorative object is appearance. In objects of utility such as electrical parts, laboratory testing equipments, etc., the union of glass and metal must be perfectly prepared to very exacting standards. The following are procedures for preparing some of the more popular metal-to-glass seals that are used in the preparation of utilitarian objects.

METAL SEALED INTO GLASS

Tungsten in Pyrex Brand Glasses

The linear coefficient of expansion of tungsten, which is approximately 45×10^{-7} is somewhat higher than that of Pyrex Brand No. 7740 Glass, which is 32.5×10^{-7}, or of Pyrex Brand No. 7720 (Nonex) Glass, 36×10^{-7}. The match in the latter case is better than the figures indicate. Either glass adheres to the metal. Tungsten oxidizes readily in a flame necessary for working these glasses; therefore, prolonged heating is to be avoided.

Although it is not a common practice commercially, tungsten wire may be sealed directly into Pyrex Brand Glass No. 7740 under favorable conditions. Such a joint will be suitable for vacuum work if made properly from wire of 50 mils or less, and if the service requirements are not severe.

1. The tungsten wire is heated to a dull red, and immediately brought into contact with a stick of potassium nitrite (KNO_2). This dissolves the tungsten oxide resulting from the heating. Wash free of the nitrite, and dry. This leaves the surface of the wire clear. (The wire is reoxidized slightly in the flame.)

2. A close-fitting sleeve of glass tubing 12 to 18mm in length, and with a wall thickness of about 1/2mm, is slipped over the wire. This is heated progressively from one end to the other so that the glass shrinks to the metal-driving the gas out ahead through the open end. The metal in contact with the glass appears larger; therefore the progress of the seal can be followed easily. This procedure also avoids the trapping of air between the glass and the metal.

3. This bead may be enlarged at the end or at the middle so that it can be sealed into a tube. A hole is made in the tube sufficiently large to admit the beaded wire without permitting the enlarged part of the bead to pass. Seal in this bead with a sharp flame in the same manner as in making a ring seal.

4. If larger wire is to be used for vacuum, use Pyrex Brand Glass No. 7720 for beading. This glass contains lead and must therefore be worked in an oxidizing flame. It is widely used for tungsten seals commercially.

The procedure is similar to that used for Pyrex Brand Glass No. 7740. The wire is beaded with Pyrex Brand Glass No. 7720, and the bead is sealed into Pyrex Brand Glass No. 7740. Some glass blowers prefer to

Fig. 19-6. Glass solder sealed to the inner rim of a steel bezel. (Courtesy: General Electric Co.)

seal the bead into a short length of Pyrex Brand Glass No. 7720 tubing, and in turn seal this to Pyrex Brand Glass No. 7740 in a straight seal.

Pyrex Brand Glass No. 7991 can be used for tungsten seals in place of Glass No. 7720. Its main advantage is that it is a non-lead glass and an oxidizing flame is not required in working. It also has a higher softening point which allows for higher baking-out temperatures in vacuum tubes or similar items.

Tungsten seals may vary in color but generally a good seal is indicated by what might be termed copper bronze or golden amber color.

The tungsten wire used should be free from any longitudinal cracks or pores, since the presence of these would make a vacuum-tight seal impossible. To avoid cracking large tungsten wire when straightening it, it should be bent while it is red hot.

Platinum in Pyrex Brand Glasses

Platinum, having a coefficient of expansion of approximately 90×10^{-7}, can be sealed very easily into a number of soft glasses. When sealed into Pyrex Brand Glasses, the platinum must be very thin, so that it will give with the glass rather than tear away from it; and if made, it should not be intended for use in vacuum work.

Two methods are applicable. One is to use a very light-walled platinum tube. The glass is fused about the outside of the tube for a distance of five to fifteen millimeters. If the platinum is to carry current, its capacity can be increased by passing a wire through the tube and welding it at one end.

The other method is useful in making electrodes. Fine wire is flattened so that it is very thin, and the edges are filed to a sharp "V" shape. This flat portion is sealed into the end of a glass tube, the end of which has been shrunk to a diameter slightly greater than the flattened portion of the wire. In making the seal, the heated glass is pressed with tweezers to assure good contact with the metal.

Copper In Pyrex Brand Glasses

Copper can be sealed to Pyrex Brand Glasses in the same manner as platinum except that copper has a much lower melting point and oxidizes in the flame; therefore, very careful control of the heating operation is required. With good technique copper seals of almost any size can be made.

Copper ribbon which is to be sealed into glass should have its edges filed to a sharp angle. The heated glass will shrink about and adhere to the metal. To assure good contact the glass can be pinched about the metal by using tweezers. If wire is to be sealed into glass a short section should be flattened so as to be comparatively wide and thin. The flat section can then be sealed

into the glass in the same manner as copper ribbon.

Copper discs can also be sealed to glass. Two glass tubes are flared (90%) and the copper is sealed flat between the flares. Care should be taken so that the glass on one side does not run over the edge to make contact with glass on the opposite side. After the seal is completed one of the glass tubes can be cut off at the flare to leave essentially a glass ring on the copper.

The following types of seal are used in the sealing of **copper tubing** to glass:

1. seal having the glass on the inside of the copper tube;
2. seal having the glass on the outside and the inside of the copper tube; and
3. seal having the glass on the outside of the copper.

The copper tubing is prepared by **machining,** to a knife edge, the **end to be sealed.** This knife edge should be formed by a gradual taper of a length somewhat greater than the length of the glass section which will contact the metal. The seal having the glass on the inside must be carefully worked to avoid softening the glass without burning the copper, but from the standpoint of freedom from strain in the seal this type is better.

The part of the copper which is sealed to the glass must be cleaned free of grease, finger marks, etc. Borax, although not necessary, may be used as a flux. If used, it should be applied uniformly and in light coating. The completed copper-to-glass seal will have a layer of cuprous oxide between the metal and glass which accounts for the crimson color.

Molybdenum-Fernico-Kovar
In Pyrex Brand Glasses

Glass seals to the above metals can be made by using special glass, as intermediates between the metal and Chemical Glass No. 7740. The size and shape of the metal and glass tubes have a bearing on which particular intermediate glasses to use, as does also the service to which the seal is to be subjected. The skill of the glass blower will also determine to some extent the type of seal which is most applicable.

For small size metal and glass tubes (10 mm or smaller metal), (15 mm or smaller glass tube), the seal is made as follows:

Metal to Glass No. 7040 to Glass No. 7720 to Glass No. 7740.

All sizes of metal and glass may be sealed as follows: Metal to Glass No. 7052 to Glass No. 7050 to Glass No. 7070 to Glass No. 7740.

Glass No. 7040, which has higher electrical resistivity, may be substituted for Glass No. 7052 in the above.

Glass No. 7052 has greater chemical stability than Glass No. 7040.

If the final glass in the series is to be Glass No. 7720, then the seals are made as follows:

Metal to Glass No. 7040 to Glass No. 7720.

Metal to Glass No. 7052 to Glass No. 7050 to Glass No. 7720.

In making these seals the metal must be cleaned carefully and the alloys should be heated in H2 atmosphere at 900° C. before making the seals. For more complete information as to using these alloys the user should contact the following suppliers:

Kovar-Stupakoff Ceramic & Manufacturing Company, Latrobe, Pa.

Fernico-General Electric Co., Schenectady, N.Y.

Fig. 19-7. Glass metallizing apparatus. (Courtesy: General Electric Co.)

BONDING METAL TO OUTSIDE OF GLASS (METALIZING)

There are a few different methods that can be used to join metal permanently to the surface of glass. One of the best methods is a process utilizing titanium hydride that was discovered by Floyd C. Kelly and further developed by Ralph J. Bondley and Lawrence J. Hogue, all scientists at General Electric, Schenectady, N.Y. Figure 19-6 shows a piece of glass that was solder-sealed to the inner rim of a steel bezel by this method.

The Titanium Hydride Metalizing Process, which is applicable to inorganic materials such as glass, ceramics, carbon and most metals, incorporates the use of titanium hydride, a solder, high temperatures and an atmosphere free of oxygen, nitrogen and water vapor. The titanium hydride, which is purchased as a finely divided powder, is made into a thin slurry by mixing with acetone then, with the aid of a small brush, is painted on the glass and metal pieces at the areas where they are to be joined. The parts are placed on a small **pedestal,** and the **solder** metal is positioned in **contact** with the titanium hydride painted areas. The assembly is then **covered** with a glass bell jar, ceramic closed-end tube or gas-tight metal container, the system purged of its air, then heated to 500° -600° centigrade (Fig. 19-6 and 19-7).

It was found that a **vacuum** can be used for the atmosphere if the system is tight and can be pumped down to a pressure of 10^{-3} mm Hg (1 micron) or better. If no vacuum facilities are available then you can use an **inert atmosphere** of Helium or Argon. Nitrogen, which is also considered an inert gas, cannot be used because of the great activity of the titanium metal, which combines directly with nitrogen at high temperature to form the nitride.

The heating generally presents the greatest problem but **induction heating** has been used with excellent success. A thin tantalum-metal susceptor **heat-shield** made in the form of a cylinder is placed **around the parts** to be joined and is **heated** by positioning the water-cooled **work-coil** from a 550 kc oscillator around the **outside** of the glass (pyrex) bell jar opposite the heat shield. This method of heating is clean, fast and easy to assemble. If no induction heater is available the parts may be heated with **resistance-wire** heaters or by placing the atmosphere-controlled set-up in a **high-temperature furnace.**

This mechanism of metalizing is dependent upon the high activity of titanium (or zirconium) in its pure surface state. This is achieved when the titanium hydride is heated to 500° centigrade where it undergoes dissociation to liberate nascent hydrogen, which is an excellent cleaner and scavenger at high temperature, and pure titanium reacts with the solder to form a compound that is able to "wet" the surface of the glass in the areas where the titanium is present. This allows selective metalizing to be performed, since the solder will not adhere to parts which were not painted with the hydride.

Normally, a soft solder is used which is the eutectic mixture of lead and silver (2-1/2% silver) for ease of application. Since this solder melts at about 300° centigrade, some provision must be made to keep the molten solder in contact with the hydride. A soft soldered joint will withstand reheating and cooling without danger of breakage due to mechanical stress or strain since the solder is flexible. Metals having high vapor pressures such as cadmium and zinc (or brass) must be avoided since they vaporize at these temperatures and plate-out on the cooler areas within the bell jar.

The description of the process may appear to be somewhat complicated but in essence it is relatively simple to operate if the proper type of equipment is available. Size limitations of the parts are imposed only by the ability to heat uniformly and rapidly enough.

In metalizing with the titanium hydride process (active metals method) there is an intermolecular-reaction bonding between the glass and solder, resulting in a vacuum-tight seal that is essentially impossible to remove without damage to the glass and which maintains its good sealing characteristics up to the softening point of the solder. Tensile tests always result in a glass fracture when the seal is pulled apart.

Other Methods of Joining Metal to the Surface of Glass which, although simpler to produce, do not gen-

erally result in seals that are as realiably vacuum-tight or temperature-durable as those made with titanium hydride. For example: Glass may be metalized by painting the surface with a silver paint which can be either air-dried or baked. This method is known as the **conductive coating process.** This bonding is not vacuum-tight since it depends largely on the use of an organic binder. **Platinum salt** solutions and compounds can be **painted** on glass and fired to reduce the surface to platinum which can then be copper-plated and tinned with solder. The bonding strength of this type of metalizing is not strong because of the thinness of the platinum sublayer.

Another method consists of using **low melting solder alloys** which usually contain indium or gallium and are characterized by their ability to "wet" a clean hot glass surface. The cohesive forces of this type of metalizing are weak and temperature sensitive and although the

Fig. 19-8. Glass metallizing apparatus. (Courtesy: General Electric Co.)

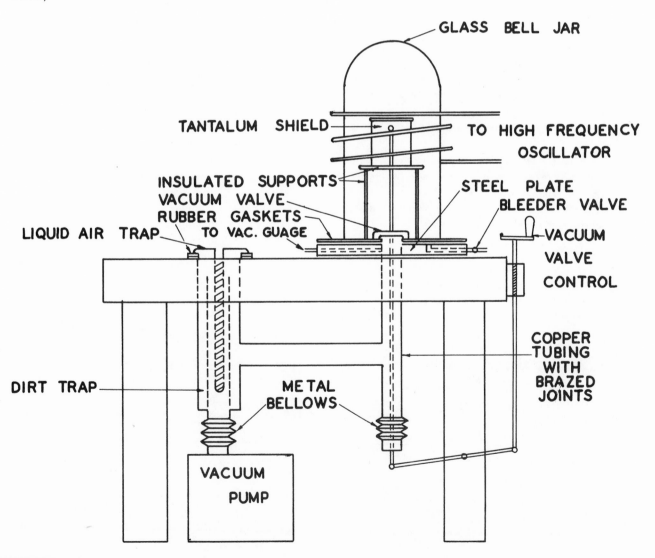

metal wets the glass there is little bonding.

There is also a **fusion method** of joining glass to metal that involves the use of special glass compositions to make direct glass-to-metal seals in an oxidizing atmosphere. In this method the molten glass combines with the previously formed oxide on the metal to produce a relatively strong bond. Seals of this type, however, are dependent on obtaining a close match of the coefficients of thermal expansion between the glass and metal, since the joint is composed of rigid materials and both parts must expand and contract at about the same rate throughout all temperature variations to prevent stresses and strains that would break the seal.

A modification of the fusion method for small glass beads takes advantage of a mismatch in the expansion characteristics between the glass and metal and results in added mechanical compression around the periphery of the glass bead to help restrain the effect of stresses. This is known as a **compression seal** and is used in the making of small lead-through terminals. The disadvantage of the method is that the seals are vibration and shock sensitive.

There are several vendors of metalized products on the market. Most of the major ceramic companies now sell glass and ceramic sealed bushings and terminals as do several smaller companies which specialize in making seals only.

MENDING BROKEN GLASS

Broken glass can be mended so it can again be useful. It can never be mended to the point where the break is invisible, but with care, the break can be made to appear so slight as to be unobjectionable. When the mend is properly made with certain cements, the glassware can be used for hot liquids and washed in hot water and soap suds.

1. The mending procedure starts at the time of the break; at this time salvage all the pieces you can. With a strong detergent clean all the pieces to be joined and then dry them thoroughly.

2. To increase the holding quality of the cement remove the sheen from the edges of the glass to be joined. This can be done by applying glass frosting cream (see page 65) to the surfaces for about five minutes. Care should be exercised not to frost the exposed surfaces of the glass.

3. Plan how the break will be reassembled to make sure all the pieces will fit. When only one piece is involved in the break you merely have to be sure that the piece can be slipped back into place. When several pieces are involved, you must make sure that the next piece will fit after a previous one is cemented into place. Generally, this requires that you work from the inside of the break out toward the edge.

Before the broken glass is cemented certain decisions must be made. If the glass is to **hold hot liquids** or to be washed in hot soap and water, use "Dow Corning Glass and Ceramic Adhesive" (see page 137) or "Kleins #200 Cement" (see page 139). If **transparency** is a consideration, the two previously mentioned cements may be used as well as "Duco" cement (page 137). When mending **opaque** glass where heat and transparency are not considerations, any of the cold adhesives previously listed (see page 137) can be used. If the cement requires mixing, do so just before it is ready to be used. Certain cements do not require mixing and can be applied directly from the tube.

4. Apply the cement with a stiff brush or knife as previously recommended, or as recommended on the container by the manufacturer. The cement should be

Fig. 19-9. When the break consists of one piece cover the entire surface of both pieces with adhesive. (Courtesy: Dow Corning Corp.)

Fig. 19-10. Save all pieces possible at the time of the break. (Courtesy: Dow Corning Corp.)

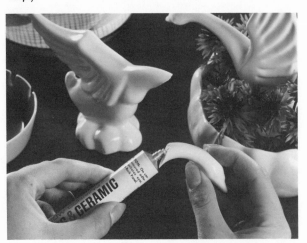

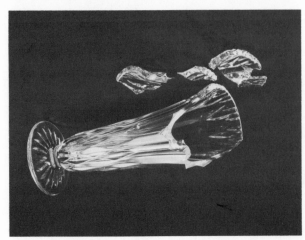

Fig. 19-11. Apply a thin coat of adhesive. (Courtesy: Down Corning Corp.)

Fig. 19-13. A carefully executed glass mend is almost invisible. (Courtesy: Dow Corning Corp.)

Fig. 19-12.Use a sand box to position the glass to keep the pieces from shifting. (Courtesy: Dow Corning Corp.)

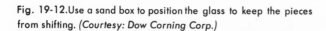

applied in a thin coat to both surfaces to be joined. (A thick coat encourages the piece to slip out of place.)

If the break involves only one piece (Fig. 19-9) apply the cement to the entire edge to be joined. However, if the break includes more than one piece (Fig. 19-10), do not cover the entire edge to be joined, but leave the edge uncoated for about 1/8 inch back from the corner that is to butt up to the next piece to be cemented (Fig. 19-11). This prevents the cement from oozing out, and forming a lump in the corner when the piece is pressed in place. Such a lump will prevent the next piece from fitting properly into place.

5. Balance the broken piece so the cement-coated edges are in tight contact and perfect alignment.

. A **sand box** can be used to make the balancing easy (Fig. 19-12). This consists of a box slightly larger than the glass object to be mended, filled with about four inches of sand. By **packing the sand around** the broken object it can be positioned at any angle so the broken piece can be balanced in alignment without difficulty.

The alignment of the crack can be checked by running the end of your nail across it. Masking and "Scotch" tape can be used, if need be, to keep the piece from slipping.

6. Allow the cement to cure completely (Fig. 19-13). The time required will depend on the cement used. In a break requiring the placement of two or more pieces,

make sure the **cement** is **cured** on each piece before proceeding to the next. Where force-drying with an oven or heat lamp is used, make sure the glass object is cooled completely before filling it with a liquid or food. The excess dried cement can be cut away with an "X-acto" knife or razor blade.

Correcting Faulty Alignment

Should a piece of broken glass be accidentallly cemented out of alignment, proceed to correct it as follows:

1. Carefully re-break the pieces apart.

2. Remove the hardened cement. Use the thinning material recommended by the manufacturer, then wash with a strong detergent. If no thinner is recommended, cut away as much of the hardened cement as you can, soak the edges with lacquer thinner, then wash with a strong detergent.

3. Re-cement the pieces, following the procedures recommended for the original union.

Chapter 20
Making stained glass panels and windows

There is an elegance and glory in stained glass work that cannot be duplicated in any other form. The early history of stained glass work is lost in antiquity; however, it is known that the twelfth and thirteenth centuries in Europe were the "Golden Age of Stained Glass". Following this period it diminished in popularity until very recent time. Following the Second World War a contemporary form of stained glass work has gained popularity and wide acceptance.

WHAT IS STAINED GLASS WORK?

Stained Glass work is perhaps best described as a decorative composition, constructed of hundreds, even thousands of pieces of irregularly cut white and colored glass, bound together by strips of grooved leads, soldered at the joints, to form a planned design (Fig. 20-1).

With the exception of a stain painted and fired to produce a yellow tone in white glass, the only pigment used is a reddish brown or black powdered oxide to delineate features and form, drapery and pattern. The pigment is rendered permanent by fusing in the surface of the glass at a high temperature.

THE GLASS USED

The stained glass used in this craft is known as antique or pot metal glass. Other types, such as Flemish and Cathedral, are used on account of their low price, but they have none of the interesting qualities that are to be found in the antiques.

The color in glass is in the substance itself. While the mass of colorless glass is still in its molten state, various chemicals are added to it in the pot. Copper oxide,

Fig. 20-1. (Courtesy: Immerman & Sons)

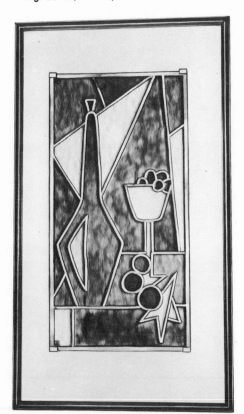

Fig. 20-2. Cartoon being prepared in a commercial studio. (Courtesy: The Payne-Spiers Studios)

Fig. 20-3. Student prepares cartoon. (Courtesy: Stephens College, Columbia, Mo.)

under different conditions produces ruby, blue and green, while cobalt is the principal base of fine pure blues. Green is also obtained from chromium and iron oxide. Golden glass is sometimes colored with uranium, cadmium sulfide, or titanium, and there are fine celenium yellows as well as vermilions. Ruby is also colored with noble metal, gold. This method of staining or dyeing glass is the same as that used in the Middle Ages. For this reason it is called antique glass or pot-metal, from the pot in which it is made.

The less expensive type of glass used is machine drawn and comes in sheets approximately 32" x 84" in length.

The more expensive mouth-blown type, of which there is practically an unlimited color variety, runs anywhere from 15" x 23" to 23" x 35". This latter type is made by dipping a long tube into a molten mass of glass and collecting a lump on the end. This lump is formed into an elongated bubble by blowing down the tube. While still plastic the cylinder is cut down one side and allowed to settle on a flat stone. In settling, the underside of the glass becomes relatively flat. The upper surface of the glass retains undulations from blowing which make interesting variations in the color. The imperfections found throughout the glass are often a part of its glory.

Flashed glass is made by dipping a ball of molten white glass into molten colored glass which, when blown and flattened, results in a less intense color because it will be white on one side and colored on the other. The use of flashed glass enables the craftsman to obtain certain effects not otherwise possible.

HOW A STAINED GLASS PANEL IS MADE

Stained glass craft is practiced in America today in virtually the same manner as it was in the Middle Ages, although some of the tools—notably the glass cutter and the soldering iron—have been improved for rapid and more skillful handling.

The steps in the production of stain glass panels are as follows:

1. **Design.** The design is usually a small-scale study of the panel, intended to convey an impression of the

Fig. 20-4.

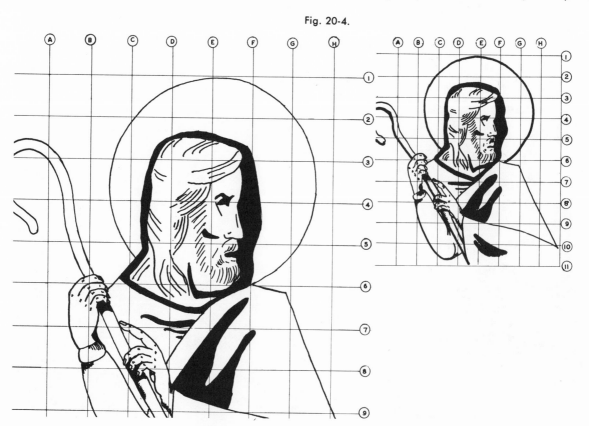

Fig. 20-5. Cutline drawing being prepared in a commercial studio. *(Courtesy: The Payne-Spiers Studios)*

Fig. 20-7. Patterns being checked in a commercial studio. *(Courtesy: The Payne-Spiers Studios)*

stained glass panels wherein diamonds, squares, rectangles and triangles are used are extremely effective (Fig. 20-1). American Indian designs with their squared-off corners work out to very attractive designs.

Some general rules of design to remember are these: Try to pattern dark glass next to light and light on dark. The various shapes, which go together to form the

color and light of the full-sized panel. Before attempting a design the student should first read through this entire section and become familiar with the entire process involved in making the panel. He should realize that he is working with three basic materials—stained glass, lead separators, and colors fired on the surface of the glass. The design should be kept within the limitations of these materials.

The novice should not attempt an elaborate design as may be found in the local church window. Contemporary

Fig. 20-8. The student prepares her patterns. *(Courtesy: Stephens College, Columbia, Mo.)*

Fig. 20-6. Student prepares her cutline drawing. *(Courtesy: Stephens College, Columbia, Mo.)*

whole panel, must not run contrary to the character of glass. Shapes which are long and narrow are difficult to cut and fire, and make for weakness in the finished window. The position for the leads is of great importance, too, for they not only connect the glass components but, when well-designed, greatly strengthen the finished product.

2. **Cartoon** (Fig. 20-2 and 20-3). The cartoon is an enlargement—generally in black and white—designed to the full-size of the finished panel desired. The suggestions of the design are usually developed further in the cartoon.

The design may be enlarged to cartoon size by any of the customary enlarging methods, such as pantograph, photography etc. It is usually done by "squaring up". (Fig. 20-4). Small squares are drawn on the design. An equal number, though proportionately larger squares, are drawn on the cartoon paper. The figure in the design is transferred to the cartoon paper a square at a time until it is completed. Either a heavy lead pencil, charcoal or water-color is used as the medium for the cartoon. If the panel is a small one, the cartoon can be drawn on a piece of heavy paper right on a table. When large panels or windows are being made, the cartoon is drawn on special cartoon paper or on several layers of heavy paper that have been glued together. Such large sheets are usually pinned to the wall at a convenient height for working.

The finished cartoon should clearly indicate the leadlines, and the shapes for the individual pieces of glass. The layout for the individual glass shapes should be drawn lightly with a pencil at first so that alterations and corrections can be made.

When preparing the leadlines in the cartoon a few simple rules should be born in mind. The lead should not attempt to compete with the glass in the formation of the pattern. Lead formations that are objectionable from both an aesthetic as well as strength point of view are those wherein the leads form an X in a prominent part of the panel. Or where the lead closely crowds an important detail in the panel. Or where two leads join to the outside lead in the form of a K. On large panels to be used for windows the leads should be carried through from side to side every ten to twelve inches. These leads can serve to support reinforcing rods that are later attached.

3. **Cutline and Pattern.** From the cartoon, the CUTLINE and PATTERN drawings are made. The modern cutline drawing (Fig. 20-5 and 20-6) is a careful, exact tracing of the leadlines of the cartoon on heavy paper. The leadlines are the outlines of the shapes for patterns to which the glass is to be cut. The color of each piece of glass is indicated. This drawing serves as guide or reference for the subsequent placing and binding with

Fig. 20-9. The glass worker selects and cuts the glass. *(Courtesy: The Payne-Spiers Studios)*

Fig. 20-10. A student traces the pattern on the glass. *(Courtesy: Stephens College, Columbia, Mo.)*

Fig. 20-11. The student waxes-up her panel. *(Courtesy: Stephens College, Columbia, Mo.)*

lead of the many pieces of glass (Fig. 20-7 and 20-8).

The pattern-drawing, usually on heavy paper, is a carbon copy of the cutline drawing. It is cut along the black or lead lines with double-bladed scissors or knife which, as it passes through the middle of the black lines, simultaneously cuts away a narrow strip of paper, thus allowing sufficient space (approximately 1/16") between segments of glass for the core of the grooved lead. This core is the supporting wall between the upper and lower flanges of the lead, which is something like a miniature girder or like the letter H lying on its side.

4. **Cutting** (Fig. 20-9 and 20-10). The pattern is placed on a piece of the desired colored glass and, with a steel wheel cutter, the glass is shaped to the pattern. (See Chapter 3 for details on cutting). As mentioned before, the two faces of the glass will be found to vary considerably. The cut should always be made on the flatter surface. All pieces to be cut from the same colored glass should be cut at the same time. In this way, the patterns can be layed out on the glass so that the most economical cuts can be made.

The pattern may be held on top of the glass and traced with the cutter or the glass placed over the cutline and the cutter guided around the shape observed through the glass. The preference of the person doing the cutting will determine the method used.

5. **Waxing Up** (Fig. 20-11). When all the stained glass pieces are cut they are then placed on a sheet of clear glass and secured in their correct positions by drops of beeswax at the corners. The clear glass is laid over the cartoon which is used as a guide in the placement of each piece of stained glass. If all the stained glass pieces are properly cut and placed, there will be an even margin separating each unit which will represent the core of the lead. For small panels, plain window

Fig. 20-12.

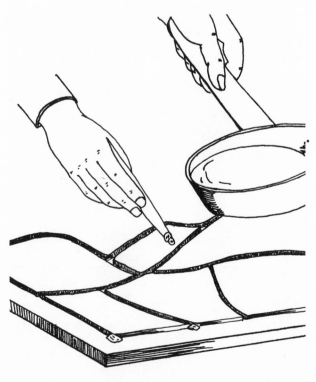

Fig. 20-13. The student makes corrections in the waxed-up panel. *(Courtesy: Stephens College, Columbia, Mo.)*

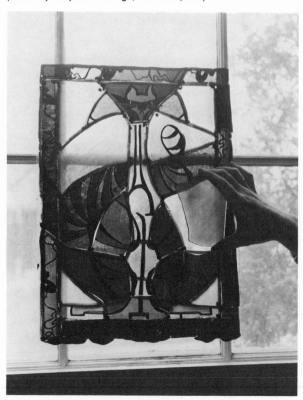

glass may be used for mounting, but for large panels plate glass is preferred.

This waxing process must be carried out accurately and with a degree of speed. The accuracy concerns the placement of the beeswax. It must be placed against the edges of the corner of the stained glass and never allowed to touch the face of the glass (Fig. 5-12). Wax on the face of glass will make it difficult to paint on the glass in the following process. It must be done speedily so that the wax will not require too frequent reheating.

The beeswax, which comes in the form of a brick, can be warmed up in a can or pan until it is molten. A pointed piece of glass or metal may be used to transfer the molten beeswax from the pan to the spot on the glass desired. Sufficient molten wax may be picked up on the point to drop quickly onto the required position.

Plasticine may be used instead of wax for fixing the glass. It is a much slower process but has one advantage over the wax for it allows frequent removal and replacement of the stained glass shapes without the need to take the panel down from the vertical position.

When the waxing is completed the panel can be raised to a vertical position for close examination and further work (Fig. 20-13). To get a truer picture of how the completed panel will look, black lines are painted on the back of the mounting glass where the lead will eventually go as well as a black border. Black poster paint is used for this purpose.

6. **Painting** (Fig. 20-14; 20-15 and 20-16). After the glass

has all been cut, the painting of line work takes place. The painting is done on each required piece with special vitrifiable paint. This may be done directly onto the panel as a whole as it stands waxed upon an easel, or separate pieces of glass may be removed from the frame and traced accurately over the cartoon. Painting while the panel is being held up to the light approximates the conditions in which the panel will eventually be seen. If the panel is to be used as a window, the painting should be done on the side that will be inside the building. It should be the flatter side of the glass.

The paint used for stained glass work is an oxide in powder form which is mixed with water, painted on, then fired into the glass so that it fuses and becomes part of the basic material. The colors used by the stained glass worker consist of about four or five shades of brown and two or three of black. These few colors are composed mainly of oxides of iron and, when applied strongly, completely prevent light passing through the glass. When the paint is applied to the glass it sticks rather securely allowing it to be worked with several different tools and techniques.

To prepare the dry pigment for painting you would first shake some out on a pallet—a clean sheet of glass will serve well for this purpose. Use the pigment liberally to avoid running short. Add water and thoroughly

Fig. 20-15. Painting ornaments and figures. (Courtesy: The Payne-Spiers Studios)

Fig. 20-14. Painting of flesh parts. (Courtesy: The Payne-Spiers Studios)

Fig. 20-16. The student paints the glass in preparation for firing. *(Courtesy: Stephens College, Columbia, Mo.)*

Fig. 20-18. Leading-up a stained glass panel in a commercial studio. *(Courtesy: The Payne-Spiers Studios)*

mix with a pallet knife to a creamy consistency. Add a few drops of gum arabic to prevent the pigment from being rubbed off the glass. Too much gum in the mixture may cause "frying." Mix well until the gum is evenly distributed and the paint is smooth and satin-like in

quality.

The paint should be applied fluently and smoothly. Thick and retouched work will result in frying. The main lines of the design are first lightly painted on. They may be traced directly from the cartoon or sketched free hand on the glass. To fill in large areas several techniques familiar to art work may be used. It may be stippled or matted to inject interesting light variations in the painted area. Dots, diamonds, lines, stars, etc. may be drawn in the paint to add additional variation and sparkle. These latter variations are known as "stick lights" and may be done with sharpened sticks, metal or bone.

Silver Staining is a coloring process used to obtain pale yellows and deep oranges. The color is obtained in a powder form and mixed as indicated above. Unlike

Fig. 20-17. Loading painted glass in the kiln for firing. *(Courtesy: Stephens College, Columbia, Mo.)*

Fig. 20-19. Students leading-up modern stained glass panels. *(Courtesy: Stephens College, Columbia, Mo.)*

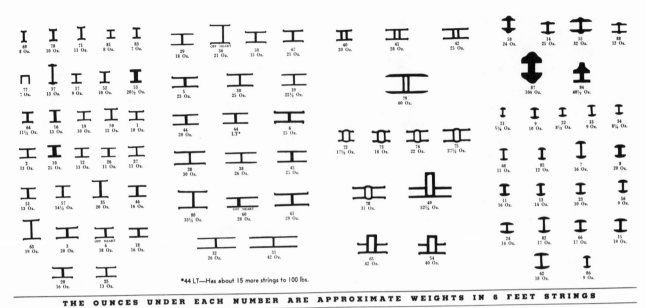

THE OUNCES UNDER EACH NUMBER ARE APPROXIMATE WEIGHTS IN 6 FEET STRINGS

*44 LT—Has about 15 more strings to 100 lbs.

Fig. 20-20. Typical came-lead shapes. (Courtesy: White Metal Rolling & Stamping Corp.)

Fig. 20-21.

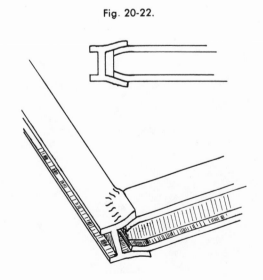

Fig. 20-22.

the previous mentioned colors, silver stain is applied to the undulated or outside of the glass. Silver stain used on a specially prepared white flasked glass called "Kelp" produces some very brilliant effects, but colored glass will also take the stain.

7. **Firing** (Fig. 20-17). The separate painted pieces of glass are fired in the kiln at least once and perhaps several times to fuse the paint and glass. In the school a regular pottery kiln may be used for the firing. In stained glass shops a specially designed kiln called a "lehr" is used. Each kiln requires different handling according to its special construction, but in every case the temperature should be raised slowly and after the required heat has been reached the glass should be cooled slowly. The ultimate heat to which the kiln is fired

Fig. 20-23.

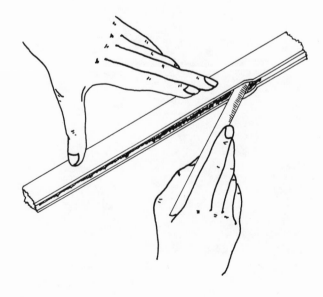

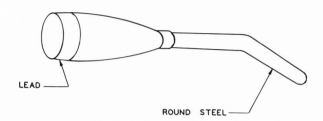

LEAD

ROUND STEEL

Fig. 20-24 Lathekin.

is determined by the directions accompanying the pigment used.

The pieces of glass to be fired are placed on iron trays which are completely covered with some heat-resisting powder such as silica, flint, or plaster of Paris. The surface of the powder should be rendered perfectly flat and smooth with a roller or knife blade.

The pieces of painted glass should be firmly nested in the powder with the painted sides uppermost. Care must be taken not to touch the painted surfaces while this is being done. Each piece should be checked to make sure it is flat in the powder and that there are no

air-holes trapped beneath. Air-holes may cause breaks or mounds to form in firing. The edges of the pieces of glass should not touch each other as they might fuse together. The tray should be carefully slipped into the kiln so that no pieces are shifted or upset.

'Frying' refers to a defective condition that might arise in firing. Sometimes the paint will not fuse into the glass but, instead, will dry into little bubbles that are easily wiped off. This condition is called 'frying' and when it occurs the glass must be cleaned and the painting and firing process repeated.

8. **Leading Up** (Fig. 20-18 and 20-19). The separate pieces of glass are joined together by strips of girder-shaped lead called "came" (sometimes called "calm"). It is started with the cutline drawing being spread out on a bench and strips of wood lath being nailed down along an edge and end of the drawing to form a right angle. A border of about 1/4 inch should be left between the drawing and inside edge of the lath.

Before commencing work the lead should be stretched to strengthen and straighten it. This can be done by placing one end in the vise and pulling on the other end with a pair of pliers. After the lead has stretched a little it usually breaks off at the vise or the pliers.

Came leads (Fig. 20-20) come in dozens of different

Fig. 20-25.

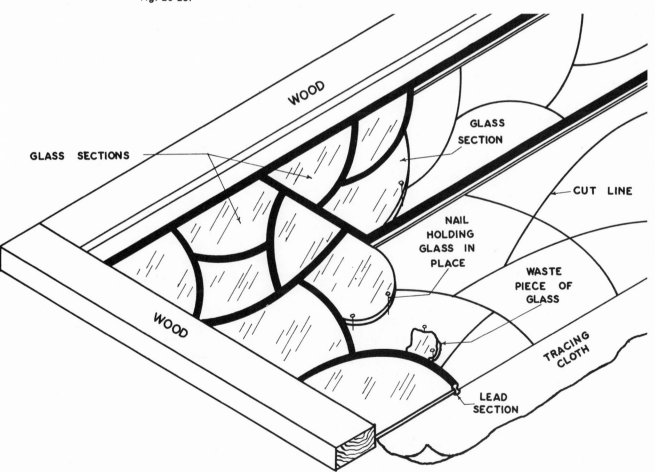

GLASS SECTIONS

WOOD

GLASS SECTION

CUT LINE

NAIL HOLDING GLASS IN PLACE

WASTE PIECE OF GLASS

WOOD

TRACING CLOTH

LEAD SECTION

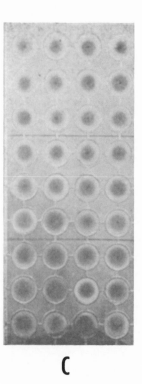

Fig. 20-26. A. Leaded Diamonds Cathedral Glass Random Colors.
 B. Stained Glass Figure Window.
 C. Leaded Rondels.

sizes. For most panels 1/2-inch came lead is used for the outside edge while 3/8-inch lead is used for all the inside. An old scraper sharpened to a fine edge (Fig. 20-21) may be used to cut the lead. It is done by holding the scraper vertically and pressing straight down.

Cut two lengths of 1/2-inch lead about one inch longer than the panel each way and place them inside the laths. Lap joint them at the corner (Fig. 20-22). Fit the first piece of glass tightly into the right angle corner formed by the lead. A strip of narrow lead should next be fitted around the remaining exposed edge or edges of this first piece. The procedure is to cut off a piece of lead a little longer than required and shaping one end to butt up to the outside lead (all inside joints are butted). The lead should then be bent around the glass, marked off to fit, and cut to length. Should the cutting press together the lips of the lead it should be reopened with the lathekin (Fig. 20-23 and 20-24). A second piece of glass may now be fitted and held in position for

leading by small nails lightly tapped in the bench. The back end of the handle of the lathekin or knife is weighted with lead to serve as a hammer for tapping nails. A lead is now fitted around the exposed edges of this piece. This procedure is continued until all pieces are leaded in place (Fig. 20-25).

When all the glass pieces are leaded in place, another two wide came leads are fitted along the open edge and end. These leads are lap jointed to the first two wide calms and to each other. Wood lath is nailed around the remaining sides of the panel. Finally, the corners of the panel are checked for squareness.

Large panels use extra reinforcing. One method of injecting this extra strength is to use steel-coved calm (calm with steel wires running through center) at regular intervals throughout the panel. Another is to attach saddle bars (thin steel rods) to the panel by means of wires soldered to the panel. Large panels are sometimes made in sections and later connected by wiring.

9. **Copper (Tiffany) Foil Method**. In the Tiffany Foil Method, copper foil is used in place of lead between the individual pieces of glass in the construction of stained glass objects. The edge of each piece of glass is wrapped with copper foil, then placed over a pattern and soldered together.

(1)

(2)

(3)

(4)

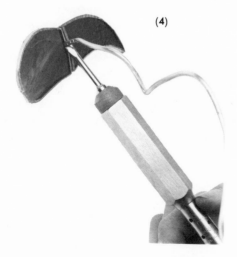

Fig. 20-27. The steps in preparing a stained glass panel the copper (Tiffany) foil method are illustrated above:

1. Press foil to the edge of a cut piece of stained glass. 2. Fold the foil over glass edges firmly. 3. Using a burnishing tool smooth the foil to the glass. 4. Arrange glass pieces into desired pattern, then bead solder over foil edge to hold together. *(Courtesy: C.R. Laurence Co., Inc.)*

The copper foil used comes in three thicknesses: Soft (.001"), Medium (.0015"), and Hard (.002"). It can be had in "Adhesive-Backed" precut widths of 3/16, 7/32, 1/4, 5/16, 3/8, and 1/2 of an inch. The roll usually contains thirty-six (36) yards. On rare occasions, some craftspersons buy their foil by the yard in six inch widths and cut it themselves to the desired widths.

When using the adhesive-backed precut foil (which is the type that is best for the novice) the process is quite simple. Check the edges of each piece of glass that will make up the stained glass panel or object to make sure it is clean, that there is no oil or grease present. Select foil with a width that will cover the edge of the glass and overlap each surface a distance about equal to the thickness of the edge of the glass. The distance the foil laps over on the surface of the glass can be varied depending on the effect desired. The thinner the strip on the surface the more delicate the final effect will be. However, the overlap of the surface should not be made so thin that the object may fall apart. Thicker foil should be used for the thicker, heavier glasses.

Proceed by pressing the copper foil firmly on to the edge and surface of the glass as you work around it. Your fingers are best for this purpose as it presses the foil into the many variations to be found in the glass surfaces.

When all the pieces of glass have the foil pressed on them, place each piece in its proper position on the cartoon. Hold the pieces properly in place with nails as previously described. Make sure everything is properly positioned before proceeding to solder.

The difference between using lead came and copper foil is essentially size and appearance. The flexibility of copper foil makes it easy to use on small pieces of glass and on bent surfaces. Large pieces of glass are held in place better with lead came. Copper foil usually imparts a more delicate, lacy appearance. Lead came, on the other hand, produces a heavier, bolder appearance. Both materials are sometimes used on the same stained glass object.

10. **Soldering**. The soldering should be done with a soldering copper. The size of the tip should vary with the size of the surface to be soldered. For copper foil, a one-fourth inch semichisel point, on a well-tinned soldering copper will do nicely. For soldering thicker lead came, a larger tip could be used.

All joints to be soldered must be **perfectly clean**, coated with a paste flux (such as Nokorode) or a liquid flux (such as Ruby Fluid), and fitted tightly. The solder customarily used consists of 60% tin and 40% lead and is in tape form.

Fig. 20-29. Foil burnishing tool. (Courtesy: C.R. Laurence Co., Inc.)

Fig. 20-28. "Adhesive-backed" pre-cut copper foil that is used in preparing stained glass work the Tiffany method. (Courtesy: C.R. Laurence Co., Inc.)

Each joint must be soldered. **The soldering copper must be left to rest on each joint just long enough to get the solder into the joint and no longer**. Both sides of the panel or object should be soldered. Care should be exercised in turning over the panel; when only one side has been soldered it is quite fragile.

On copper foil work all joints between glass pieces must be "beaded." This consists of running a second layer of solder over all the previously soldered joints to give them a rounded appearance. To do this, the heat of the soldering copper must be lowered (by means of a rheostat or similar device) so that the solder is plastic, but not a runny liquid.

11. **Cementing**. The panel is finally cemented on both sides to make it firm and water-tight. The cement used for this process is very special in nature and can be purchased ready-prepared or can be mixed in the shop. The ready-prepared cement is supplied in two parts, a powder and a liquid, which are mixed together as used. A cement can be made up by adding enough boiled linseed oil and turpentine to equal parts of whiting and plaster of Paris to make a thick paste. A little lamp black and red lead are then added. The mixture may be stored in an airtight container.

The window to be cemented is laid flat on the bench and cement is placed upon it. The cement is then vigorously rubbed in under the lead throughout the panel with a stiff-bristled scrubbing brush. Following this a stick is used to press down the edges of all leads. When one side is completed, the window is turned over and the process is repeated on the other side.

The process is completed by scrubbing both sides of the panel with whiting then resting the panel on edge to dry.

12. **Fitting**. The use to which the finished stained glass panel is going to be put will determine the final fitting it will receive. If it will be used as a window, it must be framed as any ordinary window is framed, in wood or metal. If it is to be a part of a lamp or room separator it can be delicately framed in wood or metal whichever blends into the general design. To make the panel into a mobile a wire can be run in the opening of the outer calm and twisted into a loop at the top for hanging. As an overlay for an existing window it can be cemented or tacked in place without any additional framing.

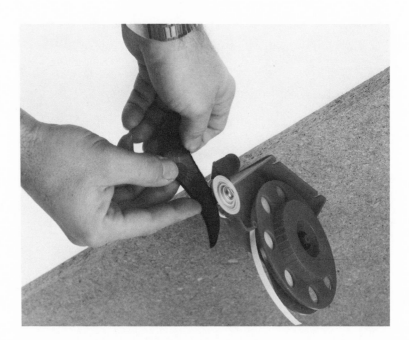

Fig. 20-30. Wrap foiling machine. Dispenses, centers, and crimps foil tape onto all glass shapes. *(Courtesy: C.R. Laurence Co., Inc.)*

Chapter 21
Simulated stained glass

A glass activity that is becoming increasingly popular is simulated stained glasswork. Unlike regular stained glass panels, this one is produced on a single piece of textured glass. Lead stripping is glued to the glass with a pressure-sensitive adhesive and the spaces between the leads are painted with glass stains of varying transparency.

Very attractive wall panels, mobiles, pictures, room dividers, and lamp shades, as illustrated in figure 21-1, may be made with these materials.

Materials Needed

All the following materials may be secured from local craft shops.

Textured Glass (Fig. 21-2). There are several textured designs on glass that can be used for this purpose. The variation in texture on the glass will produce a variation in the effect of the finished panel, therefore the glass used should be chosen in terms of the final effect desired. In general, a light texture is to be preferred.

Glass Stain. These are transparent colors and are available in almost unlimited shades. All stains manufactured by the same concern are compatible and may be blended to obtain other shades. Extender may be

Fig. 21-1.

Fig. 21-2. Attractive textured glasses that can be used for simulated stained glass work. *(Courtesy: American-Saint Gobin Corp.)*

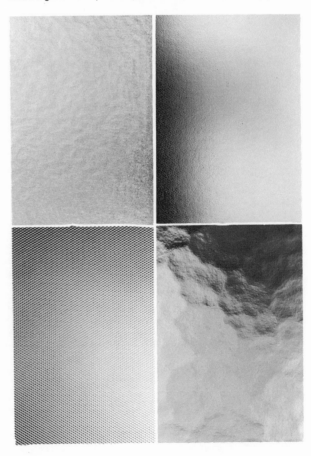

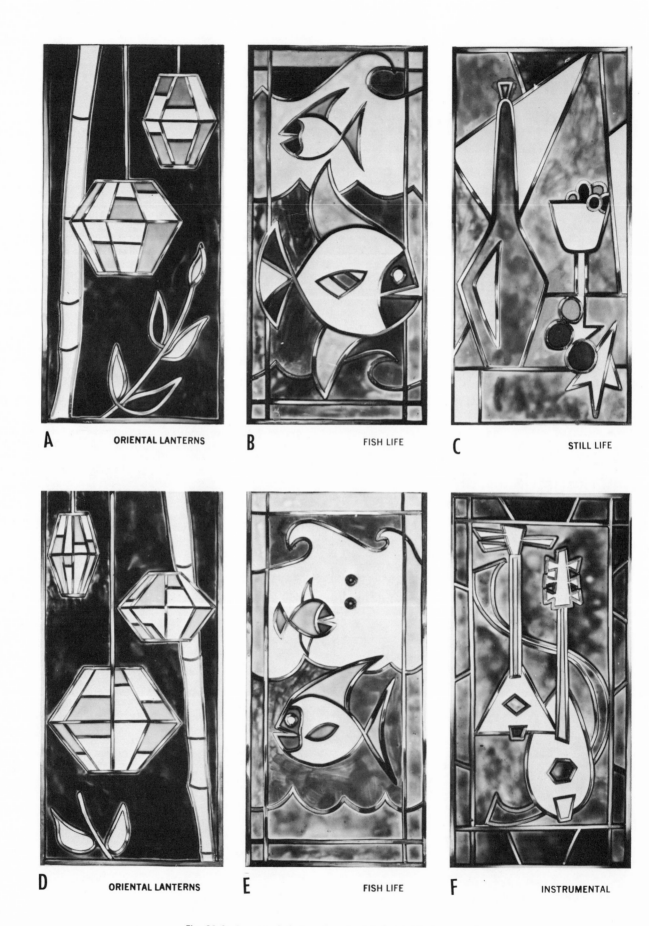

A ORIENTAL LANTERNS B FISH LIFE C STILL LIFE

D ORIENTAL LANTERNS E FISH LIFE F INSTRUMENTAL

Fig. 21-3. Suggested designs for simulated stained glass panels.
(Courtesy: Immerman & Sons)

G   CALYPSO DANCER

H   MEXICAN SIESTA

I   CATS

J   CALYPSO DRUMMER

K   MEXICAN SIESTA

L   CATS

used to lighten colors and obtain tints such as changing red to pink, etc.

Thinner. This is used to thin the glass stain and to clean brushes.

Lead Tape. This dead soft lead tape is available in widths of 1/8" and 1/4" and up to 100 feet in length.

Lead Adhesive. A pressure-sensitive adhesive which when brushed onto lead tape and permitted to dry thoroughly converts lead into a pressure-sensitive tape which will adhere wherever pressed into place.

To construct a simulated stained glass panel the following operations are involved:

1. **Design.** The design should be worked out on paper at a convenient size and proportion capable of being scaled up to the full size of the proposed panel. It should include the general color scheme.

Simulated stained glasswork is especially effective when produced in contemporary designs such as shown in Figure 21-3 A through L as well as in religious designs. (Both kinds of designs, printed on clear film in full size, may be purchased in craft shops.)

2. **Cartoon.** The approved design is next enlarged to full size (See page 152).

3. **Cutting Glass.** A piece of textured glass is cut to the size of the finished panel desired.

4. **Applying Adhesive to Lead Strips.** Start with wide border lead and cut off two pieces equal to the length and two equal to the width of the panel. (Lay balance aside to be used if hanging procedure as shown in Fig. 21-4 is desired.) Cut narrow outline lead into 24-inch lengths. (Shorter lengths can be cut from these as needed.) Straighten lead strips and flatten out on paper surface preparatory to applying adhesive (Fig. 21-4-1). The lead can be easily flattened with the small roller.

Apply adhesive with a brush (Fig. 21-4-2). If brush is not stiff enough, trim end slightly to leave a stiff stubble. Cover entire lead surface liberally and permit to dry for 15 to 20 minutes.

Lead can then be treated the same as any self-stick tape.

Note: Adhesive-treated lead can be stored indefinitely if kept clean and dry. KEEP FINGERS OFF ADHESIVE AT ALL TIMES.

5. **Applying Lead to Border.** Determine top of glass (SMOOTH SIDE). This is side you must use for staining

Fig. 21-4. (Illustrations 1 through 14 *courtesy: Immerman & Sons*)

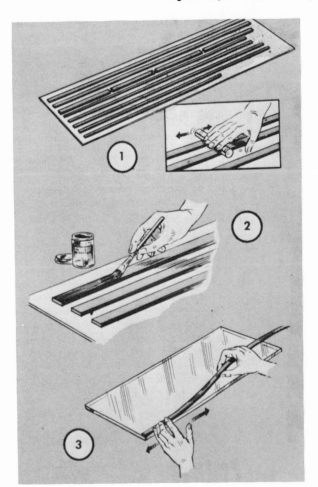

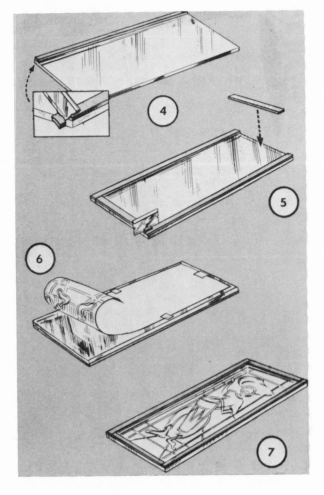

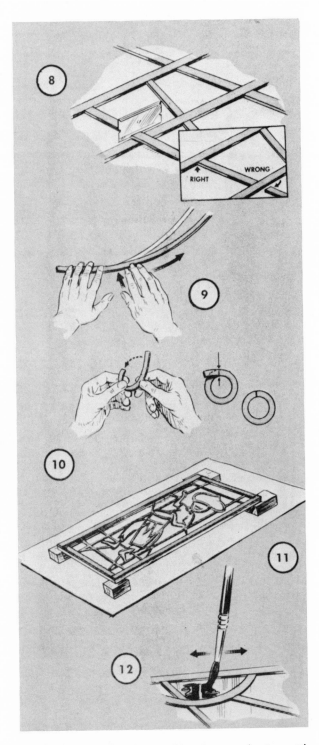

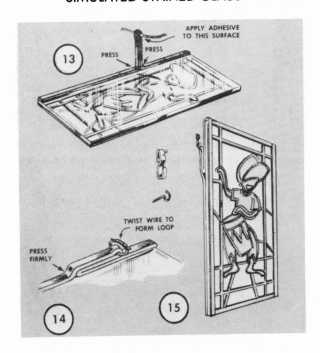

and applying lead. Thoroughly clean and wipe with alcohol to assure a surface free of fingerprints and moisture.

Starting at bottom corner, apply lead to edge of glass (Fig. 21-4-3) starting flush with the bottom and keeping flush with the edge. Press lead in place firmly. Roll with wooden roller.

Cut off excess lead with razor blade (Fig. 21-4-4). Repeat on opposite border.

Apply top and bottom pieces to front (smooth side) of glass, cutting as indicated (Fig. 21-4-5).

6. **Fasten Cartoon Drawing to Glass.** Using masking or scotch tape, fasten the cartoon to the underside of the glass so that design shows through the glass (Figures 21-4-6 and 21-4-7).

7. **Applying Outline Lead.** Position and adhere lead onto glass carefully following pattern designs. Work with lead strips only slightly longer than needed. Cut lead carefully to fit as tightly as possible (Fig. 21-4-8). Accurate joints and pressing lead down into a tight bond with the glass will prevent colors from running into each other. When following curves (Fig. 21-4-9), lay flat lead strip in starting position. Press down and hold. With other hand press lead to conform to pattern. Advance both hands along the strip as you work. Use roller for added pressure and flattening. Small circles and curves can be formed of "unadhesived" lead (Fig. 21-4-10 and adhesive applied after the shape is finished.

Note: If lead will not adhere securely, the glass may be moist or fingerprinted and should be wiped again with alcohol. IF ADHESIVE HAS ACCIDENTALLY ADHERED TO TOP SURFACE OF LEAD, RUB IT OFF WITH FINGERS OR WIPE GENTLY WITH NAPHTHA BEFORE USING STAIN.

8. **Staining Glass.** Remove paper pattern from glass. Work over a white surface and raise the glass about 1/2 inch. This will help you see and control the color intensities (Fig. 21-4-11). For even flow of color, be sure glass is level. Your first application of color will indicate where additional propping is necessary. Use a wide brush for large areas . . . a narrow brush for small areas and a toothpick or matchstick to spread stain into corners or along edges.

Shake stains well before using.

Staining is done by flowing colors . . . NOT brushing them. Dip your brush in stain and transfer a liberal amount to the glass surface. Hold the brush vertically permitting the stain to flow off. Stain will flow over the glass surface and blend. By putting more or less stain on the glass and spreading slightly with your brush (Fig. 21-4-12) you can control color depth. More stain will darken the color intensity, less stain will give a lighter color. Perfection in color evenness IS NOT NECESSARY. Differences in intensity help achieve the authentic stained glass look.

Before stain dries and hardens, remove any you may have brushed onto lead by wiping or if necessary using brush cleaner.

Spreading stain along edge of lead will help seal lead to glass.

CLEAN BRUSHES WELL BEFORE CHANGING COLORS. Squeeze excess stain from the brush with a paper towel or tissue. Then dip it into brush cleaner and repeat. To minimize brush cleaning, use lightest colors first. COVER ALL AREAS REQUIRING THE SAME COLOR BEFORE CHANGING TO ANOTHER STAIN.

9. **Hanging Panel.** To hang panel on wall use 1/4" border tape previously set aside. Cut in half and apply adhesive as before . . . and permit to dry.

a. When panel stains are completely dry turn over and, working from top, firmly press down 6 inches of lead to back of glass (Fig. 21-4-13). Form a 2-inch loop. Insert a 4-inch length of picture wire in loop and continue to apply border lead down to bottom of glass and cut off excess lead. Apply adhesive to top face of loop. Permit to dry and press loop down firmly (Fig. 21-4-14).

b. Repeat on opposite border.

c. To hang panel on wall, use picture hangers or screw eyes (Fig. 21-4-15). Hook previously inserted picture wire over hanger. Adjust by making wire loop larger or smaller.

10. **Additional Staining and Leading Pointers.** Stains can be applied faster by using an eye dropper or small squeeze bottle (depending on size of the area) and then spreading them with a brush. Use plastic bottles for squirt application. Do not store stains in plastic.

Small objects can be dipped and drip-dried.

When staining large free-standing panels it is easier to judge intensity of color by working over a light.

Jars and similar round objects can be flow-stained. Apply stain to top and permit them to flow freely to bottom . . . adding additional stain wherever flow stops. This must be executed rapidly and evenly to avoid streaking. Drippings may be caught in a container and re-used.

Lead may be used as channels to break up large vertical areas.

Lead may be painted gold, silver or any other color desired.

When decorating glass with adhesived lead only, brush clear extender along both lead edges to seal.

Stencil designs may be created with scotch or masking tape.

Unusual designs can be achieved by spattering or stippling stain using sponges, tooth brushes, steel wool, cloth pads and other applicators. Swirling one color over or through another produces interesting marbleized effects.

Removing Glass Stain Color. Color can be removed from flat glass surfaces by scraping it with a razor blade. For cut or textured surfaces use thinner or soak with very hot water or in a dishwasher.

Test Work Surface. When applying Stained Glass colors to materials other than glass, pre-test with stain before proceeding with the project.

Caution: Stained surfaces are not "scratchproof" and should be handled with normal care. Clean stained objects only with warm water and mild soap.

Glass Stains should be used with normal precautions: contains acrylic resin. In case of contact with eyes or repeated contact with skin flush well with water.

Variations

Concerning Leading: There are simulated stained glass kits on the market, such as "Great Glass," that furnish a plastic lead that comes in a sealed tube. When the lead is to be applied, the stem on top of the tube is cut or punctured. When the stem is opened, the tube is pressed and the ejected lead is made to follow the leading lines on the cartoon. The leading technique recommended by the makers of "Great Glass" is as follows:

* Hold the tube like a pencil, with the tip about 1/4-inch above the work. Support elbow on work surface.

* Squeeze tube with a constant even pressure, following the pattern line with a slow continuous motion.

* Stop squeezing at the end of a line and touch tip down on surface of glass. Pull tail of leading back over the leaded line.

* Work from top of design, downward. Be careful not to put sleeves or hands in wet leaded lines.

* *Lead will spread slightly and flatten as it dries.* Squeeze the lead tube with this in mind.

* If leading has spilled, clean it quickly with soap and water.

* Let leading dry for 24 hours after entire pattern is leaded.

Note: You may benefit from practicing leading before actually leading your project. To practice leading technique, place a sheet of plastic food wrap over project pattern and practice technique. Throw plastic sheet away when practice is completed.

When the plastic lead is used it must be finished with a

pewter silver or gold metallic wax. It is applied with your finger, to the dried leading lines. This will give the leading the look of real solder. This step must be done **before** coloring glass with the glass stain paint. If any silver wax gets onto the surface of the glass, it can be removed with lacquer thinner.

Concerning Glass Stain Application: In some of the simulated stained glass kits on the market the glass stain is of a consistency that enables it to be applied with an eye-dropper. The following technique should be used:

* With project lying flat (so glass stain won't run to one side) apply liquid glass stain to each section with the eye-dropper, following color notations on pattern. The leading creates a shallow "receptacle" to hold the paint until it dries. "Puddle" or drip the paint on, then spread paint around evenly over the section using the eye-dropper. Make sure the paint goes to the edges of each section, touching the leading; in fact, slightly overfill the section, since the paint shrinks slightly while drying. However, it should not be on top of the leaded lines.

* If you should drop or spill some of the wrong color onto an uncolored section, it can be removed with lacquer thinner.

* The glass stain colors can be lightened or darkened, and instructions to do this are included on the paint jar or can.

* Let panel dry thoroughly (lying flat) overnight.

Chapter 22
Making slab glass panels

Slab glass panel work (also called Faceted Glass) is similar to stained glasswork in general structure. They differ in that a cement is used to hold the colored glass pieces in place in the slab glass panels, while lead is used in the other. Unlike the delicateness of stained glasswork, slab glass panels have a massivness and ruggedness that gives it a particular appeal of its own. Its use has become increasingly popular in architectural forms.

There are two types of slab glass panels that can be constructed: One is made with thin colored glass pieces, the same as that used in stained glass panels. The other uses glass pieces that are approximately an inch thick.

Thin Glass Slab Panel

To make a slab glass panel using thin glass you would proceed as follows:

1. **Design.** Panels made in this way are usually of simple, formal design and rely upon geometrical shapes to show a rich pattern of color. The geometrical shapes used sometimes have symbolic meaning. For example, in Figure 22-1, the cross is a symbol of passion and atonement. The triangle, with its sides and angles equal, expresses the idea of unity and equality. The circles, having no beginning or end confirm the eternal nature of the world without end.

Figure 22-2 is an example of slab glass work without symbolism. The Byzantine mosaics and the Early Christian mosaics in Rome used such flat, decorative designs. Figure 22-3 depicts a "Nativity Scene" similar to that

Fig. 22-1. Geometrical shapes used in a slab glass panel. *(Courtesy: LaCrosse Glass Co.)*

Fig. 22-2. An example of a slab glass panel without symbolism. *(Courtesy: LaCrosse Glass Co.)*

frequently seen in regular stained glass windows.

The colors used in slab glass panels have significance. **Blue**, the color of sky, symbolizes Heaven and heavenly love. It is the color of truth, because blue always appears in the sky after the clouds have dispelled, suggesting the unveiling of truth. **Green**, the color of vegetation and of spring, and therefore symbolizes the triumph of spring over winter, or of life over death. **Purple** has always been associated with royalty and is the accepted sign of imperial power. It is also the color of sorrow and penitence. **Red** is the color of blood, symbolic of both love and hate. **Violet** symbolizes love and truth or passion and suffering. **Golden Yellow** is the emblem of the sun and of divinity.

2. **Cartoon.** The cartoon is a full size drawing of the design similar to that prepared in stained glasswork (see page 152). The slab glass cartoons differ from the stained glass in that a great deal more room is left between the glass components. The distance allowed between the

Fig. 22-3. A slab glass panel that depicts a Nativity scene. (Courtesy: Sauereisen Cement Co.)

glass pieces will depend upon the size of the panel and thickness of the cement.

In the cartoon each piece of glass should be shown 1/4" larger all around than it calls for on the design. This is the part of the glass that will be covered with the cement. The design should be so arranged that reasonably wide strips of cement will separate the glass shapes even after the 1/4" overlap has been allowed for.

3. **Cutline and Pattern.** The cutline and pattern drawings are, as in stained glasswork, carbon copies or tracings of the cartoon (see page 152). The pattern drawing is cut apart to use as a guide for cutting the glass pieces.

4. **Cutting.** Using the shapes from the pattern drawing the individual pieces of glass are cut to size (see page 153). The 1/4" allowance must be left on each piece of glass.

5. **Form Preparation.** The form is the housing into which the cement is poured (Fig. 22-4). A flat sheet of waterproof plywood or 1/4" Plexiglass, larger in size than the finished panel, is placed on the bench. Over this is placed the cutline drawing.

The pieces of glass are now prepared for placement in the form (Fig. 22-4). This is done by rolling out a slab of clay equal in thickness to the thickness of the panel desired (from 1/2" to 1" is customary). Cut from this slab, designs equal in shape to each piece of glass in the panel except that each piece of clay is cut 1/4" smaller all around than its comparable shaped glass. The clay forms are pressed on the backs of each matching piece of glass and will cover the glass except for a 1/4" border that will protrude all around. The clay must not be allowed to dry or it will cease to grip the glass and will not resist the pressure of the cement.

The pieces of glass with their clay backing are now glued in their proper place on the cutline drawing, or a thin sheet of polyethylene plastic may be placed over the drawing and the pieces of glass placed on it. If the latter method is used, the face of the glass against the plastic should be coated with a thin film of oil or grease to prevent the cement from sticking to them. The edges of the glass must be kept from oil or grease, however, to allow this cement to bind to the glass.

The side boards should now be put in place. They can be made of pine or any other inexpensive wood and are made into a form, the inside measurement of which determines the final shape and size of the panel. The height of the side boards should be equal to the thickness of the panel desired. The side boards are nailed or clamped in place on the cutline drawing. Inside surfaces of the side boards must be oiled so the cement will not adhere to them and to facilitate their removal when the cement slab has hardened. Side boards may be made with a tongue and groove design which will be reproduced in the sides of the cement panel and will allow the panels to be interlocked, when erected.

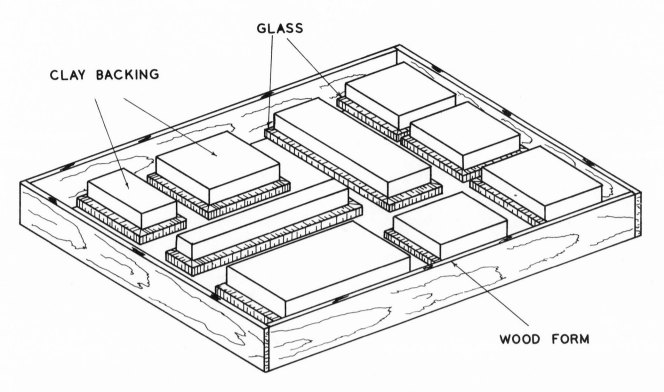

CLAY BACKING

GLASS

WOOD FORM

Fig. 22-4. A prepared slab glass mold.

6. **Cementing.** A special cement is prepared and poured into the form around the glass and the clay backings.

The word "cement" as used in this section does not refer to the ordinary type that is used to make concrete, as this type would shrink and crack in such thin slabs as used in this work. For cement, in this instance, you can use plaster of Paris and keen cement. Both

Fig. 22-5.

CEMENT

GLASS →

OPENING

EXTERIOR FACE→

INTERIOR FACE

CEMENT

work rather well, but produce a finished panel that is quite fragile, until encased in a firm framing.

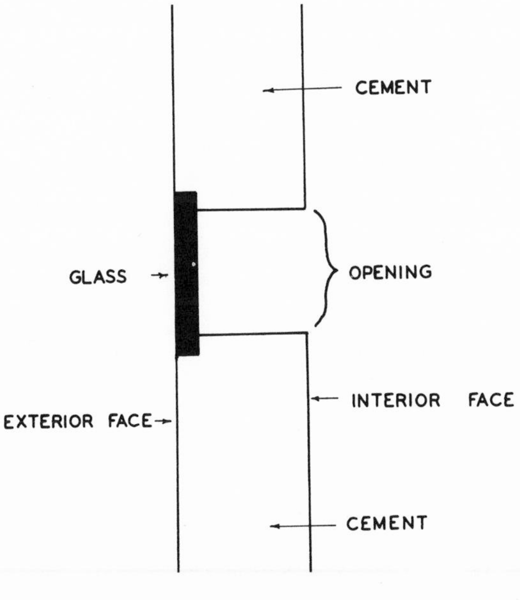

The best cement for slab work is a special preparation called Sanereisen No. 54 Pour-lay Cement produced by the Sanereisen Cement Company of Pittsburgh, Pennsylvania. This gray-colored cement consists of two parts, a powder Filler and a liquid Binder, which are mixed together as used. They may be mixed either by hand or in a mixer. The standard mixture is 4 parts· filler powder to 1 part Binder liquid by weight, however, this is not a critical mix and can be varied slightly when either a thinner or heavier consistency is desired for ease of application.

After the cement is thoroughly mixed the pan or bucket should be vibrated until all the air bubbles have been eliminated, so the cement is free of bubbles which may cause weak spots in the panel.

A pouring pitcher is then filled with the cement mixture and the cement is poured into the form around the pieces of glass to approximately 1/2 the desired final thickness. Care should be taken to avoid creating bubbles and the cement should be spread uniformly over the entire form.

Metal reinforcing rods (usually cold, rolled galvanized steel rods 1/8" by 1/4") should be laid in position on top of the cement. These rods should be clean, dry and free of oil, grease, or alkali. When the reinforcing rods are in place the remainder of the panel depth should be poured with the cement.

Fig. 22-6. A slab glass mold being poured. *(Courtesy: Sauereisen Cement Co.)*

contact with the glass the more difficult its removal.

The cement may be colored if desired, by adding an oilless base coloring pigment to the mixture.

7. **Finishing.** When the cement has hardened sufficiently, the clay backing on each piece of glass should be removed leaving the glass exposed (Fig. 22-5). The side boards are then carefully removed from around the panel, if they are not intended to be used as a permanent framing.

The finished panel may be coated with transparent waterproofing compounds such as silicone-base products to reduce water absorption.

Thick Glass Slab Panels (Fig. 22-6)

The construction of a THICK glass slab panel is identical to that described above for making a THIN glass one with only two exceptions:

1. The glass used in the thick glass panel is usually 1 inch in thickness while the thickness of the glass in the thin glass panel is only 1/16 to 3/16 inches. This thick glass cannot be accurately cut with an ordinary glass cutter and therefore must be shaped with a diamond impregnated saw. A novice wishing to use thick glass in a slab panel would most likely have to order the shapes he wanted already cut to size from the supplier.

2. The other exception can be found in the preceding Section 5 (Form Preparation). All references in this section to a clay backing for the thin glass is not needed in working with thick glass as the glass itself is sufficiently thick to hold back the cement. It is customary, when using 1" thick glass, to pour the cement only 3/4" thick to permit full light transmission at the lower edges of the glass.

Any excess cement which becomes smeared on the glass should be removed by wiping with a damp sponge or cloth, before the cement hardens. When the pouring is completed, wash the tools and equipment with warm water to remove the cement before it hardens.

The poured panels should be permitted to set undisturbed at room temperature for at least 24 hours, although, if desired, the side boards may be removed at the end of 8 hours. After the cement has set up completely hard the baseboard should be removed and the underside cleaned. Paper may be removed with a wire brush or steel wool. Excess cement may be removed with a sharp tool. This cleaning should be done as soon as possible since the longer the cement sets in

Chapter 23
Fiber glass construction methods and materials

Fiber glass reinforced plastics is a relatively new family of production materials whose use has grown tremendously in recent years. "Fiberglass" has become almost synonymous with such well-known products as boats, sports cars, translucent building panels, vaulting poles, fishing rods and missiles.

The reasons for selection of fiber glass reinforced plastics vary considerably by end use, but in every case one or more of the basic properties of these materials have dictated their use—design flexibility, light weight, high strength, corrosion resistance, electrical properties, and low tooling investment.

Fiber glass reinforced plastics—FRP for short—are composed of several component materials which are combined in first part fabrication. In order to evaluate FRP from an end-use standpoint, it is necessary to have an understanding of the laminate properties and costs afforded by the various fabrication processes available and the combination of glass reinforcement and resin systems which can be selected.

Fiber Glass Reinforced Plastics Is
A Combination Of Materials

Fiber glass reinforced plastics is a combination of glass fibers (which furnish the reinforcement), and resin (that forms the combination into a solid mass) (Fig. 23-1). It is this combination of glass and resin that make up a family of production material referred to by different names, such as, **fiber glass reinforced plastics, FRP, fiber glass laminated plastics, fiber glass laminates, laminates,** and other similar terms.

Fiber glass materials (reinforcements) are supplied as continuous strands, fabrics, mats, chopped strands, and other forms. More than ninety per cent of all reinforced plastics use glass. Somewhat less than ten per cent use sisal, cotton, jute, asbestos, synthetic fibers, and metallic fibers.

Many types of resins are used to give a wide range of chemical and thermal performance. Polyester resins are used in about eighty-five per cent of all reinforced plastics. Epoxy resins are most often selected where high performance is a must. Acrylics, melamines, phenolics, silicones, nylon, polystyrene and others are used also.

FOUR PRINCIPLES IN USING
FIBER GLASS REINFORCED PLASTICS

1. MECHANICAL STRENGTH DEPENDS ON COM-

Fig. 23-1. FIBER GLASS REINFORCED PLASTICS (FRP) Glass Fiber Reinforcement + Plastics (Resin+Additive) = FRP (Laminate)

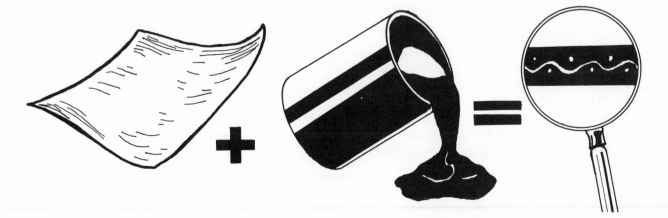

BINED EFFECT OF AMOUNT OF FIBER GLASS REINFORCEMENT AND ARRANGEMENT OF GLASS STRANDS IN FINISHED OBJECT.

2. CHEMICAL, ELECTRICAL AND THERMAL PERFORMANCE RESULT FROM CHOICE AND FORMULATION OF RESIN MIX MATERIALS.

3. MATERIALS SELECTION—PLUS DESIGN AND PRODUCTION REQUIREMENTS—DETERMINE WHICH PROCESS IS USED.

4. TOTAL VALUE RECEIVED—OR ECONOMICAL COST AND PERFORMANCE COMBINED—RESULTS FROM GOOD DESIGN BASED ON JUDICIOUS SELECTION OF RAW MATERIALS AND PROCESSES.

First Principle

Amount of glass. Strength of the finished object is directly related to the amount of glass in the finished object. As shown by the straight line graph of Figure 23-2, strength increases directly in relation to the amount of glass.

A part containing 80 per cent glass and 20 per cent resin is almost four times stronger than a part containing opposite amounts of these two materials.

Arrangement of glass. Equally important, strength is related to the arrangement of glass in the finished object.

Consider three cases: (1) when all glass strands are laid parallel to each other, (2) when half the strands are laid at right angles to the other half, and (3) when the strands are arranged in a random manner.

When all the strands are laid parallel to each other, maximum strength results in one direction. This strength is supplied for end uses such as rocket motor housings, golf clubs and fishing rods.

When half the strands are laid at right angles to the other half, strength is highest in those two directions. Although strength is less than with parallel arrangement, it is still considerable. This strength is supplied for application in boats, airplane wing tips, and swimming pools.

When glass fibers arranged in a random manner, strength is no longer concentrated in one or two directions. Safety helmets, chairs, electrical parts, luggage and machine housings utilize this strength. This random arrangement results in equal strength in all directions. This condition is called isotropic.

Arrangement and amount of glass are related. There is a relationship between the way the glass is arranged and the amount of glass that can be loaded in a given object. This relationship is analogous to stuffing a shoe box full of objects: the neater the arrangement, the more objects that can be placed in a given volume.

By placing continuous strands next to each other in a parallel arrangement, more glass can be placed in a given volume. Glass loadings range from 45 to 90 per cent.

DIRECTIONAL STRENGTH PATTERNS

One Direction (Uni-directional)   Two Directions (Bi-directional)   All Directions (Isotropic)

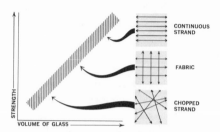

Fig. 23-2. *(Courtesy: Owens-Corning Fiberglas Corp.)*

When half the strands are placed at right angles to the other half, glass loadings range from 55 to 75 per cent.

The relationship of amount of glass, strength characteristics, and arrangement of glass, is shown in Figure 23-2. Note that continuous parallel strands give the highest strength range, bi-directional arrangement gives a middle strength range, and random arrangement gives the lower strength range.

Reinforcements are sold in forms which permit the designer to utilize this directionality to maximum advantage. The basic forms of glass are continuous strand, fabric, woven roving, chopped strand, reinforcing mat, and surfacing mat.

Continuous strand gives unidirectional reinforcement.

Fabric essentially reinforces the object in two directions.

Woven roving gives high strength and is lower in cost than conventional glass fabrics.

Chopped strands give a random reinforcement.

Reinforcing mats are lower in cost than fabric and give random reinforcement.

Surfacing mat gives virtually no reinforcement but gives a decorative and smoother surface finish.

Second Principle

The major resins used in fiber glass-reinforced plastics vary in resistance to corrosion and heat.

Formulation of the resin mix has a similar but less produced effect. By varying ingredients such as filler, pigment and catalyst system, each resin mix can be made to vary in performance.

Resin also helps prevent abrasion of the glass fibers by maintaining the position of the fibers and keeping them separated.

Polyester resins are used in approximately 85 per cent

of all fiber glass-reinforced plastics because they are economical. Other resins in use are epoxies, phenolics, silicones, melamines, acrylics, polyester modified with acrylics. Some thermoplastic resins—nylon, polystyrene polycarbonate and fluorocarbons—are reinforced with fiber glass.

Third Principle

Processes vary in ability to utilize different arrangements of glass, different amounts of glass, different resins. A given combination of raw materials, required to meet performance criteria in a given application, narrows the choice of processes to those which can successfully and economically form the raw material into a completed part.

Process determined by design and production needs. Production flexibility of a process is often the single most important economic factor. If a large number of parts are to be made from one mold, for example, the lowest total cost is probably achieved by using presses and molds and automating materials handling where justified. Conversely, if only a few parts are required, a process minimizing investment in molds and other equipment would be the logical choice. Continuous forming, being used for bench slats, FRP paneling etc. can greatly improve the economics of large volume items.

Fourth Principle

Economical cost and performance result from good design based on judicious selection of both raw materials and process. Proper materials must be combined in a process or processes so that potential performance is realized at an economical cost of manufacturing. Design of the part must take advantage of the material and turn potential limitations into advantages.

FIBER GLASS MATERIALS (REINFORCEMENTS)

How Is It Made? (Fig. 23-3). Fiber glass is glass in fiber form. Fibers are made from two types of glass—"E" and "C" glass. Both types are similar, however, each is

Fig. 23-3. *(Courtesy: Owens-Corning Fiberglas Corp.)*

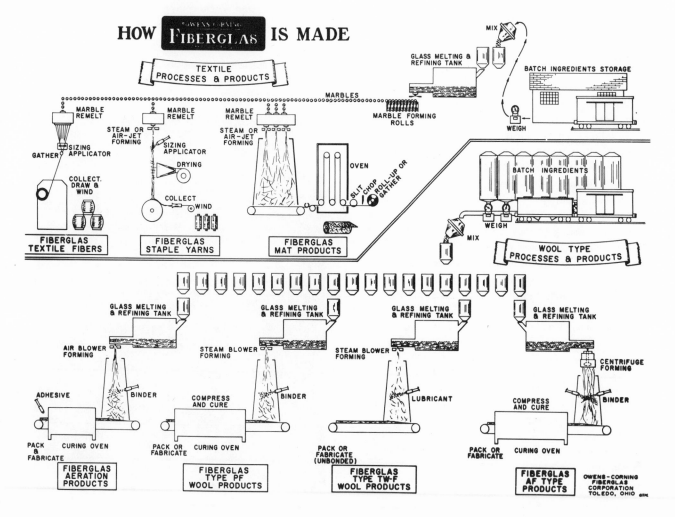

Fig. 23-4. Fiber glass yarn. *(Courtesy: Owens-Corning Fiberglass Corp.)*

better suited for different applications. "E" glass provides superior electrical characteristics and high heat resistance-properties which are particularly suited for electrical insulation applications. "C" glass has superior resistance to corrosive action of most chemicals and is widely used for applications where such resistance is required, i.e., chemical filtration.

There are two basic forms of fiber glass textile fibers—continuous filament and staple fiber. Almost all of the continuous filaments are made from E glass, and staple fibers from C glass.

Both forms of fiber glass begin with the same manufacturing process. Various ingredients are mixed to make a batch of a specified formulation. The batch is then fed into a furnace where it is melted at high temperatures to form glass. Fine, precisely controlled **filaments** are attenuated—drawn or pulled rapidly—from the molten glass. These filaments may be sized and then wound on packages ready for further fabrication.

The difference between continuous filament and staple filaments is the manner in which they are drawn from the furnace.

A continuous filament is an individual fiber of a continuous filament strand. A strand is composed of many continuous, fine filaments—from 51 to 408—depending upon the specific requirements. These continuous filaments are drawn from the furnace at a speed of more than two miles a minute.

A staple fiber is an individual fiber—8 to 15 inches

long. It is formed by jets of air which pull the glass filaments from the furnace and gather them into a strand on a revolving vacuum drum.

Continuous filament and staple fibers can be fabricated into yarns and cords through conventional twisting, plying and cabling operations.

Fiber glass reinforcements are supplied in several basic forms. These forms allow for flexibility in cost, strength and choice of process. Many variations of the basic forms have been developed over the years to meet performance and economic needs which vary over a wide range. Research is constantly conducted to develop new forms and applications.

Variations of each form occur in the manufacture of the basic glass fiber. Most fibers are made by mechanically drawing a filament from a stream of molten glass. During the early stages of processing, a chemical treatment is applied to the surface of the glass fiber.

A number of fibers, or filaments, are formed simultaneously. These "filaments" are collected into a bundle known as a "strand" at a gathering device where the chemical surface treatment is applied. Below the gathering device, the strand is wound into a forming package. The forming package is a delicate intermediary from which shippable forms of fibrous glass are produced as **continuous strand, yarn, roving, chopped strands, fabric,**

Fig. 23-5. A magnified comparison of the Continuous Filament Yarn (top) and the Staple Fiber Yarn (bottom). *(Courtesy: Owens-Corning Fiberglas Corp.)*

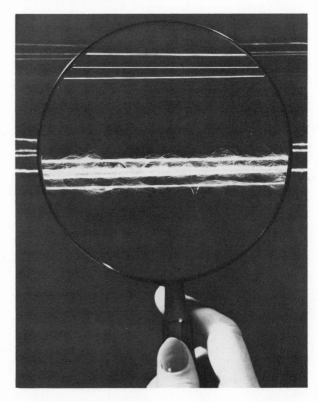

Fig. 23-6. Fiber glass continuous and spun roving. *(Courtesy: Owens Corning Fiberglas Corp.)*

reinforcing mat, surfacing mat, and woven roving.

The following shows the most commonly used reinforcements:

Continuous strand is supplied in the form of yarn and roving. Yarn (Fig. 23-4) is supplied as twisted single-end strands, on tubes. Roving is supplied as untwisted multistrands, wound together on a spool. It is also made in the form of "spun roving" by looping a continuous single strand many times upon itself and holding it together with a slight twist. Yarn and continuous roving are used in high strength parts (Fig. 23-5). Laminates of glass contents as high as 90 per cent are possible.

Spun roving (Fig. 23-6) is the lowest cost glass rein-

forcing material available. Performance of the end product ranges from low to medium strength, depending upon the process used. Glass content of these parts may range up to 45 per cent.

Fabrics (Fig. 23-7) are woven from yarns of various twist and ply construction into a wide range of types, weights and widths. Fabrics are selected by number of factors, but primarily thickness and weight. Weights vary from 2-1/2 to 40 ounces a square yard. Thickness of fabrics varies .003 inches to .045 inches. These same yarns are also fabricated into non-woven fabrics similar to cloth. Both materials come close to duplicating strength properties achieved in the use of continuous parallel strands. Maximum glass content is from 65 to 75 per cent.

Continuous or spun rovings (Fig. 23-8) are woven into coarse, heavy drapeable fabrics called "woven roving." They give high strengths to a part and are lower in cost than conventional fabrics. The weight of woven roving varies from 15 to 27 ounces a square yard. Thickness varies from 0.035 to 0.048 inches. Woven roving is used mainly in the manufacture of large structural objects such as boats and swimming pools and to make plastic tooling for the metal-stamping industry. A newer form of woven roving, called "woven spun roving," produces laminates with superior inter-laminar shear strength. In addition, woven spun roving improves adhesion to other materials and drapeability.

Chopped strands (Fig. 23-9) are made from continuous strands normally chopped into 1/4, 1/2, 1, 1-1/2 or 2 inch lengths. Spun roving is normally cut into 1/4 to 1/2 inch lengths, although it is also used in 1 inch lengths. Milled fibers are hammermilled into 1/32 or 1/8 inch lengths are used for low strengths. Strands of 1-2 inches, from either continuous or spun roving, are used to fabricate large medium strength parts of uniform cross section. Low to moderate strength parts, which require complex cross-sectional walls, utilize these chopped fibers of 1/4 and 1/2 inch.

Reinforcing mats (Fig. 23-10) are made of either chopped

Fabric

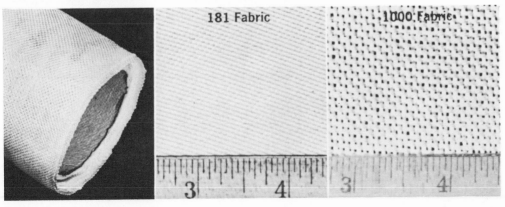

Fig. 23-7. Fabrics are woven from yarns of various twist and ply construction into a wide range of types, weights and widths. *(Courtesy: Owens-Corning Fiberglas Corp.)*

Woven Rovings

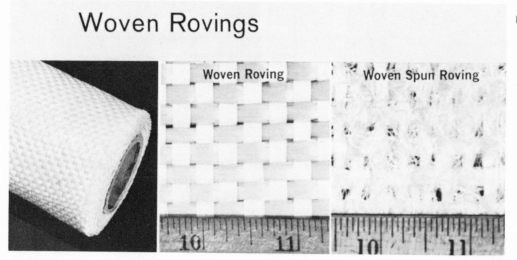

Woven Roving

Woven Spun Roving

10 11 10 11

Fig. 23-8. Continuous or spun rovings are woven into coarse, heavy drapeable fabrics called "woven roving". (Courtesy: Owens-Corning Fiberglas Corp.)

strands or continuous strands laid down in a swirl pattern. Strands are held together by adhesive resinous binders or mechanically bound by "neddling." Lower in cost than woven materials, mats are slightly more expensive than bulk chopped strands or roving. Chopped and continous strand mat weight varies from 3/4 to 3 ounces a square foot. Mechanically bonded or needled mat weight varies from 2 to 10 ounces a square foot. Reinforcing mats are used for medium strength parts with uniform cross-section.

Surfacing mat (Fig. 23-11) is often used (with other reinforcements) for appearance and weathering. It covers irregularities by drawing a slight excess of resin to the surface next to the mold. This resin-richness compensates for resin shrinkage and forms a smooth surface. The reinforcing value of surfacing mat is lower than other forms because glass filaments are not designed for strength. Cutting down on the diameter of the reinforcing strands in mat or roving, or using fine-yarn fabric, also helps reduce surface irregularities. Surfacing mats are available from .010 to .030 inches.

RESINS USED FOR FIBER GLASS REINFORCED PLASTICS

Resins used with FRP contribute mechanical strength, determine electrical, chemical and thermal performance, and prevent abrasion of fibers by keeping them separated. By varying resin ingredients and their treatment, many different materials are produced for a wide range of end uses.

Thermosetting and thermoplastic are two main classes of resins. Thermosetting resins become hard when heated and further heating will not soften them—the action is irreversible. Thermoplastic resins become soft when heated and hard when cooled—the action is repeatable.

Thermosetting resins undergo a chemical change called polymerization—the linking together of "monomers" to form "polymers." Time and temperature accomplish this effect.

Thermoplastic resins are heated to fluid form before they are injected into the mold. The part can be removed from the mold when sufficiently cooled.

Polyester resin—a thermosetting type—is the work horse of the industry, accounting for approximately 85 per cent of resin materials used. Other thermosetting resins in use are epoxies, phenolics, silicones, melamines, acrylics, and polyester resins modified with acrylics. Other resins are being developed or modified continually to meet the needs of the industry.

Some thermoplastic resins used by the industry with glass reinforcement added include polystyrene, nylon, polycarbonates, and acetals. In many cases the addition of the glass reinforcement doubles mechanical properties and greatly improves dimensional stability.

Various resins, resin characteristics and typical uses are shown in Table I.

Polyesters offer the advantage of a balance of good mechanical, chemical and electrical properties, dimen-

Fig. 23-9. Fiber glass chopped strands. (Courtesy: Owens-Corning Fiberglas Corp.)

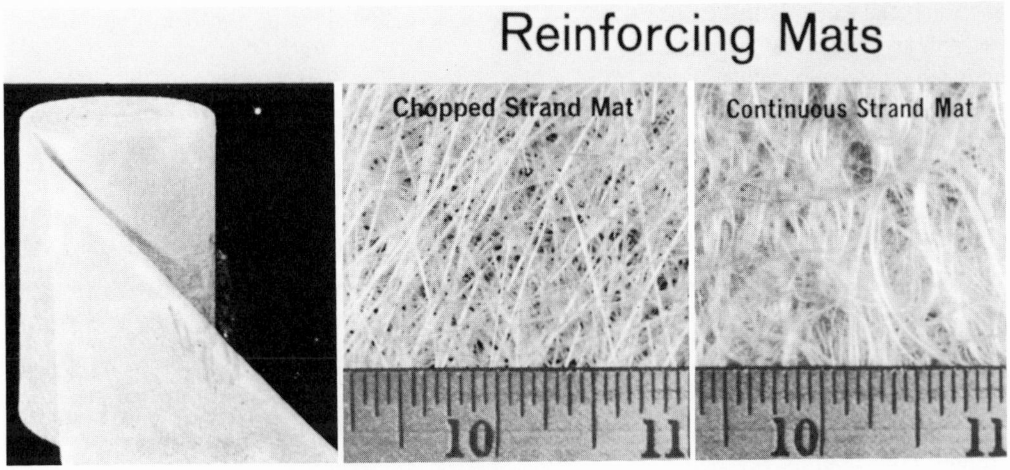

Reinforcing Mats

Chopped Strand Mat   Continuous Strand Mat

Fig. 23-10. Reinforcing mats are made of either chopped strands or continuous strands laid down in a swirl pattern. *(Courtesy: Owens-Corning Fiberglas Corp.)*

sional stability, low cost, and ease of handling. A selection guide for polyester resins, Table II will guide you in selecting, from the many types of polyesters available, the particular type best suited for your end use.

Much of the versatility of polyester resins results from a wide selection of raw materials and basic processing methods available to meet product characteristics that may be required.

When resin, monomer, and catalyst are mixed together, a cross-linking reaction starts. In this reaction,

TABLE — RESINS

	Resin	Characteristics	Typical Uses
THERMOSETS	Polyesters	Simplest and most versatile production techniques of all thermosets, good electrical properties, good mechanical properties and chemical resistance (especially acids).	Used in all markets. Typical applications are: boats, auto and truck bodies, building panels, pipes and ducts, furniture, appliance components.
	Epoxies	Excellent mechanical properties, dimensional stability, chemical resistance (especially alkali), low water absorption, self-extinguishing, good adhesion to other materials, low cure shrinkage, good abrasion resistance.	Plastic tooling, electrical and electronic applications, aircraft and missile components, adhesives, piping, process equipment, storage tanks.
	Phenolics	Good acid resistance, good electrical properties (except arc resistance), high heat resistance.	High pressure electrical grade laminates, printed circuit boards, transmission parts, high temperature applications e.g.: ablative missile nose cones.
	Silicones	Highest heat resistance (950°F), low water absorption, excellent dielectric properties, excellent arc resistance.	Heat barriers in jet and rocket engines, military high temperature applications, electronic equipment.
	Acrylic	Weather resistant, good gloss.	Building and glazing panels.
	Melamine	Good arc, fire, and heat resistance, high impact strength.	Electrical apparatus.
	Diallyl phthalate (DAP)	Good electrical insulation values in arcing atmospheres.	Ignition parts, coil forms, connector insulators.
THERMOPLASTICS	Polystyrene	Low cost, moderate heat distortion point (220°F), good dimensional stability, good stiffness and impact strength.	Large injection molded parts, automotive applications, blower wheels.
	Polyamide (Nylon)	High heat distortion point (498°F), low water absorption, low elongation, good impact strength, good tensile and flexural strength.	Self-lubricating gears, bearings, washers.
	Polycarbonates	Self extinguishing, high dielectric strength, high strengths and stiffness.	Electrical high performance components.
	Styrene-acrylonitrile	Good solvent resistance, good long term strength and appearance.	Battery cases.
	Fluorocarbons (Teflon)	High heat distortion (500°F), self lubricating.	Gasketing, encapsulation of electronic components, bushings and bearings.

Table I. *(Courtesy: Owens-Corning Fiberglas Corp.)*

the unsaturated polyester resin connects with the monomer to form a thermoset polyester resin. When fully cured, thermoset resins have higher temperature resistance than most thermoplastics.

Completion of this reaction is dependent both on the formulation and on the time and temperature balance selected in the design of the formulation. Pressure is not necessarily required for curing, but does effect other factors such as surface smoothness, density, and other process considerations.

The cure itself proceeds in two distinct stages. The first is the formation of a soft gel. Immediately after gelation, the cure proceeds rapidly with considerable evolution of heat which must be properly controlled. This evolution of heat is called an exothermic reaction. Complete cure is obtained without liberation of volatile materials. This last consideration coupled with low pressures accounts in large measure for the simplicity of the fiber glass reinforced plastic molding operation.

Fig. 23-11. Fiber glass surfacing mat. *(Courtesy: Owens-Corning Fiberglas Corp.)*

OTHER MATERIALS USED IN FRP RESIN MIX

The resin mix consists of materials other than the glass reinforcement. These ingredients include fillers, monomer, catalyst, activators, inhibitors, pigments, and mold release.

Fillers. Inorganic fillers such as clay, talc, calcium carbonate, and calcium silicate are used for economic

TABLE — POLYESTER RESINS

	Polyester	Characteristics	Typical Uses
CLASSIFIED BY Characteristic of Cured Resin	General purpose	Rigid moldings.	Trays, boats, tanks, boxes, luggage, seating.
	Flexible resins and semi-rigid resins	Tough, good impact resistance, high flexural strength, low flexural modulus.	Vibration damping: machine covers and guards, safety helmets, electronic part encapsulation, gel coats, patching compounds, auto bodies, boats.
	Light stable and weather resistant	Resistant to weather and ultraviolet degradation.	Structural panels, skylighting, glazing.
	Chemical resistant	Highest chemical resistance of polyester group, excellent acid resistance, fair in alkalies.	Corrosion resistant applications such as pipe, tanks, ducts, fume stacks.
	Flame resistant	Self-extinguishing, rigid.	Building panels (interior), electrical components, fuel tanks.
	High heat distortion	Service up to 500°F., rigid.	Aircraft parts.
CLASSIFIED BY Processing Characteristics	Hot strength	Fast rate of cure, "hot" moldings easily removed from die.	Containers, trays, housings.
	Low Exotherm	Void-free thick laminates, low heat generated during cure.	Encapsulating electronic components, electrical premix parts — switchgear.
	Extended pot life	Void-free and uniform, long flow time in mold before gel.	Large complex moldings.
	Air dry	Cures tack free at room temperature.	Pools, boats, tanks.
	Thixotropic	Resists flow or drainage when applied to vertical surfaces.	Boats, pools, tank linings.

Table II. *(Courtesy: Owens-Corning Fiberglas Corp.)*

and performance reasons in fiber glass-reinforced plastics. Economics, surface appearance, strength, resistance to environment, and moldability can be improved by proper selection and use (Fig. 23-12).

Monomer. The resin as supplied by the resin manufacturer contains monomer. Additional monomers, such as styrene or vinyl toluene may be added to the resin in the molder's shop as directed by the resin manufacturer to vary the viscosity and lower the cost of the resultant resin mix. The addition of excessive amounts of styrene monomer lowers the weathering performance of the molding.

Catalysts. Organic peroxides are widely used (to catalyze polyester resins) due to convenience, cost, speed and control of action. Polymerization of polyester resins can also be achieved by exposure to radiation, ultraviolet light, and heat.

Some factors regulating the catalyst choice are:

1. Desired temperature, allowable time and/or operation.
2. Type of monomer or mixture of monomers employed.
3. Desired pot life.
4. Desired gel and cure times.
5. Influence of sunlight and weathering on product performance.
6. Any other specifically required properties of the finished end product influenced by the catalyst choice.

The most commonly used organic peroxides are benzoyl peroxide, methyl ethyl ketone peroxide, and lauroyl peroxide.

Epoxy hardening agents take part in the chemical reaction. The two major classes, amines and amides, have different effects on performance.

Activators and Inhibitors. Activators—or accelerators—promote the action of the polyester catalyst to reduce the required processing time. Inhibitors are added to the resin to give control over the cure cycle and to impart adequate shelf life to the material. Both of these materials are normally added or specified by the resin supplier.

Pigments. Reinforced-plastic resins can be colored to nearly any desired shade. This flexibility is one reason for the growth of the reinforced-plastics industry (Fig. 23-13). Choice of pigments affects the difference in reflected and transmitted color, clarity of the resin mix, reactions between colorants and other additives such as catalysts, and performance of the end product in terms of color fastness and resistance to heat.

Mold Release. Easy release of the part from the mold after completion of the curing cycle is normally effected by treating the mold with waxes or silicones to lubricate the surface. At other times such substances as zinc stearate are added to the resin mix in order to facilitate the removal of parts. This is common practice on parts where the draft is low or where removal is difficult.

Other Additives. Other substances are added to the resin mix either by the resin supplier or the molder to impart special performance qualities. Typical of these substances are ultra-violet absorbers. They are added to resin mixes subject to exposure to ultra-violet rays in natural sunlight or fluorescent light.

Flame-proofing substances, such as antimony oxide or chlorinated waxes, are added to give a fire retardant effect. In this case the effect of these additives on outdoor weathering performance must be carefully considered.

SELECTING A MOLDING PROCESS (Fig. 23-14)

Many processes are available to produce the desired combination of design performance and economics. Each process has its own usefulness for combining different kinds and amounts of glass and resin.

The third basic principle states that "Materials determine process. But process defines design and production flexibility." To aid you in selecting a process, the many basic processes can be considered broadly in two classes: open molds and closed molds.

Open Molds (Fig. 23-14). Open Molds are single cavity molds, either male or female, used in processes which require little or no pressure. The principal characteristics of the molded object are:

1. Only one side is finished.
2. Complex shapes may be formed.
3. Large objects may be formed.

The principle characteristics of the open mold processes are:

1. Low investment due to need for only one mold and lack of other equipment, mold materials can be plaster, wood, metal, or reinforced plastics (ab-

Fig. 23-12. Fiber glass, filler and resin give high durability and low cost in these trays. *(Courtesy: Owens-Corning Fiberglas Corp.)*

Fig. 2-13. Pigments add color to this boat made of fiber glass reinforced plastic. *(Courtesy: Owens-Corning Fiberglas Corp.)*

sence of high pressure).
2. Relatively high labor cost.
3. Speedy mold production and easy design changes.
4. Relatively slow production of parts.

Closed Molds (Fig. 23-14). Closed molds are two-piece male and female molds, usually made of metal. The principal characteristics of the molded object are:
1. Controlled surface finish.
2. Two sides finished.
3. Excellent reproduction of detail from part to part.
Principal process characteristics:
1. Low labor cost.
2. Efficient use of raw materials.
3. Highest production rate.
4. Higher equipment and mold cost.

Open or Closed Molds?

In most cases, the first step is to determine the combination of the glass and resin to be processed. The second step is to determine size and configuration of object. The third step is to decide on the quantity required. These decisions place limitations on processes that will accomplish the desired results.

Determine Glass and Resin Combination to be Processed

Different processes vary in their ability to use different forms of glass reinforcement and to achieve proper distribution of the desired amount of glass reinforcement. Some processes can only use short chopped fibers. Others can only use continuous strands. Some processes can only achieve a low percentage of glass reinforcement. In others, as much as 90 per cent of the entire laminate can be glass fibers.

Consider Part Size and Configuration

With uncomplicated shapes and up to a certain size, both open and closed processes can cope equally well with these problems. With large parts, however, a process usually requiring presses and other preparatory equipment might well be eliminated due to physical or economic reasons, so open molds are chosen.

The degree of surface finish required, on both the inside and outside surfaces, is also evaluated. Open processes give a different surface finish on two sides. The mold side is the smooth side.

Determine Parts Needed in a Given Time

This determination permits a choice between closed mold high speed processes resulting in lower cost per part but requiring more investment in molds and equipment, and open mold processes with relatively lower investment but slower production rates, which usually mean higher labor costs. If a large number of parts is required, it is always advisable to give serious consideration to the "closed mold" process that produces the part in the least amount of time with the least amount of labor.

Open and closed mold processes have many variations to give a range of performance and economic characteristics. The variations are explained in the following pages, along with a summary of the advantages and limitations of each, the raw materials used, production rates that can be expected from the processes, and examples of parts now being produced by each process.

A chart is presented at the conclusion of these process descriptions for the reader's convenience. Processes described and explained in the next few pages are:

Open Molding
  A. Hand Lay-Up
  B. Spray-Up
  C. Encapsulation

Fig. 23-14. Open Mold. *(Courtesy: Owens-Corning Fiberglas Corp.)*

D. Filament Winding
E. Centrifugal Casting
F. Continuous Pultrusion
Closed Molding
  A. Matched-Die Molding
    1. Preform
    2. Mat
    3. Fabric
    4. Premix/Molding Compound
  B. Injection Molding
  C. Continuous Laminating

Open-Mold Processes (Hand Lay-Up)

This open-mold process is the oldest and simplest. A cavity or positive shaped mold is made of easily worked materials, such as wood, plaster or reinforced plastics. Fiber glass and resin are placed in the mold manually; squeegees or rollers work in resin to remove air. Layers of materials are added for thickness. For a high quality surface, gel coat is brushed or sprayed on the part prior to lay-up. Variations of hand lay-up are shown below.

Matched-Die Molding

Preparing for Molding—(Preforming): Mat and fabric are tailored for the mold by standard textile cutting

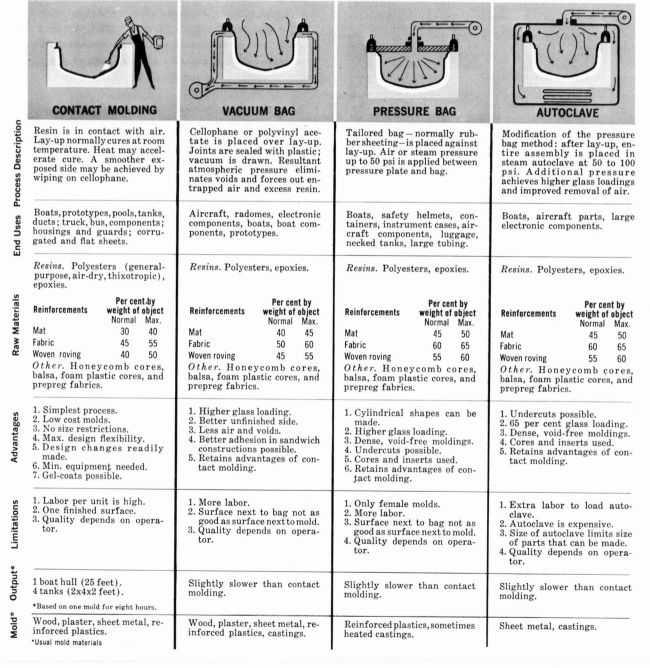

	CONTACT MOLDING	VACUUM BAG	PRESSURE BAG	AUTOCLAVE
Process Description	Resin is in contact with air. Lay-up normally cures at room temperature. Heat may accelerate cure. A smoother exposed side may be achieved by wiping on cellophane.	Cellophane or polyvinyl acetate is placed over lay-up. Joints are sealed with plastic; vacuum is drawn. Resultant atmospheric pressure eliminates voids and forces out entrapped air and excess resin.	Tailored bag—normally rubber sheeting—is placed against lay-up. Air or steam pressure up to 50 psi is applied between pressure plate and bag.	Modification of the pressure bag method: after lay-up, entire assembly is placed in steam autoclave at 50 to 100 psi. Additional pressure achieves higher glass loadings and improved removal of air.
End Uses	Boats, prototypes, pools, tanks, ducts; truck, bus, components; housings and guards; corrugated and flat sheets.	Aircraft, radomes, electronic components, boats, boat components, prototypes.	Boats, safety helmets, containers, instrument cases, aircraft components, luggage, necked tanks, large tubing.	Boats, aircraft parts, large electronic components.

Raw Materials	CONTACT MOLDING		VACUUM BAG		PRESSURE BAG		AUTOCLAVE	
	Resins. Polyesters (general-purpose, air-dry, thixotropic), epoxies.		*Resins.* Polyesters, epoxies.		*Resins.* Polyesters, epoxies.		*Resins.* Polyesters, epoxies.	
	Reinforcements	Per cent by weight of object — Normal / Max.	Reinforcements	Per cent by weight of object — Normal / Max.	Reinforcements	Per cent by weight of object — Normal / Max.	Reinforcements	Per cent by weight of object — Normal / Max.
	Mat	30 / 40	Mat	40 / 45	Mat	45 / 50	Mat	45 / 50
	Fabric	45 / 55	Fabric	50 / 60	Fabric	60 / 65	Fabric	60 / 65
	Woven roving	40 / 50	Woven roving	45 / 55	Woven roving	55 / 60	Woven roving	55 / 60
	Other. Honeycomb cores, balsa, foam plastic cores, and prepreg fabrics.		*Other.* Honeycomb cores, balsa, foam plastic cores, and prepreg fabrics.		*Other.* Honeycomb cores, balsa, foam plastic cores, and prepreg fabrics.		*Other.* Honeycomb cores, balsa, foam plastic cores, and prepreg fabrics.	

Advantages	1. Simplest process. 2. Low cost molds. 3. No size restrictions. 4. Max. design flexibility. 5. Design changes readily made. 6. Min. equipment needed. 7. Gel-coats possible.	1. Higher glass loading. 2. Better unfinished side. 3. Less air and voids. 4. Better adhesion in sandwich constructions possible. 5. Retains advantages of contact molding.	1. Cylindrical shapes can be made. 2. Higher glass loading. 3. Dense, void-free moldings. 4. Undercuts possible. 5. Cores and inserts used. 6. Retains advantages of contact molding.	1. Undercuts possible. 2. 65 per cent glass loading. 3. Dense, void-free moldings. 4. Cores and inserts used. 5. Retains advantages of contact molding.
Limitations	1. Labor per unit is high. 2. One finished surface. 3. Quality depends on operator.	1. More labor. 2. Surface next to bag not as good as surface next to mold. 3. Quality depends on operator.	1. Only female molds. 2. More labor. 3. Surface next to bag not as good as surface next to mold. 4. Quality depends on operator.	1. Extra labor to load autoclave. 2. Autoclave is expensive. 3. Size of autoclave limits size of parts that can be made. 4. Quality depends on operator.
Output*	1 boat hull (25 feet). 4 tanks (2x4x2 feet). *Based on one mold for eight hours.	Slightly slower than contact molding.	Slightly slower than contact molding.	Slightly slower than contact molding.
Mold*	Wood, plaster, sheet metal, reinforced plastics. *Usual mold materials	Wood, plaster, sheet metal, reinforced plastics, castings.	Reinforced plastics, sometimes heated castings.	Sheet metal, castings.

Table III. *(Courtesy: Owens-Corning Fiberglas Corp.)*

Spray-up

Encapsulation

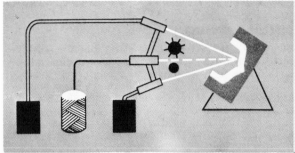

Process Description

Fiberglas and resin are simultaneously deposited in a mold. Roving is fed through a chopper and ejected into a resin stream, which is directed at the mold by either of two spray systems: (1) A gun carries resin premixed with catalyst, another gun carries resin premixed with accelerator. (2) In a second system, ingredients are fed into a single gun mixing chamber ahead of the spray nozzle. By either method the resin mix precoats the strands and the merged spray is directed into the mold by the operator. The glass-resin mix is rolled by hand to remove air, lay down the fibers, and smooth the surface. Curing is similar to hand lay-up.

Milled fibers or short chopped strands are combined with catalyzed resin and poured into open molds to cast terminal blocks, electrical casings, and electronic components and similar objects. The fibers decrease shrinkage and crazing, and increase useful temperature range of the resin system. Cure is at room temperature. A post-cure of 30 minutes at 200°F. is normal.

End Uses

Boats, display signs, prototypes, tank linings, large integral moldings, truck roofs, roofing.

Embedding of coils, windings, transformers, chokes, resistors, transistors, diodes and other electrical parts.

Raw Materials

Resins. Polyesters (same as contact molding), epoxies.

Reinforcements	Per cent by weight of object	
	Normal	Maximum
Continuous roving	30	40 to 50

Other. Core materials or inserts can be easily incorporated.

Resins. Polyesters (low exotherm), epoxies.

Reinforcements	Per cent by weight of object	
	Normal	Maximum
Milled fibers	5	15

Advantages

1. Portable inexpensive equipment, total investment low.
2. Uses lowest cost form of reinforcement (roving).
3. Complex shapes with reverse curves can be formed with minimum waste.
4. Labor for complex shapes is less than with hand lay-up molding due to portability.
5. Low cost molds (same as hand lay-up).
6. Permits on-site fabrication.

1. Simple process, low tooling cost.
2. Can be automated.
3. High materials utilization.
4. Inserts of any size, shape, material or number may be used.

Limitations

1. Surface away from mold is not finished.
2. Difficult process to control.
   a. High materials usage.
   b. Laminate thickness.
3. Labor costs comparable to hand lay-up on simple shapes.

1. Slow process because a slow cure is required.
2. Limited to minimum reinforcement.

Output*

2 boats up to about 28 feet.
1 boat from 28 to 41 feet.
3 to 4 tanks, 2 x 4 x 2 feet.

*Based on one mold for eight hours.

3 to 4 batches or "gangs" of terminal blocks or electronic parts.

Mold*

Wood, plaster, reinforced polyester and epoxy, sheet metal.

Silicone, rubber, plaster, wood, FRP, glass, metals.

*Usual mold materials

Filament Winding

Centrifugal Casting

Continuous Pultrusion

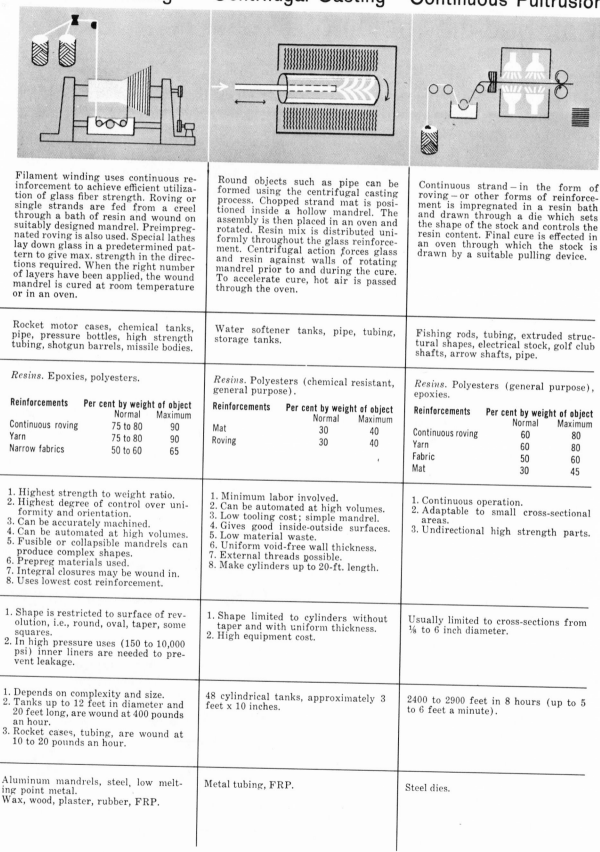

Filament winding uses continuous reinforcement to achieve efficient utilization of glass fiber strength. Roving or single strands are fed from a creel through a bath of resin and wound on suitably designed mandrel. Preimpregnated roving is also used. Special lathes lay down glass in a predetermined pattern to give max. strength in the directions required. When the right number of layers have been applied, the wound mandrel is cured at room temperature or in an oven.

Round objects such as pipe can be formed using the centrifugal casting process. Chopped strand mat is positioned inside a hollow mandrel. The assembly is then placed in an oven and rotated. Resin mix is distributed uniformly throughout the glass reinforcement. Centrifugal action forces glass and resin against walls of rotating mandrel prior to and during the cure. To accelerate cure, hot air is passed through the oven.

Continuous strand – in the form of roving – or other forms of reinforcement is impregnated in a resin bath and drawn through a die which sets the shape of the stock and controls the resin content. Final cure is effected in an oven through which the stock is drawn by a suitable pulling device.

Rocket motor cases, chemical tanks, pipe, pressure bottles, high strength tubing, shotgun barrels, missile bodies.

Water softener tanks, pipe, tubing, storage tanks.

Fishing rods, tubing, extruded structural shapes, electrical stock, golf club shafts, arrow shafts, pipe.

Resins. Epoxies, polyesters.

Reinforcements	Per cent by weight of object	
	Normal	Maximum
Continuous roving	75 to 80	90
Yarn	75 to 80	90
Narrow fabrics	50 to 60	65

Resins. Polyesters (chemical resistant, general purpose).

Reinforcements	Per cent by weight of object	
	Normal	Maximum
Mat	30	40
Roving	30	40

Resins. Polyesters (general purpose), epoxies.

Reinforcements	Per cent by weight of object	
	Normal	Maximum
Continuous roving	60	80
Yarn	60	80
Fabric	50	60
Mat	30	45

---

1. Highest strength to weight ratio.
2. Highest degree of control over uniformity and orientation.
3. Can be accurately machined.
4. Can be automated at high volumes.
5. Fusible or collapsible mandrels can produce complex shapes.
6. Prepreg materials used.
7. Integral closures may be wound in.
8. Uses lowest cost reinforcement.

1. Minimum labor involved.
2. Can be automated at high volumes.
3. Low tooling cost; simple mandrel.
4. Gives good inside-outside surfaces.
5. Low material waste.
6. Uniform void-free wall thickness.
7. External threads possible.
8. Make cylinders up to 20-ft. length.

1. Continuous operation.
2. Adaptable to small cross-sectional areas.
3. Undirectional high strength parts.

---

1. Shape is restricted to surface of revolution, i.e., round, oval, taper, some squares.
2. In high pressure uses (150 to 10,000 psi) inner liners are needed to prevent leakage.

1. Shape limited to cylinders without taper and with uniform thickness.
2. High equipment cost.

Usually limited to cross-sections from ⅛ to 6 inch diameter.

---

1. Depends on complexity and size.
2. Tanks up to 12 feet in diameter and 20 feet long, are wound at 400 pounds an hour.
3. Rocket cases, tubing, are wound at 10 to 20 pounds an hour.

48 cylindrical tanks, approximately 3 feet x 10 inches.

2400 to 2900 feet in 8 hours (up to 5 to 6 feet a minute).

Aluminum mandrels, steel, low melting point metal.
Wax, wood, plaster, rubber, FRP.

Metal tubing, FRP.

Steel dies.

Table V. *(Courtesy: Owens-Corning Fiberglas Corp.)*

techniques. When the shape of the object is too complex to use mat or fabric, a preform is used. A preform is a mat of chopped strands bonded together in the shape of the end product. Economic use of spun roving and continuous roving (low cost reinforcements), characterize this process. Two basic preforming methods are air preformed and water slurry. Two methods of air preforming are directed fiber and plenium chamber.

Molding Process. The mass production method for manufacturing fiber glass-reinforced plastics is matched-die molding (Fig. 23-15). Mat, fabric, or preform reinforcement is combined with a resin mix at the press, either just prior to, or just after, placing in the mold. Heated metal molds form and cure the part at 100 to 300 psi. Molding temperatures usually range from 225° F. to 300° F. depending on thickness, size and shape, cure cycles range from slightly under a minute to five minutes.

High Pressure Laminating, a variation, uses fabrics and mat as reinforcement, utilizes molding pressures to 3,000 psi and curing temperatures to 350° F. Resins are normally phenolics, melamines, silicones and epoxies. Reinforcements are impregnated with resins prior to molding. Thick laminates up to 6 inches are made in this manner.

Rubber Plunger Molding uses preforms, mat, and fabric as raw materials in a variation of the matched die molding process. A closed mold consists of a heated metal female mold, or outer half, and a rubber plunger male mold. The female mold, is sometimes made from an aluminum casting, since the pressures involved are low. Mold cost is reduced by use of this process.

Some advantages of matched-die molding, such as high glass loadings, are gained. Modest undercuts are possible. The process is limited to producing objects which a pressure bag can form. The surface next to the rubber plunger is not as smooth as the surface achieved by a polished metal mold. The system is not practical for use with large molds since the strength of the plunger will not stand the pressures involved.

DESIGN CONSIDERATIONS

To successfully design FRP units requires a logical sequence of considerations. The steps usually followed in this effort are: (1) Determine essential performance and economic requirements; (2) Make initial sketches; (3) Determine materials and process; (4) Make detailed drawing; (5) Determine decorations; (6) Prototype; (7) Build mold.

Determine Performance and Economic Requirements. Over-all performance of a part in service depends largely on thoroughly anticipating and analyzing the performance requirements in terms of physical, mechanical, and chemical criteria. Investigate completely environmental factors, such as probable operating atmosphere and degree of expected abuse. Study carefully, also, the job that the part must do. Economics, volume of production, and comparative costs must also be investigated.

2. Make Initial Sketches. At this point the designer makes his initial decisions on the shape and construction that the part will take in order to best meet the performance and economic requirements. It is important at this point to keep in mind the basic advantages of fiber glass-reinforced plastics and the unique freedom of design allowed the designer.

3. Select Materials and Process. At this point the designer can make a final decision on materials and process, keeping in mind the first three principles:

A. Strength results from arrangement and amount of glass.

B. Chemical, electrical, and thermal performance result from resin.

C. Materials selection—plus design and production needs—determine process.

4. Make Detailed Drawings: In making these drawings —as with any other material—there are certain Design Rules that a designer must follow to get the most out of the individual material. The special considerations that the user of FRP must follow are shown in Table X. At this point, the thickness of a part should be determined, since both performance and cost depend on this decision.

This is especially true in the case of large flat surfaces. Selection of thickness will effect the cost significantly

Fig. 23-15. Matched die mold. *(Courtesy: Owens-Corning Fiberglas Corp.)*

since the raw materials used normally constitute over 50 per cent of the total manufacturing cost. Thickness of typical parts is shown in Table X. Use this guide.

5. Determine Decorations. The materials used as decorations in FRP may effect the strength, thickness, flexibility and shape of the object. Certain decorative materials cannot be bent or shaped to conform to sharp or successive bends and curves. Discoloration is another consideration for decorations. Some decorative materials fade when subjected to the effects of the resins or the heat required in the formation of the FRP. Others discolor the FRP product.

The decorative material should be tested on small pieces of the same FRP material in which it will be used. The sample should be subjected to the same forming conditions as the final piece will be.

6. Prototype. A prototype can now be constructed for use in testing and evaluating the design as near as possible. In the case of parts to be molded with metal molds, a compromise can be made to employ hand lay-up methods. If this is done, advice should be sought from experienced prototypers to assure that mechanical performance is duplicated.

Exact duplication of the finished surface is not possible in this case. Samples of other moldings already in production can be studied and material and methods

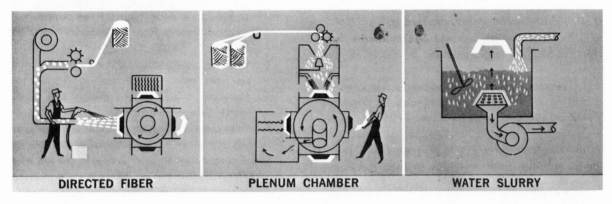

	DIRECTED FIBER	PLENUM CHAMBER	WATER SLURRY
Process Description	Roving is cut into 1 to 2 inch lengths of chopped strand which are blown through a flexible hose onto a rotating preform screen. Suction holds them in place while a binder is sprayed on the preform and cured in an oven. The operator controls both deposition of chopped strands and binder.	Roving is fed into a cutter on top of plenum chamber. Chopped strands are directed onto a spinning fiber distributor to separate chopped strands and distribute strands uniformly in plenum chamber. Falling strands are sucked onto preform screen. Resinous binder is sprayed on. Preform is positioned in a curing oven. New screen is indexed in plenum chamber for repeat cycle.	Chopped strands are pre-impregnated with pigmented polyester resin and blended with cellulosic fiber in a water slurry. Water is exhausted through a contoured, perforated screen and glass fibers and cellulosic material are deposited on the surface. The wet preform is transferred to an oven where hot air is sucked through the preform. When dry, the preform is sufficiently strong to be handled and molded.

Raw Materials

Reinforcements	Per cent by weight of object Normal	Maximum
Continuous roving	35	50
Spun	35	50

Other. Preform binders (polyester and acrylic).

Reinforcements	Per cent by weight of object Normal	Maximum
Continuous roving	35	50
Spun roving	35	50

Other. Preform binders (polyester and acrylic).

Reinforcements	Per cent by weight of object Normal	Maximum
Chopped strands	35	50
Chopped spun	35	50

Other. Cellulose fibers used as preform binders.

Output*

Safety helmet: 960 (120 per hour).
Chair: 480 (60 per hour).
17-ft. boat: 40 (5 per hour).

Safety helmet: 960 (120 per hour).
Chair: 480 (60 per hour).

Safety helmet: 960 (120 per hour).
Chair: 480 (60 per hour).

*Based on one preform machine for 8 hours.

Comparison of Processes for Preparing Preforms

	Directed Fiber	Plenum Chamber	Water Slurry
Direct labor	Most	Moderate	Least
Waste	Most	Moderate	Least
Uniformity	Least	Moderate	Most
Direct labor skill	Most	Moderate	Moderate
Set-up time and cost	Lowest	Medium	Highest
Cost of preform screens	Least	Moderate	Most
Inserts in preforming	Easy	Impractical	Impractical
Wall thickness	Controllable by operator	Moderate changes	More abrupt changes than air plenum
Production change flexibility	Maximum	Medium	Minimum
Inside radii (inches)	⅛	⅛	1/32
Walls (thick to thin ratio)	2:1	1.5:1	3:1

Table VI. *(Courtesy: Owens-Corning Fiberglas Corp.)*

duplicated to insure that the anticipated surface finish is achieved.

7. Build Mold. After completion of tests on the prototype, modifications are made in design on the detailed drawing. The mold is then built. Remember that on an open-mold the surface of the FRP that is against the mold will be smooth while the other will not.

In a matched-die mold both sides will be smooth.

SUGGESTED PROCEDURES FOR SCHOOL AND HOME WORKSHOP

Having read the preceding to understand the nature of the product and its fabrication methods, you are now ready to prepare a FRP product of your own. It is suggested that your first effort be the hand lay-up method in an open contact mold. When you have mastered this technique you can proceed to the closed molds and the more complex techniques.

The Open Mold process is discussed on page 183. This type of mold is not subjected to any pressure and therefore can be made from wood, plaster, keen cement, fired clay, metal, reinforced plastic, etc. Any material that can be shaped and whose surface can be given a very smooth finish is acceptable (Fig. 23-16). Small dishes, trays, etc. may be turned out on the wood lathe. Free

Fig. 23-16. An open mold made of wire and plaster with steel bracing. *(Courtesy: Owens-Corning Fiberglas Corp.)*

form design may be molded of plaster, clay, etc. Large molds may be built up of a few different materials, such as making a large wood skeleton frame then covering it with chicken or hardware wire and burlap or canvas then coating it with plaster for a surface.

To Prepare The Mold Surface you must first polish it

		PREFORM MOLDING	MAT MOLDING	FABRIC MOLDING
End Uses		Seating, machinery housings, safety helmets, engine shrouds, automotive parts, truck cab parts, luggage, washbaskets, trays, materials handling containers.	Machinery, cabinets, trays, electrical flat sheet, printed circuit boards, caskets.	Aircraft, missile components; electrical laminates for insulators, breakers, buss bars.

Raw Materials		*Resins.* Polyesters (general purpose, flexible, semi-rigid, chemical resistant, flame resistant, high heat distortion, extended pot life).
		
		Reinforcements Per cent by weight of object
		                              Normal Max.
		Chopped strand preforms 30 to 40 50
		
		Other. Surfacing mat.
			Resins. Polyesters (same as preform), phenolics, melamines, silicones, epoxies.
			
			Reinforcements Per cent by weight of object
			                            Normal Max.
			Chopped strand mat 30 to 40 50
			
			Other. Surfacing mat.
				Resins. Polyesters (same as preform), phenolics, melamines, silicones, epoxies.
				
				Reinforcements Per cent by weight of object
				                          Normal Max.
				Fabrics 55 to 65 75
				Prepreg fabrics 55 to 65 75
				Non-woven prepregs 55 to 65 75

Advantages		1. Most economical on complex shapes. 2. Uses lowest cost reinforcements. 3. Easily automated. 4. Can work to closer wall thickness tolerance than mat or fabric as preform thickness can be varied over a continuous range. 5. Imbed inserts possible, both decorative and functional.	1. Most economical on large flat or simple shapes. 2. Uniform wall section easily molded.	1. Higher physical strengths than preform or mat laminates due to higher glass contents possible. 2. Laminates over 6 inches thick are possible with prepreg fabrics. 3. Thin laminates to .010 inch are possible.
Limitations		1. Maximum molded wall thickness is ¼ inch. 2. Largest size part made is 17 feet.	1. Economically limited to simple shapes. 2. Wall thickness limited to 0.25 inch.	1. Relative high cost of fabrics. 2. Economically limited to simple shapes.
Output*		Safety helmet: 248 (31 per hour). Chair: 112 (14 per hour). 17 ft. boat: 24 (3 per hour). Auto underbody: 40 (5 per hour). *Based on one mold for eight hours.	Tray: 200 (25 per hour). 4 ft. x 8 ft. panel: 160 (20 per hour).	4 ft. x 8 ft. electrical panel: 160 (20 per hour). 0.5-inch-thick laminates: 8 to 16 (1 to 2 per hour).
Mold*		Cast steel, forged steel, cast aluminum, cast iron.	Cast steel, forged steel, cast aluminum, cast iron.	Cast steel, forged steel, cast aluminum, cast iron.

*Usual mold materials

Table VII. *(Courtesy: Owens-Corning Fiberglas Corp.)*

Premix/Molding Compound

Injection Molding

Continuous Laminating

Process Description

Prior to molding, glass reinforcement, usually chopped spun roving, is thoroughly mixed with resin, pigment, filler, and catalyst. The premixed material can be extruded into a rope-like form for easy handling or may be used in bulk form.

The premix is formed into accurately weighed charges and placed in the mold cavity under heat and pressure. Amount of pressure varies from 100 to 1500 psi. Length of cycle depends on cure temperature, resin, and wall thickness. Cure temperatures range from 225°F. to 300°F. Time varies from 30 seconds to 5 minutes.

Transfer molding is a high speed modification of compression molding utilizing premix/molding compounds. The premix is preheated in a chamber (called a pot), then forced into the hot mold cavity for curing. This process is used to mold small, complicated parts incorporating many delicate inserts.

This high production process is designed for use with thermoplastic materials. The glass and resin molding compound is introduced into a heating chamber where it softens. This mass is then injected into a mold cavity that is kept at a temperature below the softening point of the resin. The part then cools and solidifies.

In continuous laminating, fabric or mat is passed through a resin dip and brought together between cellophane covering sheets; the lay-up is passed through a heating zone and the resin is cured. Laminate thickness and resin content are controlled by squeeze rolls as the various plies are brought together.

End Uses

Electrical switchgear, laundry tubs, trays, housings, automotive and appliance components, impeller blades, pump housings.

Small gears, automotive instrument panels, coil forms, clutch parts, bobbins.

Construction panels (corrugated and flat), glazing panels, electrical insulation sheets.

Raw Materials

Resins. Polyester (general purpose, chemical resistant, flame resistant, high heat distortion, hot strength, low exotherm, extended pot life), epoxy.
Molding Compounds. Phenolic, polyester, silicone, melamine, alkyd, diallyl phthalates, epoxy.

Reinforcements	Per cent by weight of object	
	Normal	Maximum
Chopped strands	10 to 30	45*
Chopped spun roving	10 to 30	45

*Exception: 60 per cent with some molding compounds.
Fillers. Up to 65 per cent by weight.

Resins. Polystyrene, nylon, polycarbonate, teflon, styrene-acrylonitrile.

Reinforcements	Per cent by weight of object	
	Normal	Max.
Chopped roving	40	50
Milled fibers	10	20

Resins. Polyester (weather resistant, flame resistant), acrylic.

Reinforcements	Per cent by weight of object	
	Normal	Max.
Mat	25	35
Fabric	35	45

Other. Surfacing mat is used to improve weather resistance.

Advantages

1. Unlimited part and cross-sectional configuration.
2. Can mold inserts and attachments.
3. Low cost mass-production using low cost molding materials — 14¢/lb. and up. High performance molding compounds are usually more expensive.
4. Parts weighing 200 pounds have been made.
5. Molding to close tolerances possible.
6. Direct labor is low.
7. Molded in holes and threads.

1. Automated process used on high production runs.
2. Low direct labor.
3. High reproducibility.
4. Complex details easily molded.
5. Very small, delicate precision parts possible.

1. No limit to length of panels produced.
2. Automated process.
3. Low tooling cost.
4. Wide variety of surface textures.
5. Many different shapes.
6. Uniform wall thickness.

Limitations

1. Strengths generally lower than preform.
2. Knit lines may cause weakness.

1. Part size limited.
2. High priced molds.

1. 3/32-in. max. wall thickness.
2. Uneconomical for short runs.

Output*

Small motor housings: 240.
Laundry tubs: 80.

Fiberglas-reinforced thermoplastics give molding cycles 25 to 50% faster than with unreinforced plastics.

Part Weight	Mold Cycle
Under 8 ounces	15 to 30 seconds
8 to 16 ounces	30 to 60 seconds
Over 16 ounces	60 to 90 seconds

Construction panels, 4 feet wide (8 to 10 feet a minute).

*Based on one mold for eight hours.

Mold*

Steel, cast iron.

Steel.

Steel.

*Usual mold materials

Table VIII. *(Courtesy: Owens-Corning Fiberglas Corp.)*

until it is perfectly smooth (scratches, dents, and other defects in the mold will be duplicated in the FRP. The surface is then sealed. The sealer used will depend on the material the mold is made of. For dry wood, plaster and other porous materials you can use lacquer, shellac, varnish, enamel, paint, commercial sealers or plastic resin. Plastic resin is the best. A few coats of sealer must be applied so the final coat can be buffed without cutting through to the original mold. Metal molds can be highly polished, plated, or given a lacquered and buffed finish.

The Mold Release is a layer of materials that are applied to the finished mold surface to make it easy to remove the shaped and hardened FRP. This consists of two coats of hard paste wax (automobile or floor wax will do) that is applied and buffed with a clean soft cloth to the finished mold surface, followed by a coat of a material that is called mold release (Fig. 23-17).

	Process	Resin	Fiberglas	Normal Per Cent Fiberglas By Weight	Maximum Filler Per Cent By Weight	Molding Temperature °F	Molding Pressure psi	Molding Cycles	Finished Surfaces	Size of Products To Date
OPEN MOLD	Hand Lay-up									
	Contact Molding	Polyester Epoxy	mat fabric woven roving	30 45 40	none	70 to 110	0	30 min. to 1 day	1	small prototypes to 1 piece boat hulls, 80 feet long
	Vacuum Bag	Polyester Epoxy	mat fabric woven roving	40 50 45	none	70 to 110	12 to 14	30 min. to 1 day	1	same as above
	Pressure Bag	Polyester Epoxy	mat fabric woven roving	45 60 55	none	70 to 220	50	30 min. to 1 day	2	same as above
	Autoclave	Polyester Epoxy	mat fabric woven roving	45 60 55	none	70 to 250	50 to 100	30 min. to 1 day	2	only limited by size of autoclave
	Spray-up	Polyester Epoxy	continuous roving	30	none	70 to 110	0	30 min. to 1 day	1	boats to 30 feet, only limited by size of mold
	Encapsulation	Epoxy Polyester	milled fibers	5	none	70 to 110	0	120 min. to 1 day	1	very small electronic parts
	Filament Winding	Epoxy Polyester	continuous roving yarn narrow fabrics	75 75 50	none	70 to 150	wrapping tension	10 lb. to 400 lb./hr.	1	12 feet by 20 feet tanks to small tubes
	Centrifugal Casting	Polyester	mat continuous roving	30 30	none	180 to 200	20 to 40	10 to 20 min.	2	tubes up to 20 feet long
	Continuous Pultrusion	Polyester Epoxy	continuous roving yarn fabric mat	60 60 50 30	none	180 to 220	0	up to 5 ft. to 6 ft./min.	1 or 2	1 inch wall-thickness, up to 6 inches diameter, any length extrudable shape
CLOSED MOLD	Matched-die Molding									
	Preform: directed fiber, plenum chamber, water slurry	Polyester	chopped strands	30	40	225 to 300	100 to 300	1 to 5 min.	2	from safety helmets to 17-foot boat hulls (directed fiber); from safety helmets to chairs (plenum chamber and water slurry)
	Mat	Polyester Phenolic Melamine Silicone Epoxy	mat	30	40	225 to 350	100 to 3,000	1 to 10 min.	2	from small trays to 4 feet by 8 feet panels
	Fabric	Polyester Phenolic Melamine Silicone Epoxy	fabrics, prepreg woven and non-wovens	60	none	225 to 350	100 to 3,000	1 to 30 min.	2	panels up to 5 inches thick, 4 feet by 8 feet
	Compression Molding of Premix/ Molding Compound	Polyester Epoxy All molding compounds	continuous roving	25 25	65	250 to 350	250 to 2,000	0.5 to 5 min.	2	housings; electrical switch gear to laundry tubs; objects to 200 pounds
	Transfer Molding of Premix/ Molding Compound	Polyester Epoxy All molding compounds	chopped strands chopped roving	25 25	65	250 to 350	1,000 to 5,000	0.5 to 5 min.	2	electronic components
	Injection Molding	Thermoplastics: nylon, Teflon, polystyrene etc.	continuous roving milled fibers	40 10	5	275 to 750	10,000 to 25,000	0.25 to 1.5 min.	2	under 0.5 pound to 2 pounds
	Continuous Laminating	Acrylic Polyester	mat fabric	25 35	none	190 to 250	5 to 20	40 sq. ft. to 50 sq. ft./min.	2	4 foot-wide panels, unlimited length

Table IX. PROCESS COMPARISON GUIDE. (*Courtesy: Owens-Corning Fiberglas Corp.*)

TABLE   — DESIGN RULES

		Spray-up, Hand lay-up (contact)	Hand lay-up (Pressure Bag)	Filament Winding	Continuous Pultrusion	Matched-die molding Premix, Molding compound	Preform, Mat
Minimum Inside Radius (inches)		¼	½	⅛	NA	1/32	⅛
Molded-in Holes		Large	Large	NR	NA	Yes	Yes
Trimmed-in Mold		No	No	Yes	Yes	Yes	Yes
Built-in Cores		Yes	Yes	Yes	NA	Yes	Yes
Undercuts		Yes	Yes	No	No	Yes	No
Minimum Draft Recommended (degrees)		0°	5°	2°-3°	NA	1°	1°
Minimum Practical Thickness (inches)		.060	.060	.010	.037	.060	.030
Maximum Practical Thickness (inches)		.500	1	3	1	1	.250
Normal Thickness Variation (inches)		±.020	±.020	±.010	±.005	±.002	±.008
Maximum Thickness Buildup		As Desired	As Desired	As Desired	NA	As Desired	2 to 1 Max.
Corrugated Sections		Yes	Yes	Circumferential Only	In Longitudinal Direction	Yes	Yes
Metal Inserts		Yes	Yes	Yes	No	Yes	Yes
Surfacing Mat		Yes	Yes	Yes	Yes	No	Yes
Maximum Size Part to Date (Sq. ft.)		3000	2000	1500	NA	25	200
Limiting Size Factor		Mold Size	Bag Size	Lathe Bed Length & Swing	Pull Capacity	Press cap., Flow	Press Dimens.
Metal Edge Stiffness		Yes	NR	Yes	No	Yes	Yes
Bosses		Yes	NR	No	No	Yes	Yes
Fins		Yes	Yes	No	NR	Yes	NR
Molded in Labels		Yes	Yes	Yes	Yes	No	Yes
Raised Numbers		Yes	Yes	No	No	Yes	Yes
Gel Coat Surface		Yes	Yes	Yes	No	No	Yes
Shape Limitations		None	Flexibility of Bag	Surface of Revolution	Constant Cross Section	Moldable	Moldable
Translucency		Yes	Yes	Yes	Yes	No	Yes
Finished Surfaces		One	One	One	Two	Two	Two
Strength Orientation		Random	Ply Orient.	Depend on Wind	Directional	Random	Random
Typical Glass Ldg. % by Wt.		20-30	45-65	75-90	25-70	10-35	25-40

NR — Not Recommended   **NA** — Not Applicable

Table X. DESIGN RULES. *(Courtesy: Owens-Corning Fiberglas Corp.)*

Fig. 23-17. Mold release is applied on the finished mold. (Courtesy: Owens-Corning Fiberglas Corp.)

Mold release material comes in either water-soluble film form or paste type. The water soluble type may be applied with a rag, brush, or sprayer. The paste type is wiped on with a clean soft rag. The water soluble type can be completely washed off the mold with water. Make sure the mold release is thoroughly dried before proceeding to the application of resin.

Notes Concerning Resin. Heat is required to convert liquid resin to a solid (cured). This heat may be applied from the outside, as is the case with heated metal molds, or may be generated internally (exothermic heat), by the addition of chemicals. The chemicals added to the resin to produce this internal heat are called accelerators and catalyst. Both of these must be present to cure resin at room temperature. Generally, the accelerator (cobalt) is added to the resin by the manufacturer. There are occasions, however, when you might have to add the accelerator. The catalyst (usually methyl-ethyl-ketone-MEK-peroxide) is always kept separate from the resin until a few moments before it is to be used. The resin manufacturer will indicate the amount of catalyst to be added (usually between 4 to 8 drops per ounce—1% to 2%). **Important: (1) Do not permit even the slightest amount of catalyst to get in your main supply of resin as it will cause it to harden rendering it worthless. (2) Never mix accelerator and catalyst together as they could explode.** When you must add accelerator to the resin, mix it in first thoroughly then mix in the catalyst. (3) Never mix anymore resin than is needed at the moment. (4) All additives such as colors, thinners, thickeners, etc. should be mixed in the resin before the catalyst. (5) Remove any resin, resin mix, and catalyst from skin, clothing, and tools before hardening (curing) progresses. Wash with lacquer thinner, then with soap and water. (6) **Do not throw any activated resin mix into waste paper container.** Instead, immerse in water

and discard (when cool) in unburnable trash receptacle.

Apply Gel Coat of resin to the mold on the mold separator. This first coat should be a heavy, full strength resin (about the consistency of heavy cream). Some manufacturers sell a special gel coat resin. When this is not available, regular resin used full strength will do. To prepare the resin, measure out one ounce for each square foot of mold area to be covered. If accelerator must be added (see manufacturer's instructions) mix it in at this time. Now add the catalyst according to instructions and mix thoroughly (Fig. 23-18). Apply an even thick coat (approximately .025") of gel coat with a brush (Fig. 23-19). Allow it to cure completely. Lightly sand the cured gel coat surface with 6/0 or medium steel wool before applying additional coats of resin. The gel coat may also be applied with a special spray gun (Fig. 23-20)

Apply Second Resin Coat over the cured and sanded gel coat. This coat is to be mixed according to the manufacturer's directions. Apply an even full coat. This may be done with a brush, spray gun, or roller with stiff carpet-material texture. **The material following is to be applied while the resin is still in liquid form.**

Apply Decorative Materials at this time if desired. The decoration applied at this time will be permanently embedded in the FRP. If a cloth material is used for the

Fig. 23-18. The catalyst is carefully added to the resin. (Courtesy: Owens-Corning Fiberglas Corp.)

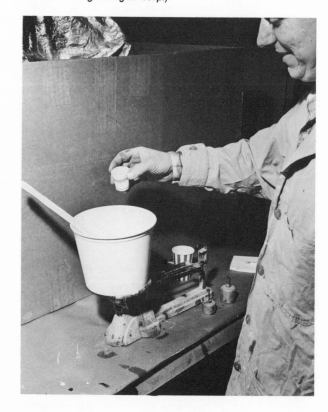

Fig. 23-19. The gel coat is applied with a brush. (Courtesy: Owens-Corning Fiberglas Corp.)

Fig. 23-20. The gel coat may also be applied with a spray gun. (Courtesy: Owens-Corning Fiberglas Corp.)

decoration it should be cut to shape and then pressed into the liquid resin a little at a time making sure it is wrinkle-free and no air bubbles are trapped underneath.

Apply Third Resin Coat At This Time If Needed. If cloth was applied as a decorative material then this resin coat must be applied over it. If such a decoration was not used, then it is not necessary to apply this third resin coat at this time, but should be applied over the first layer of glass material.

This resin coat should be mixed the same as the second and generously poured over the entire material. A stiff brush, roller or squeegee should be used to evenly distribute the resin, press the material down firmly, and work out all trapped air.

Apply First Layer Of Fiber Glass Material. This could be glass fabric, woven desired. Cut the glass material larger than the mold leaving sufficient overhang so the material can be grabbed to pull it flat where needed. The material can be cut with a large scissors or sharp tin snips or regular industrial cutters where large quantities

Fig. 23-21. Large quantities of fiber glass cloth may be cut with regular industrial cutters. *(Courtesy: Owens-Corning Fiberglas Corp.)*

are used (Fig. 23-21). It should be cut beforehand ready to be put in place as soon as the resin has been applied before it.

Starting from one side of the mold, lower the glass material onto the mold a little at a time brushing it down with a stiff brush and using a stippling motion. Work across the mold making sure the material is tight against the mold and there is no trapped air. Pull out all wrinkles. If a wrinkle cannot be removed by pulling, lift the material, cut out the excess that is folded over and press the material down into the resin again. Mats may be applied in sections (Fig. 23-22).

Apply Next Resin Coat. This will be the third or fourth coat depending on whether one was applied over a decorative material or not. This resin coat should be

mixed and applied as described for the third resin coat (Fig. 23-23).

Apply Second Layer Of Fiber Glass Material. This layer of glass material is worked into the resin in the same manner as the first one (Fig. 23-24). By placing this layer of glass material at an angle to the first one you increase the strength of FRP. This glass material does not have to be the same as the first one.

Apply Another Coat Of Resin on top of this second layer of glass material. Mix and apply this coat the same as described for the third coat of resin (Fig. 23-25).

Apply Additional Layers Of Fiber Glass Material And Resin As Desired. There is practically no limit as to the number of layers of fiber glass material that can be laminated as described. Each layer of fiber glass material must be saturated with the resin then pressed firmly in place with no air trapped beneath it. In actual practice, the number of fiber glass layers that are applied are determined by the strength and/or stiffness required. Each layer that is added adds to both. It should be remembered that some fiber glass materials add more strength to the final laminate than others. By using these stronger materials the same total strength may be had with fewer layers in the laminate.

If desired laminate is very thick, layers should be applied in stages of no greater than 1/8 inch each. The chemical reaction in the resin generates a great deal of heat and too thick a laminate can result in damage to the mold as well as degradation to the laminate.

One-fifth of 1% (.2%) of a dark green pigment added to the resin will cause voids or air bubbles to show up through the fiber. Before leaving the laminate to cure, make a last check to see that the fiber has not pulled out of corners or indentations of the mold. Squeeze excess resin off rather than let it puddle in the mold.

Pre-wetting is a variation on hand lay-up that is espe-

Fig. 23-22. Mats may be applied in sections. *(Courtesy: Owens-Corning Fiberglas Corp.)*

Fig. 23-23. The second resin coat is applied. *(Courtesy: Owens-Corning Fiberglas Corp.)*

cially useful on perpendicular or inclined surfaces of the mold when getting the resin evenly distributed would be difficult. In pre-wetting the cloth or mat is saturated with activated resin away from the mold on a wetting table and then rolled onto a cardboard or wooden spool and unrolled on the mold. From this point on the procedure is the same as described for hand lay-up on the mold. Care must be taken with this type of operation, especially with fiber glass material, since the fibers are held together with a polyester bonding material that is soluble in polyester resin. If the piece is too large, it may be impossible to get it into the mold before the mat can no longer be handled. It is usually best in this type of operation to use mat in conjunction with woven roving or cloth to permit handling.

Gun-wetting is a refinement of hand lay-up in that cloth and mat is placed on the mold and the resin sprayed on. This system reduces the pot life problems, wastage and spillage. Either a two-pot gun or a catalyst injector gun may be used.

Cellophane Sheeting placed over the final layer of the laminate while it is being squeezed and rolled, will give the surface a smoother than usual finish. The cellophane, when used, is left in place over the final surface of the laminate until it is cured and then it is peeled off. If the surface of the laminate opposite to the mold is not going to be seen, then there is no point in using the cellophane sheeting as described.

Checking Thickness of a laminate is important. A sliding disc on a pointed steel rod is one cheap, easily-made gauge. The disc is preset at the desired laminate thickness. Then the point, coated with wax and wiped smooth, is inserted through the uncured laminate. When the disc just touches the surface, the desired thickness has been achieved.

Cure The FRP Laminate. If the temperature and humidity conditions are within reason the FRP laminate should cure between one to three hours. The FRP laminate may require additional time for curing due to atmospheric conditions, defective resin, or improper mixing of additives to the resin. **By raising the temperature, increasing the amount of catalyst or acceler-**

Fig. 23-24. A second layer of fiber glass is added. *(Courtesy: Owens-Corning Fiberglas Corp.)*

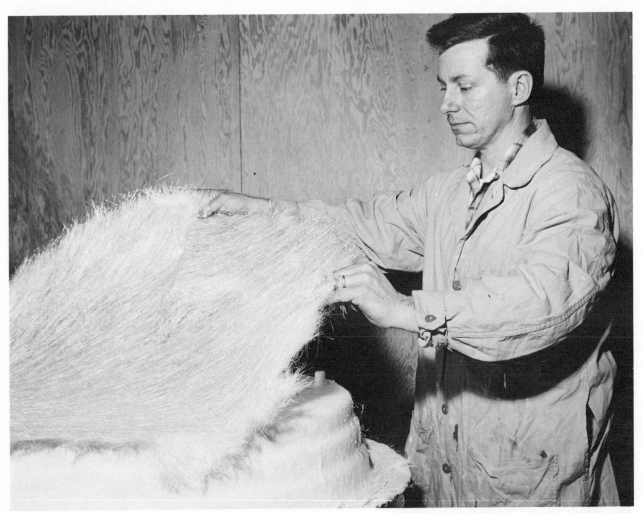

Fig. 23-25. Additional layers of fiberglass material may be applied. (Courtesy: Owens-Corning Fiberglas Corp.)

ator, or lowing the humidity you can speed up the curing process.

Remove FRP Laminate From Mold. The FRP laminate should be removed from the mold when it is cured solid but before it is fully cured. If an attempt is made to remove the FRP laminate before it is sufficiently cured the fiber glass layers may separate. If fully cured, the laminate may be too rigid and be hard to remove. There is a period just before full curing when the FRP laminate has a flexibility that permits the easiest removal (Fig. 23-26).

Use sharpened wooden sticks, putty knife, or dulled wood chisel to separate the laminate from the mold. Be very careful not to injure the laminate or the mold.

Apply Final Resin Coat. When the FRP laminate is fully cured, apply a final coat of resin on the back side.

If you plan to finish the back make this coat rather thick, if not, it can be a thin coat.

Trimming. One time-saving "trick of the trade" is to use a sharp paper-board knife to cut on or near the trim line while the resin is at a point of heavy gel before final cure. This is done while the part is still in the mold. Sample cuts can be made at short intervals to help determine the best time to do this. After ejection, a quick sanding will produce desired dimension.

Cutting off excess laminate can be done in a variety of ways; the best method depends on the size, shape and thickness of the part to be trimmed. Flat, straight-sided moldings can be run through a table saw equipped with an abrasive wheel. When the part is too large or complex for this, hand tools (preferably air powered) are available.

If very little flash is involved, a high speed grinder with 40-60 grit sanding disc can very quickly cut down and smooth the edge. Cutting of large, heavy laminates is done with saber saws, abrasive wheels, hand routers and sometimes diamond blades. The use of air or water is helpful in keeping blades cool. Figure 23-27 shows finished part after trimming.

Repairs. Repair of voids, scratches, cracks, chips or spalls can be easily done with a little practice and care. Dirt, wax, mold release, oil, etc., should be sanded out

Fig. 23-27 The finished FRP laminate after it has been trimmed. (Courtesy: Owens-Corning Fiberglas Corp.)

Fig. 23-26. The FRP laminate is removed from the mold. (Courtesy: Owens-Corning Fiberglas Corp.)

Fig. 23-28. Spray-up method of applying fiber glass and resin. (Courtesy: Owens-Corning Fiberglas Corp.)

SPRAY-UP METHOD OF APPLYING FIBER GLASS

Spray-up is a method that applies single strand fiber glass rovings and resin at the same time (Fig. 23-28). The special spray gun is capable of spraying various amounts of rovings without any resin (Fig. 23-29) or resin alone (Fig. 23-30). After each layer the fiber glass across the flaw, making each layer succeedingly larger so that the patch tapers out on all sides.

USING FIBER GLASS AS A PERMANENT COVERING

Fiber glass reinforced plastic is being increasingly used as a permanent protective coating over woods that are subjected to extended weathering, abrasion, and abuse, viz., boats, water skis, commercial furniture, etc. The entire coating process includes three basic steps: (1) Apply two foundation coats of resin; (2) bed the fiber glass cloth in the resin while wet; and (3) add finish coats of resin.

Prepare the Wood Surface so it is bare, clean, and dry. All old paint, lacquer, varnish, grease, etc. should be removed by sanding. Do not use a blow torch, electric paint remover, or liquid paint remover as they leave the surface with a residue that may disturb the FRP coating. A power sander with coarse paper (aluminum oxide of 4-16 grit and grade) is the best for this purpose. Course sandpaper will leave the surface with the desired "tooth" for taking the FRP coating, but it is important, of course, that the surface be smooth to the extent of not showing low spots where the sander may

of the area to be patched so that an absolute bond can be achieved. Using the same resin and pigment as the original laminate, fill the imperfection a little more than level full. After cure, sand the surface down level, finishing with very fine sandpaper (6/0 grit). A wool buffing pad on a high speed hand tool will usually bring the gloss up to match the surrounding surface.

The back side of a laminate can be reinforced or patched by adding layer after layer of fiber and resin across the flaw, making each layer succeedingly larger so that the patch tapers out on all sides.

Fig. 23-29. The special spray gun can be adjusted to spray the fiber glass alone. (Courtesy: Owens-Corning Fiberglas Corp.)

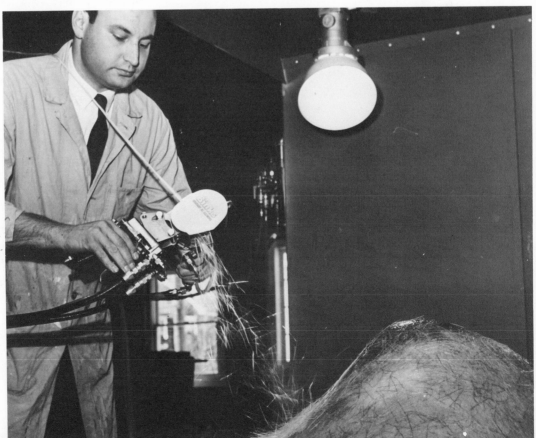

have scoured deep. Dents and gouged places will show up rather than be concealed by the covering. Before sanding, all seams, groves, dents, holes, etc. should be filled with a thick, paste-like mixture of ground fiber glass and the resin mix you have been using. Finally, brush the surface clean and inspect. Oil or grease spots can be cleaned with carbon tetrachloride or other suitable solvent.

Precut Fiber Glass material to a size that is a couple of inches over the actual size of the wood surface. Choose a width of fiber glass most economical for your job (38", 44", 50", or 60"). When it is necessary to use two or more widths of cloth to cover a surface, plan for the pieces to overlap at least three inches. **Do not attempt to make cutouts in the fiber glass material before laying, to go around permanent fittings, but make slits in the material as it is laid to accommodate these fittings.** If the pieces of fiber glass material are large, roll them up on a mailing tube or broomstick. This stores it safely and makes it convenient for applying.

Prepare and Apply the Resin. The surface is to be given two coats of resin. Measure out enough resin for up to approximately 30 square feet—it is inadvisable to mix more than a quart at a time. Add 1% catalyst and mix thoroughly. Apply immediately after mixing with a brush or roller with a stiff carpet-material texture. A second thick coat of resin is put on 15 or 20 minutes after the first. Confine your application of resin to the area you can conveniently reach with a strip of cloth, i.e., apply resin to this area, imbed the cloth, then repeat with the next area.

Imbed Fiber Glass Material in Resin. Starting from one

Fig. 23-30. Spraying resin without the fibers. *(Courtesy: Owens-Corning Fiberglas Corp.)*

side, lower the fiber glass material on to the resin a little at a time working out the bubbles and wrinkles as you go. A squeegee (avoid black rubber) is good to use for this purpose. When applying a large piece of fiber glass, the application will be easier if the resin is allowed to become tacky before laying on the material.

Apply Finish Coat of Resin. Measure out and mix the resin and catalyze as previously indicated. Brush or squeeze this resin over the material making sure it saturates the entire surface. Make sure the surface is smooth and free of bubbles. Allow the FRP to cure. If the cured surface has objectionable defects, sand out the defects, then sand the rest of the surface lightly. Apply another coat of resin as described before.

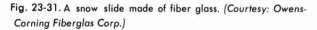

Fig. 23-31. A snow slide made of fiber glass. *(Courtesy: Owens-Corning Fiberglas Corp.)*

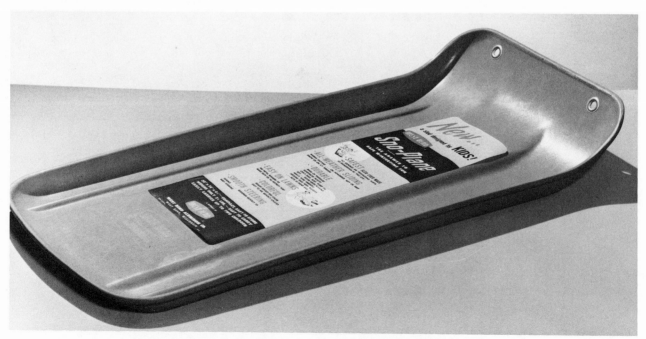

Fig. 23-32. A fiber glass tray with ferns as the decorating material. (Courtesy: Owens-Corning Fiberglas Corp.)

Trim Off Excess. After the FRP is completely cured, rough trim the excess with the tin snips. The finish trimming can be done with a disc sander.

DECORATING FIBER GLASS REINFORCED PLASTICS

There are a few different ways of decorating FRP. All the methods require that the decoration become an integral part of the laminate itself.

Over-All Color may be given to the FRP by adding a color pigment to the resin used (Fig. 23-31). The quantity of pigment used will determine the intensity of the color. Different effects can be had by adding the pigment to the resin used for different layers in the laminate, or for all layers, or for only the top layer, etc. Be sure to mix the pigment into the resin thoroughly. An interesting variation can be achieved by applying a final resin coat consisting of a few contrasting colors.

Embedded Decorations are included in the laminate under the first or second resin coat. Almost any material that can be saturated by the resin, that is not too thick, and does not have any detrimental chemical reaction, can be used for this type of decorative material, i.e., almost any woven cloth, raffia, sprinkles, designs from wallpaper, wheat strands, leaves (Fig. 23-32), tinsel, glitter, etc. The material is embedded just under the first resin coat. This method or decoration can be used with good results in the open-mold method of laminating, however, the matched-die mold gives better results.

Drawn Decorations can be applied to the first layer of material that is applied over the first coat or two of resin. The design, drawn with ordinary crayon, can be applied to the surface of the fiber glass material, or to the surface of the piece of cotton that is later embedded in the resin. Simple geometric shapes make very effective designs.

SAFETY SUGGESTIONS

Preparing an FRP laminate is a comparatively safe process if a few simple safety precautions are observed.

1. Work room should be well ventilated to eliminate fumes that can be irritating to the skin and eyes under prolonged exposure.
2. Wear rubber or plastic gloves when handling chemicals and fiber glass materials.
3. Keep catalysts away from fire or flame.
4. Never mix an accelerator and catalyst together as such a mixture can cause a rather violent reaction. (Most times the accelerator is added to the resin by the manufacturer and need not be a matter of concern.)
5. Don't throw away any unused activated resin mix into waste paper containers. Instead, immerse in water and discard (when cool) in unburnable trash receptacle.
6. Remove any resin, resin mix, and catalyst from skin, clothing, and tools before hardening (curing) progresses. Wash with lacquer thinner, then with soap and water.

FLAKE GLAS

Flake glas (Flake glass) is a comparative new material on the market that is used for the protection of metal surfaces against critical corrosion. It consists of glass flake incorporated in a specially modified bisphenol polyester resin. When applied in the recommended 35 to 40 mil coating thickness it reduces penetration by corrosive elements. In addition, Flake glas coating fills in pits in old steel, arresting corrosion and extending the service life of the metal. The metal surface should be prepared with a "good commercial" sandblast for exterior service or a white metal blast for immersion service.

Fig. 23-33. "Renaissance of the City," by international artist Robert Helsmoortel. This is a five-piece, walk-through work of art standing 33 ft. high and having a base of more than 20 ft. The sculpture weighs between 20 and 21 tons. This world's largest fiberglass sculpture is located in the plaza in front of Peachtree Center in Atlanta, Georgia (Courtesy: John Portman).

Chapter 24
Fiber glass painting

The use of fiber glass in the different art forms is becoming increasingly popular. There are a few reasons for this: (1) It can be prepared in brilliant and vibrant colors, (2) It has unusual permanence, (3) It is easy to manipulate, prepare, and correct, (4) Its cost is not excessive, (5) It lends itself to variation and new conception.

FIBER GLASS PAINTINGS

Fiber glass paintings are prepared by applying colored resins on a sheet of fiber glass laminate that serves as a backing. The fiber glass laminate sheet used for the backing may be transparent or opaque. The transparent sheet is particularly desirable for art work that would be enhanced by illumination from the rear of the painting. Such rear illumination imparts to fiber glass paintings an unusual dimension not possible with most other art mediums. Where transparency is not desired in the backing, opaque fiber glass sheets may be used. Both the transparent and opaque sheets can be secured in hundreds of different sizes and shades of various colors.

Preparing Fiber Glass Backing

The fiber glass laminate sheet used for the backing can be cut to size with a hacksaw, metal-cutting bandsaw, or (if it is quite thin) with a large pair of tin snips. The cut edge can be finished thus eliminating the need for a separate frame. To finish the edge, first sand it with fine emery cloth then polish it on a buffer. To eliminate the work of fine buffing the edge can be covered with resin. This could be the same resin that is used as the painting medium.

Preparing Outline

The art rendition to be applied to the prepared fiber glass backing can be outlined in a few different ways. The outlines can be drawn directly on the fiber glass backing with a China marking pencil (yellow is recommended) or with a soft pencil. If the fiber glass backing is transparent, the design can be prepared on paper then the transparent backing panel placed over it and the design traced on the backing.

Preparing the Colored Resins

It is recommended that polyester resins be used as the painting medium as it is the easiest to prepare and apply. Prepared color pigments (in the form of a paste) can be secured to mix with the polyester resin to render it almost any desired color. The resin may be made transparent or opaque depending on the amount of pigment added to it. Even a small amount of pigment will add a pleasant tint to the resin. For a maximum concentration it is recommended that the pigment be added at the rate of four ounces of pigment paste to one gallon of resin. The actual mixing should be done as follows: Pour into a clean container (cups, jar caps, cans, etc. will do) sufficient resin to complete all the work you will have to do with a single color. (Generally, one ounce of resin will cover a square foot of area). Add the color paste to the resin and **mix completely** to avoid any color-rich spots. Repeat this for all the colors you plan to use. The colored resin can be kept for a long time as long as it is in a tightly capped container and is not mixed with the catalyst.

Full strength heavy resin, as it comes from the can, is about equal to the consistency of a fully bodied syrup. It can be applied with a firm brush without difficulty. There are times, however, when the artist may wish to apply the resin in thin wash to produce transparent veil effect or in the form of a thick paste with the pallet knife. The consistency of the resin may be altered as follows to produce both these effects: **To Produce a Resin Wash** add styrene monomer (following manufacturer's recommendations) to the resin and mix until it is as thin as you wish it to be. **To Make a Resin Paste** add filler to the resin and mix until it is the desired thickness. There are several powdered and fibrous materials that can be used as fillers, such as whiting, China clay, pumice, mica, chalk, silicas, glass fibers, asbestos fibers, cellulose fibers, cotton flock, etc. It is advisable to thin or fill the resin before adding the color as these additives will change the intensity of color. After thorough mixing it is best to let the resin stand a few hours to allow the trapped air to escape.

Applying the Resin

Just before you are ready to apply the color, add the catalyst at the rate of about 6 drops per ounce of resin (this is approximately a 2% mix). This amount may have to be varied if the temperature and humidity are far from the standard of 75° F. and 50%. The artist must work without delay as the resin starts to set the moment the catalyst is added. Its working time may vary from half an hour to an hour depending upon the quantity and type of mixture, temperature, humidity, etc. The mixed resin may be applied with a brush, roller, spray gun, pallet knife, etc. depending on the consistency of the mix and the effect desired. (In application, the resin responds much the same as oil paint of an equal consistency.)

Corrections can be made on the fiber glass painting in the following ways: If the resin has not cured, it can be scraped off with a knife and the correct color applied. If the resin has cured to the point where it cannot be scraped off, apply the correct color over the incorrect one. If the resin has cured hard, remove the sheen from the surface to be repainted with fine emery cloth or 4/0 garnet paper before applying the additional coat of resin.

The resin applied as the painting material will cure in approximately 40 minutes to two hours or more. A humidity above 50% or a temperature below 75° F. will delay the curing. Two things may be done to speed up the curing process: (1) Add more catalyst, (2) Add a speck (1/2 drop) of cobalt napthelenate to the resin and mix thoroughly before adding the catalyst. **Never mix the cobalt napthelenate to the catalyst.** The final cured painting needs no special coating or protective materials to retain its brilliance and exquisite coloring.

Fig. 24-1. The fiber glass painting on these sculptured parts impart an almost human quality. (Courtesy: West Coast Display Mannequin)

Chapter 25
Fiber glass sculpture

FIBER GLASS SCULPTURE

Fiber glass in its different forms lends itself perfectly to the sculpturing forms of art. In its solid state it can be cut and shaped into geometric forms of all types and joined in stacked planes and angles. Fiber glass in a plastic state can be shaped and modeled into objects of intrinsic beauty limited only by the skill and imagination of the sculptor.

Plastic Form Fiber Glass Sculpture

Plastic form fiber glass sculpture is accomplished by using pliable fiber glass materials that can be molded into designs of all types, sizes, and shapes. The finished sculpture may be of the shell type or the solid body type or a combination of both. **The shell type sculpture** is made in exactly the same way as the hand lay-up method described on pages 183 and 190. A mold is made then the fiber glass laminate is formed over it and finally, the cured laminate is removed from the mold. **Solid body sculpture** is constructed in exactly the same manner as the shell type with this exception; in this type of sculpture the mold over which the fiber glass was formed is not removed but becomes a permanent part of sculpture.

Plan the Design you wish to form. This is the necessary first step as it is the determining factor in the procedure that will be followed. Once the general theme and ultimate shape of the sculpture is determined, it is then necessary to decide on the exact size it will be. It is the combination of shape and size that determines the procedures and techniques that will be employed in its construction.

It must be remembered that when a one part mold is used (as is generally the case with hand lay-up) the surface of the fiber glass laminate that will be against the mold when being formed will be smooth while the other surface will be comparatively uneven. This must be taken into consideration when planning the design for the sculpture to determine whether a male or female mold may be needed for its preparation.

Construct the Mold over which the fiber glass will be formed. This may be made from many different materials. The final decision as to material and type of construction to be used for the mold will be determined from a study of the design.

Small molds can be made of soft woods, plywood, pressed woods, cardboard, foam plastic, metal, or almost any material that will maintain its shape under the pressure and heat involved in the curing of the fiber glass laminate.

Large molds may be made from the same materials as listed above for small molds plus chicken wire, metal screening, and similar materials that can be easily bent and shaped. When these latter mentioned screenings and woven wirings are used, they must be braced with wood or metal to maintain their shape and then covered with burlap cloth to prevent sagging and draining.

When using such solid materials as woods, foam plastic, metals, etc. for the mold you shape the material to the finished design desired. For large molds, where woven wires and screening is to be used for the mold, you proceed as follows: Prepare the wood or metal to be used as supports. Form the woven wire over the supports and tack or tie it in place. Cover the woven wire with burlap cloth tieing or tacking it in place.

The Finish on the Mold is determined by the type of sculpture to be made. If the sculpture is to be of the solid type, where the mold will be a permanent part of the finished sculpture, then the surface of the mold needs no finish at all. However, if the fiber glass laminate is to be removed from the mold, making it a shell type sculpture, then the surface of the mold must be carefully finished so the cured laminate can be released. This latter mold must have its surface finished as indicated on page 190.

A mold made of woven wire may be used to form a shell type sculpture provided its surface is prepared as follows: Cover the woven wire with snugly fitting burlap cloth. Cover the burlap with at least an inch thickness of plaster-of-Paris. When the plaster is hard, file, sand, and buff the surface until it is perfectly smooth. The smoothed plaster surface is to be finished the same as wood as indicated on page 190.

Prepare Fiber Glass Material and Resin to be used to cover the mold. The different types of fiber glass materials that are available are discussed on pages 178 and 179. Choose the type of fiber glass material that is best for your need. Cut the material so it will completely cover the mold. If more than one piece of material is needed to cover the mold, make sure a sufficient amount is allowed to overlap the edges at least three inches.

If the laminate is to have several layers, cut all the pieces at this time.

Measure out the amount of resin that will be needed (see page 180). If the resin is to be colored add the coloring material at this time in the proportion as needed (see page183). Also add fillers at this time if needed (see page 182).

Apply the Fiber Glass Material and Resin to the mold in the following manner:

1. Add the catalyst to the resin at the rate of approximately 5 drops for each ounce of resin (use more or less depending on the drying speed desired—see page 194).

2. Apply a thick coat of the catalyzed resin to the mold and permit it to cure until it becomes tacky.

3. Drape the fiber glass material over the tacky resin and press it firmly against the mold. If there are undesirable overlaps or folds, raise the fiber glass material and cut out the excess material with a tin snips, then press the material back in place.

4. Saturate the fiber glass material with the catalyzed resin. Pour the resin on the material then brush it out in all directions.

Fig. 25-1. Sculptured fiber glass mannequins on display.
(Courtesy: West Coast Display Mannequin)

5. Make certain that the fiber glass material makes intimate contact with the mold. To do this, use a squeegee, roller, or brush. Start near the center and work toward the sides. Rollers work best with mat materials. Fabrics are easily worked with rollers or squeegee. If the fiber glass material spans a mold indentation it must be worked down pulling extra fibers from edges to the center. Before the resin cures, check back to see that bridging has not re-occurred.

Note: One way to greatly reduce the number of voids under the gel coat, minimizing patching later, is to lay-up the first layer of fiber and let it cure before other layers are applied. This insures that the heavier layers will not shift the first layer and cause bridging. Defects should be slit open or cut out and patched with mat materials and resin. Patching putty is also used and can be made from a combination of 1 part resin to 3 parts of fine talc or calcium carbonate (milled fiber glass fibers and resin also makes a good patching putty).

6. Lay additional layers in place and wet with catalyzed resin by pouring or brushing it on and working it in. Subsequent layers should go on immediately, repeating the process for each one, until the proper thickness has been built up. At any point, strips of fiber can be added to areas where

Fig. 25-2. A fiber glass sculpture that shows the versatility and characteristics of the material (*Courtesy: west Coast Display Mannequin*)

extra strength is needed (See pages 196 and 197 for thick laminates).

7. Apply a coat of catalyzed resin over the final layer of fiber glass material. If this surface (the one away from the mold) is not going to be seen, make this coat a thin one. If it is going to be seen, make this a thick coat.

   Note: If the surface of the laminate away from the mold is going to be exposed place a cellophane sheet over the final surface before using the roller and squeegee (see page 197)

8. Permit the laminate to cure (see page 197). If the laminate is to be removed from the mold, do so before it is completely cured (see page 198).

9. The sculpture may be completed by embellishing the surface with applications of fiber glass putty and resin paint (see page 204). Make sure the surface is roughened with sandpaper or steel wool wherever the putty and paint is to be applied.

SOLID FORM FIBER GLASS SCULPTURE

Solid form fiber glass sculptures are made from pieces of fiber glass laminates, cut to geometric shapes—squares, rectangles, triangles, circles, ellipses, etc.—then cemented together to form various designs. The geometric shapes used and the placement of them in the union of parts will be determined by the sculptor's conception. Inasmuch as there are millions of possible variations in size and form of geometric shapes and additional millions of possible variations as to plane arrangement into which they might be combined, the sculpture potential of this medium is limitless.

Materials Used for Solid Form Fiber Glass Sculpture

Fiber Glass Laminates are commonly sold in various thickness and sizes, flat and corrugated sheets. They may be had in dozens of different color variations and surface designs. In most communities they are purchased from the local hardware store or lumber yard.

Those artists who wish to prepare their own laminate sheets may do so by following the instructions for hand lay-up given on page 190 and 199. The mold can be a flat piece of 3/4 inch plywood surfaced with a hard wood such as maple or birch. The surface of the plywood should be sanded then covered with three coats of resin. Allow each coat to cure (1 to 3 hours) before applying the next. Sand the final coat with 4/0 sandpaper until the surface is smooth then finish with 320 grit wet-or-dry sandpaper. Finally, buff the surface to a high finish. To make a laminate that is smooth on both sides make up a second piece of plywood the same as the one described. When all the layers of the laminate are in place and the final layer has been rolled or squeezed to remove bubbles and air pockets, then place the second plywood piece over the laminate. Put the finished surface on the plywood against the laminate. Place weights on the second ''board,'' or place the entire combination of mold and laminates in a press if such is available. Leave the pressure on the mold until the laminate is cured sufficiently so it can be removed from the mold.

Artists who make their own laminates have the advantage of preparing the colors to their own taste as well as surface decorations and designs.

Cementing Materials used to join the pieces of fiber glass together can be the same as the non-firing materials used for cementing regular glass to surfaces. Those particularly good for this purpose are ''Dow Corning Glass and Ceramic Adhesive,'' ''Duco Cement,'' and ''Epoxy Glue.'' The two mentioned first are colorless when used in thin layers. The color of the Epoxy glue varies from brand to brand and may be a very transparent yellow to a deep gray.

Additional Materials may be included in the sculpture. Fiber glass is a good mixer and may be combined effectively with such other materials as metals, woods, plastics, regular glass, etc. These materials may be added to give the sculpture more variation and accents or to act as a bracing for the entire structure. When the materials are mixed in this way, they must be given a great deal of consideration to make sure there is no conflict or clashing between textures and finishes. Every material that is a part of the sculpture as well as the cementing material, and bracings must be carefully planned as it is the sum total of all that determines the character of the final presentation.

Suggested Procedures

The fiber glass laminate may be cut with a hacksaw or saber saw, or a metal-cutting band saw. (Do not use wood cutting tools or machines as they will be dulled). Permanent lines may be drawn on the surface with a scriber or scratch awl, guided by a straight edge or template, and metal dividers to make circles.

The edges may be smoothed with a file or hand plane. To give the edges a gloss finish, sand them with 4/0 sandpaper followed by 320 grit wet-or-dry paper and water and finally buff on a buffing wheel using rouge for the abrasive. Edges to be cemented should be left with the rough finish.

The cement should be applied as directed by the manufacturer on the tube or box. For best results do exactly as directed. Give the cement sufficient time to cure or you may damage the sculpture in handling.

It is often difficult to hold parts of the sculpture in the proper relationship while they are being cemented. This can be overcome to a large extent by using a ''sand box.'' This consists of a box (the size depends on the size of the pieces being worked on) filled with three or four inches of sand. The sand is packed around the pieces being cemented to hold them in the desired relationship to each other until the cement is cured.

Chapter 26
Repairing damaged fiber glass parts

Basically, fiber glass parts are formed from laminated sheets of fiber glass mats held together with a synthetic resin. When the liquid resin mix cures or hardens, it binds together the filaments of glass in the mats to create a solid panel. The strength of the panel is provided by the fiber glass while the resin merely acts as a bond, supplying only limited additional strength to the panel.

In general, all repairs to fiber glass parts consist of (1) removal of damaged material, (2) filling the damaged area with epoxy resin and glass (both fibrous and milled), (3) the activated resin is allowed to cure (harden), and finishing operations are performed. The procedure for repairing fiber glass covered wood is essentially the same as that described below for fiber glass panels.

In addition to fiber glass repairs being quite simple, the paint refinishing method is the same as that recommended for metal parts with the exception that temperature must be kept below 250° F. This immediately suggests the air-dry process as being the most practical. On fiber glass, air-dry enamels and lacquers are equally satisfactory.

Various manual techniques are employed depending upon the nature and extent of the damage but essentially the above system applies.

SMALL TO MODERATE HOLES (PUNCTURES)

In the first instance of damage repair described below, a hole approximately 2" x 3" and located near the center of a large panel, was repaired by the paste-patch method.

1. Grind away all loose glass fiber, crazed and fractured material and bevel the solid edges of the hole rather flatly (about 20°) on the outer surface. In addition to roughening the repair surface be certain to remove paint from the area as the resin will otherwise not adhere.

2. Mix a quantity of clear resin and catalyst (hardener) according to instruction and apply to both inner and outer surfaces plus 1-1/2" to 2-1/4" of flat surface surrounding. Saturate two layers of glass cloth on both sides with activated resin mix and apply to the inside or back surface. (See Fig. 26-1). Pressing the laminations down tightly with a sheet of wax paper or cellophane will produce a tight bond and also keep the material off the fingers.

3. Similarly saturate a single layer of glass cloth and apply to the outside making certain that this layer of glass cloth is in complete contact with inner layers entirely across opening. After the three layers of saturated glass cloth are in place (1 outside, 2 inside) a saucer-like depression should be made in them in order to increase the depth of the resin and glass paste repair material which will be troweled onto the patch.

4. To expedite curing provide external heat, preferably an infra-red lamp, no closer than twelve inches from the fiber glass surface.

   Note: 250° F. to 275° F. is the high limit for this material and to go higher is to risk material distortion. Therefore, keep close control of the external heat the only function of which is to accelerate the curing of the resin mix. This is a chemical reaction and not a drying process.

   Caution: Immerse brush and all tools (putty knife, spatula, etc.) in ordinary thinner at once

Fig. 26-1. *(Courtesy: White Motor Corp.)*

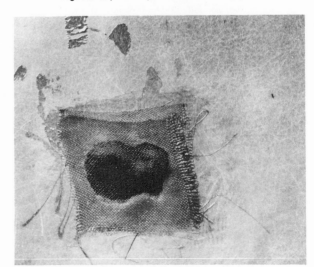

Fig. 26-2. *(Courtesy: White Motor Corp.)*

while the epoxy is still soft. Once hard, it cannot be softened. Remove soft mix from skin with thinner and wash hands thoroughly. This is important as failure to remove the wet resinous compound may cause dermititis (skin blotches and irritation) in some individuals. After original applications of glass cloth and resin are hard (cured) proceed with step 5.

5. Into a fresh batch of resin and hardener stir in enough milled glass powder to form a heavy paste. Trowel this into saucer-like depression in repair area (Fig. 26-2), leaving a sufficient mound of material to grind down smooth and flush when hard. This may be covered with cellophane or wax paper and compacted by hand pressure or a small roller may be used.

6. To expedite hardening, use external heat as in step 4.

7. When fully hard, grind smooth and flush with surrounding area with a disk sander using No. 36 grit. Finish with No. 150 emery cloth, or finer, on wood block or with a fine file (Fig. 26-3). Wipe with a cloth saturated in lacquer thinner.

SPLIT, TORN OR CRACKED PARTS

In this instance the illustrations show a fender which was repaired, having been torn apart for about 3″ in a more or less perpendicular line from the fender edge (Fig. 26-4).

1. Grind away loose and damaged material and taper edges of break in a broad V (Fig. 26-5). Clean and roughen the entire repair area with medium emery cloth or a grinding wheel in order to obtain a good bond between old and new material.

2. If parts do not align, as may be the case in a break of this nature use a bolt, nut and two large washers at the break to draw parts into place.

3. Cut two or three pieces of glass cloth allowing at least 1-1/2″ to 2″ to extend beyond the break. The extent of the damaged area will determine whether two or three layers of glass cloth will be required.

4. Mix a quantity of clear resin and catalyst according to instructions. Saturate the glass cloth and the repair area with this activated resin mix. Apply glass cloth to back side of broken part (Fig. 26-6) pressing down firmly using wax paper or cellophane to protect hands in the process.

5. Apply external heat and allow to cure.

Note: Glass cloth was applied over the washer and nut and these were left in the patch and became a permanent part of the repair. This procedure is recommended where the reverse side of the part does not show and the bulge produced by the nut and washer is therefore

Fig. 26-3. *(Courtesy: White Motor Corp.)*

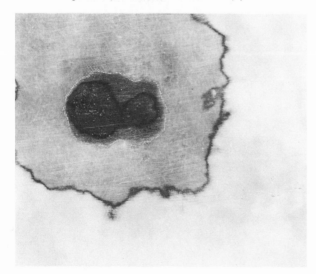

Fig. 26-4. *(Courtesy: White Motor Corp.)*

hidden.

6. Mix a fresh quantity or resin and catalyst adding a quantity of milled (powdered) glass according to manufacturer's instructions. Apply generously to top (outer) side of repair area entirely filling the prepared V-like valley. Trowel down with a putty knife or metal blade pressing the mix firmly into the repair area. When nearly hard, remove bolt and outer washer. Fill hole with resin and ground glass mix and allow to completely cure (Fig. 26-6).

7. File and sand smooth (Fig. 26-7). Wipe with a cloth saturated in lacquer thinner.

8. Allow thinner to dry. Prime and paint with air-dry finishes.

REPAIR OF LARGER DAMAGED AREA

Frequently, some damaged areas, several inches across, and too large to be repaired by the paste-patch method may be successfully and efficiently repaired by panel insertion. This is particularly practical where repair area is largely flat in nature and elaborate contours do not have to be crossed. If more than half the panel is destroyed or heavily contoured areas are involved, replace the entire section.

1. Determine a cutting line well outside actual damage area and beyond any cracks or crazing from the break. Cross as few contour differences as possible.

2. Cut along established line of repair using any of several satisfactory methods. The simplest tool for this is a broken hacksaw blade mounted in a wood handle or grip. Equally effective and much faster is a saber saw, if available.

   Caution: If cut crosses reinforcing members or underlying panels, exercise due care not to cut into this material.

3. Using as a pattern the broken piece which was cut away, lay out a repair panel 5/8" to 3/4" larger on all sides and fit it over the opening.

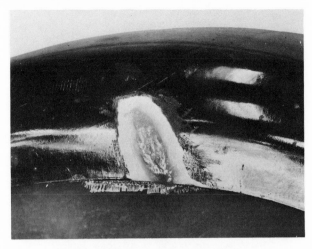

Fig. 26-6. (Courtesy: White Motor Corp.)

4. Bevel inner edges of repair panel and outer edges of original panel to approximately 20° as shown in Figure 26-8, and until insert panel fits just below flush (to provide space for resin bond.) Be certain to remove paint from the vicinity of the repair as the resin mix will otherwise not adhere satisfactorily:

5. Drill 1/8" holes for Cherry rivets spaced 2-3/4" to 3" apart. Countersink to receive rivet heads and fit panel and rivets into place to make certain repair will be in plane with surrounding surface.

6. Remove panel and rivets and prepare an adequate quantity of resin and catalyst according to directions.

7. Quickly spread activated resin on all beveled surfaces and install panel in place. Insert Cherry rivets in prepared holes and draw with Cherry rivet gun. Make certain that panel is pressed rightly into place as rivets are drawn (Fig. 26-9). Allow resin mix to cure.

Fig. 26-7. (Courtesy: White Motor Corp.)

Fig. 26-5. (Courtesy: White Motor Corp.)

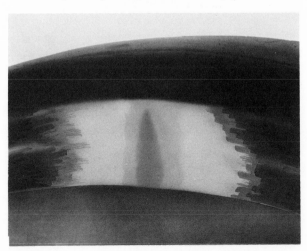

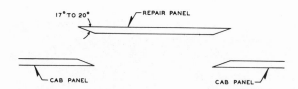

Fig. 26-8. (Courtesy: White Motor Corp.)

8. To expedite curing, provide external heat, preferably an infra-red lamp, no closer than twelve inches from the fiber glass surface.

Note: 250° F. to 275° F. is the high limit for this material and to go higher is to risk material distortion. Therefore, keep close control of the external heat the only function of which is to accelerate the curing of the resin mix. This is a chemical reaction and not a drying process.

Caution: Immerse all tools (putty knife, spatula, etc.) in ordinary thinner at once while resin is still soft. Once hard, it cannot be softened. Remove soft mix from skin with thinner and wash hands thoroughly

9. When fully hard, nip off protruding stems of Cherry rivets with side cutters and grind smooth and flush with surrounding area. Note that the fiber glass and the aluminum (in the rivets) abrades at approximately the same rate. Therefore, a smooth, flat repair should result. Use a disk sander with No. 36 grit and finish with No. 150 emery cloth, or finer, on wood block or use a fine file (Fig. 26.10).

10. Wipe with a cloth moistened with lacquer thinner to remove dust prior to prime and painting.

REPAIRING METAL PARTS WITH FIBER GLASS

Nominal damage to metal panels such as dents, deep scratch and gouge marks, and even punctures may be repaired by the use of fiber glass repair material. Fiber glass repair resins work well on aluminum, as their coefficients of expansion are nearly identical and adhesion is very satisfactory, particularly if surface is slightly roughened.

A typical repair is described below, to serve as a guide. The technique may be modified to cover a variety of damage repair.

1. Remove paint from repair area with thinner, paint

Fig. 26-9. (Courtesy: White Motor Corp.)

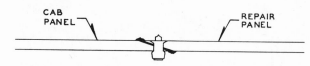

remover or sandpaper. Blow clean with compressed air.

2. Slightly roughen aluminum surface with wire brush or emery cloth to provide for better adhesion of repair material. In the case of a puncture, bend metal back into position insofar as possible in order to reduce actual opening. Blow clean with compressed air. At this stage DO NOT WASH with a solvent or cleaner.

3. Drill a few 1/8" or 3/32" holes in the metal indented area (Fig. 26-11). This will provide improved adhesion, as small "buttons" of fiber glass will form on back side of holes when repair material is troweled on.

4. Mix a moderate quantity of fiber glass repair material (resin and catalyst) in accordance with

Fig. 26-11. (Courtesy: White Motor Corp.)

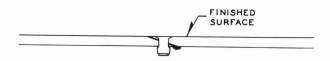

Fig. 26-10. (Courtesy: White Motor Corp.)

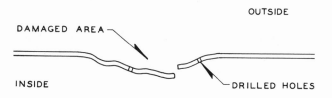

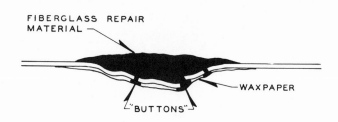

the instructions from supplier. When thoroughly mixed, trowel onto repair area, pressing firmly to force away any air pockets and to be certain that material passes through previously drilled holes. If back side of repair area is accessible, hold a sheet of wax paper or cellophane against surface to aid in formation of bonding "buttons" (see Fig. 26-11).

5. To expedite curing of repair material (resin mix) provide external heat, preferably in infra-red lamp, no closer than twelve inches from the resin.
Note: 250° F. to 275° F. is the high limit for this material and to go higher is to risk possible damage.

6. When fully hard, grind smooth and flush with surrounding area, using a disk sander and No. 36 grit, or finer. Finish with No. 150 emery cloth. Area may now be wiped with lacquer thinner or a high solvent, dried with compressed air, primed and painted (Fig. 26-11).

SUGGESTIONS FOR BETTER FIBER GLASS REPAIRS

1. Accumulate all tools, materials and replacement parts in advance, as (a) this enables you to proceed more efficiently, (b) when resin and catalyst are mixed and about to cure (harden) there is no time to hunt for tools or implements.

2. Study the repair. Think through the steps. Time spent in planning is important.

3. Use ingenuity in making repairs. Combinations of the various methods described in this section may be employed if the repair calls for it.

4. Many of the same principles true of welding also apply to fiber glass repair. (a) Increasing the lineal inches of bond will produce a stronger repair. (b) Staggering the joints in multiple-layer repairs will increase durability and strength. (c) Grinding a bevel or valley which is filled with repair material will always be stronger than a butt-joint.

5. Use only resins from a single manufacturer, make no attempt to mix with other resins.

6. Keep supplemental curing temperatures below 275° F.

7. Wherever paste-patch material (milled glass and resin mix) is used, be sure to compact it by rolling the repair with a small roller or troweling it with a putty knife or spatula. This (a) forces the fresh material firmly into the crevices of the repair and (b) removes air bubbles and pockets which would tend to produce a porous repair.

A sheet of cellophane or wax paper placed over the fresh material will keep roller or steel blade clean. This protective sheet is, in turn removed before grinding or sanding the cured repair.

8. Where Cherry or similar riveting equipment is available, it is recommended that self-plugging, flat head, aluminum Cherry rivets be used in place of self tapping screws. These rivets are left in the job and produce additional strength in the repair as well as providing considerable time saving.

9. Working with fiber glass must be a clean operation. The repair must be clean and free from grease, water and paint or a satisfactory bond is impossible.

Secondly, since the repair of fiber glass results in the production of dust (powdered glass and cured resin) it is recommended that effort be made to also keep the area as clean as possible. Although it is not toxic, the use of a respirator is suggested for those sensitive to dust.

Finally, exercise care to remove any resin, resin mix and catalyst from the skin, clothing and tools before hardening (curing) progresses. Wash with lacquer thinner, then with soap and water.

Caution: Do not throw any used activated resin mix into waste paper containers. Instead, immerse in water and discard (when cool) in unburnable trash receptacle.

Chapter 27
Glass mosaic panels and furniture

Making glass mosaics is one of the most exciting, fastest growing hobbies. You can transform pieces of glass tile into hundreds of colorful, light-catching and striking mosaic designs. As a decoration medium it can add glamour to almost any room in your home.

Glass mosaics are made by glueing or cementing small pieces of venetian glass of various colors to a surface (plywood, ceramic, plastic, concrete, plaster, etc.) in a pleasing design. The spaces between the tiles may or may not be filled with "grout" (crack filler).

MATERIALS USED

Venetian Glass Tiles are solid glass tiles made of specially-fired colored glass. Each tile is 3/4 inch square and approximately 3/16 inch thick, which may vary slightly from one color to another. They are usually available in 12" square sheets containing 15 rows of 15 tiles or 225 tiles pasted face down on paper. The glass is extremely durable, impervious to stains, acids, alkalies and unaffected by even the most extreme climate changes. The tiles may be secured in dozens of different colors and shades.

The glass used in stained glasswork and thin slab glasswork is the same as that used in glass tiles. Small pieces of glass left over from making such panels may be used for making glass mosaics. The pieces may be cut into 3/4", 1/2" or 3/8" squares and used as individual pieces, or where large areas of one color appear in a mosaic, a single piece of glass can be cut to fit the entire area then cut into squares.

To make a mosaic tile from sheets of stained glass, lay the glass on the cutting table or wad of newspaper. Clean the glass with soap and water followed by alcohol to remove grease and dust, or glass cutter will skip and leave indefinite lines. Score lines about 3/4" to 3/8" apart, use enough pressure to hear the cutter "bite" the surface. (On thick glass score lightly, and on thin glass score heavier.) Now score lines across the first set of lines. Turn the glass over and tap along the scored lines. The glass will break along the lines. To cut odd-shaped pieces, mark for cutting as shown in Figure 27-1.

To make circles or rounded edges, place the glass in the smallest of the tooth-like openings on the cutter that the glass will fit. Using a rocking back and forth motion, grind away the glass to the desired shape. Sharp edges may be dulled with an abrasive stone or melted in a kiln.

Supporting Panel is the surface to which the tiles are attached and may be made of several different materials. The one most popular for flat-surfaced objects that are used indoors is 3/4-inch fir plywood which is comparatively inexpensive and affords a flat, true surface. When hollowed or coved shapes are desired ceramics are usually used. When it is desirable to have light shine through the tiles from behind, a thick transparent plastic (such as Plexiglass) may be used.

When the glass tile panel is to be used permanently outdoors, such as part of a building, walk, concrete furniture, etc., it is best to cast the tiles directly into a supporting panel consisting of either plaster of Paris, silvercrete cement or special pour-lay cement (see page 173), colored white.

Fig. 27-1.

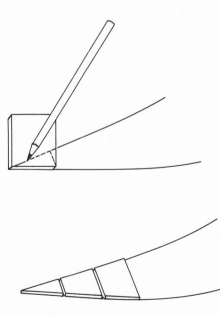

The Adhesive is used to connect the glass tiles to the supporting panel. There are three types in general use:

1. White Resin Glue is popular among the professional mosaicists to cement the tiles. It is not recommended for tabletops unless you are satisfied with the evenness of your tiles, otherwise the surface of the mosaic will be rough in texture. Do not use this glue for outdoor use. When this glue is used wait 3 or 4 days before grouting to allow for thorough drying. Advantages: dries colorless, is easily applied in squeeze bottle, may be used mixed with crushed glass mosaic for background

Fig. 27-2.

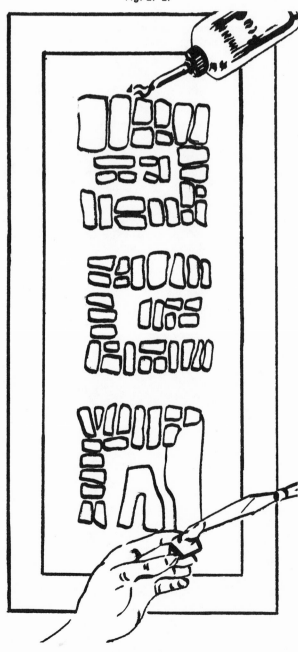

or lightly textured areas in your mosaic. Two ounces cover approximately one square foot.

2. Clear Plastic Cement like "Duco" is used where crystal-clear transparency is desired, such as when a mosaic is done on plastic or when light-colored transparent tiles are over metallic paint. Like white resin glues, it does not allow for the leveling of uneven tiles and it should not be used for outdoors. The cement is applied from a tube either to the supporting panel or directly to the back of each tile. A 1-3/4-oz. tube will cover approximately one square foot.

3. Mastic Adhesive is the tile cement used for bonding the tiles to plaster, cement-block, etc. It is recommended for outdoor use. Since mastic is a thick putty-like material it can be used to level uneven thicknesses of tile. The can containing the mastic should be kept covered and filled with water at all times when not in actual use. To clean the mastic from hands, tiles or tools, any thinner such as household cleaning fluid, kerosene, or benzine may be used. A 4-oz. can will cover 1-1/2 to 2 square feet while 1 pint will cover 6 to 8 square feet.

Grout is a filler for the space between the tiles. It is a must on tabletops to make them waterproof and sanitary but optional for wall hangings and murals unless they are to be used outside. The best kind of grout cement comes ready-mixed (except for water), free of any lime or sand, does not shrink and mixes well with dyes. It should level without difficulty. To prepare grout (which comes as a white powder) add a little water at a time and stir until it is smooth and creamy and the consistency of thick pancake batter. Any bubbles should be removed by tapping the mixing container on the floor. A 1-lb. bag covers approximately 2 to 3 square feet.

Grout Dye adds color to the grout. The dye should be

Fig. 27-3. Mastic spreader.

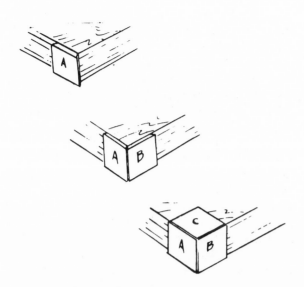

Fig. 27-4.

mixed thoroughly with the dry grout and then wet.

Silicone Polish, when applied to the finished panel, will waterproof and seal all joints as well as add a protective luster to the surface of the tiles.

PROCEDURE FOR MAKING GLASS MOSAICS

1. **Design.** The design for the mosaic should be formal and depend upon two-dimensional pattern for interest. When the design is acceptable a full-size drawing of it should be made on a sheet of heavy paper. The color of all tiles to be used should be indicated. An interest-ing stain glass effect may be injected in the mosaic by surrounding solid color areas with liquid solder or sculp metal.

Full-size prepared mosaic designs can be purchased at craft shops. These designs include tile color recommendations. The novice would do well to study these professional designs before attempting a complex design of his own.

2. **Prepare Supporting Panel.** The supporting panel is the surface to which the tiles will be attached. It should be cut to the exact finished size to take the entire design.

Paint the surface to which glass tiles are to be adhered with metallic paint. This will allow light to be reflected back through the glass, giving more color and brilliance.

3. **Transfer Design.** The full-size drawing of the design is transferred to the supporting panel by means of carbon paper.

4. **Applying Adhesive or Mastic.** When using white resin glue or clear plastic cement cover small area, working only what you can tile before drying sets in.

When using mastic you can apply directly to surface, or butter each file as shown in Figure 27-2. Mastic has longer working time than the other adhesives. On large areas the mastic can be applied with a spreader having a saw-toothed edge (Fig. 27-3). The spreader should be held at a 45-degree angle to work out an even coat. The surface of the supporting panel should be visible between the ridges of mastic.

Fig. 27-5.

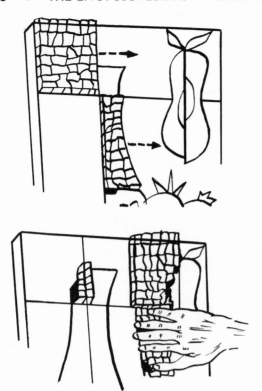

Fig. 27-6.

5. **Grouting.** This is the process of filling the cracks between tiles. You may use grout, white or colored. (See directions for coloring on dye packages.) Grouting may have to be done in two sessions with an overnight drying period in between. Before grouting, dampen tiles with water. Mix grout with cold water until all lumps are removed and it reaches whipped cream

Fig. 27-7.

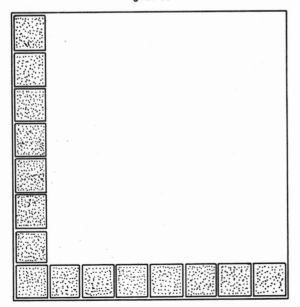

consistency. With a rubber scraper, or your hand, apply grout to a small area at a time . . . rubbing lengthwise and then cross-wise, working the grout in carefully, eliminating bubbles. Carefully wipe the excess grout from the surface without digging grout from in between tiles . . . use a damp-dry cloth. This need not be a perfect cleaning. About 15 minutes later a dusty film will appear . . . wipe then with a dry cloth.

Wait overnight for the grout to dry. If the grout is not flush with the top of the tiles, a second grouting may be necessary. The consistency of this grout should be thinner—like that of thick paint.

If regrouting is not necessary, the top of the tiles can be cleaned with a wet cloth. For difficult cleaning jobs a **muriatic acid solution** (1 part muriatic acid to 4 parts of water) can be scrubbed on with a stiff brush. Work quickly and rinse off the acid with cold water. When working with the acid wear rubber gloves to protect your hands. Acid left on too long will affect the grouting. Acid or fumes should not come in contact with metal as it will injure the finish.

TILING EDGES OF MOSAIC PANELS AND TABLES

If plywood, or a similar thick material, is used as

Fig. 27-8.

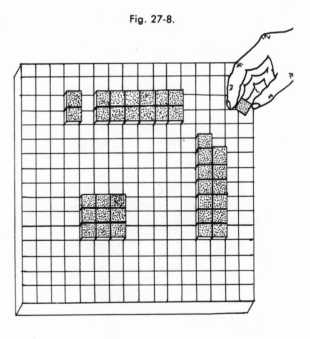

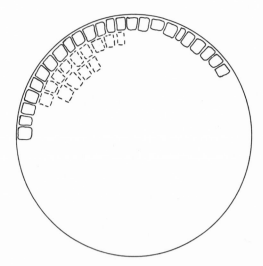

Fig. 27-9.

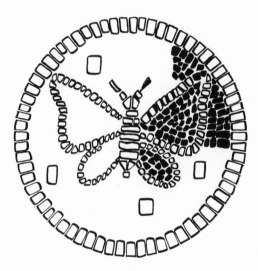

Fig. 27-10.

To tile the edge of a free-form design such as represented by the table in Figure 27-5, cement the tiles all around the edges. When the edge tiles are in place cement a row of tiles over them on the face of the panel. The end row of tiles on the face of the panel should overlap the edge tiles.

PROCEDURES FOR DIFFERENT TYPES OF DESIGNS

In the following discussions of design variations it is assumed that the design and support panel have been prepared according to the directions above and that the design has been transferred to the panel. Where the edge of the panel is to be tiled, it is assumed that this has also been done.

When prepared tiles are purchased they come mounted on paper backing. For some designs it is best to leave the tiles connected to the backing, for others it is necessary to remove them. To remove the tiles, soak the paper backing with water.

Design with Single or Cut Tiles (Direct Method). Starting with a small area in your design, cut enough tile to desired size to cover 5 or 6 square inches. Arrange them in various ways until you are satisfied with the effect. Move the cut pieces aside carefully by placing your hand firmly over all the pieces at once and sliding them away (Fig. 27-6). Apply the adhesive to the surface and replace the pieces one at a time. When using mastic, press the tiles in place with finger pressure leveling up each succeeding piece with those already set. Continue working until the design is completed and you are ready for the background.

The background can be made of whole tiles or a combination of any suggested textures. If whole tiles are used, set the adjacent edges first to establish the line spacing (Fig. 27-7). Spread the adhesive working in towards the design, cutting wherever necessary around

Fig. 27-11.

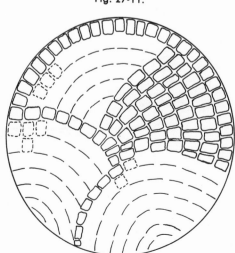

backing for the mosaic panel then the edge must be finished. The edge tiles must be cemented in place before those on the face of the panel. It is most attractive when the edge is covered with glass tiles the same as used in the border on the face of the panel. To tile the edges of a square or rectangular panel proceed as follows:

1. Sand and apply metallic paint to the edge as well as the surface of the panel.
2. Transfer the desired design to the panel.
3. Start in the corner of the panel and, working with full or half tiles, set, in the order indicated, the areas represented by A, B, and C in Figure 27-4. They should be set in such a way that, when looking straight down on tile C, tile A or B should not protrude. (Do not be disturbed by the spaces that show; they will be filled with grout later.)
4. Prepare the four corners as indicated above, then proceed to fill the spaces between the corners.

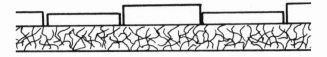

Fig. 27-13.

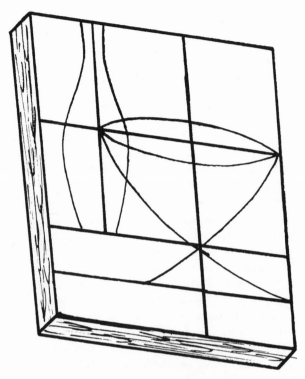

the design.

Geometric Designing. For this design there is no need to remove the tiles from the backing sheet. Decide on a design and remove only those tiles that do not belong, leaving their blank spaces (Fig. 27-8). In the blank spaces glue tiles of the desired colors.

Round Design. Start from the outer rim and work towards the center with whole, half or quarter tiles according to your choice (Fig. 27-9). Fill in the design outline and design area first, then add the background starting from outside and working in. Cut the tiles to fit around the design (Fig. 27-10). Always work according to the design requirements but attempt to maintain a flow of grout lines to establish a definite design (Fig. 27-11.

Mural Designs in Cut Tile (Direct Method). Many techniques can be used to gain contrast and texture in mosaic panels. The design in Figure 27-12 uses areas of full tiles in contrast to cut tiles. One way to work out a design is to cut each tile into quarters and then cement these quarter tiles along the outlines of the design. When the outline is complete, cut irregular shaped tile to fill inside the outline (Fig. 27-12). In mosaic wall panels or murals the surface need not be smooth and level. In most cases, roughness in the finished surface adds more interest (Fig. 27-13). If the tiles are fitted closely together grouting is not necessary . . . the contrast of color and texture is enough.

Glass Mosaics with "Cement" Backing. Gum the tiles face downward on the paper design with gum arabic or rubber cement (Fig. 27-14). It should be remembered that the finished arrangement will be the reverse of the paper design. If a direct copy of the paper design is required a tracing must be made from the original and then turned over for working.

Fig. 27-14.

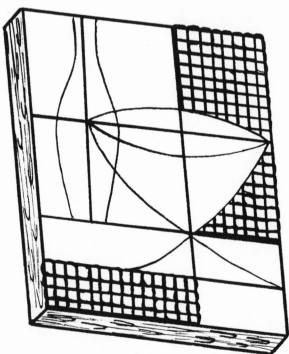

Fig. 27-12.

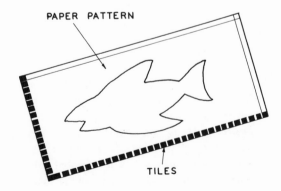

PAPER PATTERN

TILES

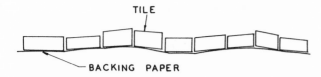

Fig. 27-15.

Note: The paper to which the glass tiles are gummed will buckle slightly as the pieces of glass are fixed in position (Fig. 27-15). No attempt should be made to avoid the unevenness as it adds a very desirable sparkle, variety and interest to the final panel.

When all the pieces have been gummed in position, a frame is fitted round the work. The frame is only temporary and need not be very strong (Fig. 27-16).

The cement or plaster is mixed and applied over the tiles in two layers. The first layer, which should fill the frame about half way, should be mixed fairly thin so it will sink into all crevices between the glass tiles and bind them together. On this layer may be placed wire lath or netting to act as reinforcement. The second layer may be a little thicker and should be applied as quickly as possible after the first so that the layers combine securely.

When the cement or plaster has set, the panel is

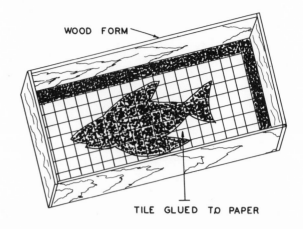

Fig. 27-16.

turned over and the paper peeled off from the front of the mosaic. The face of the tiles are scrubbed with soap and water followed by a muriatic acid solution as described above.

Applying Finish for Outdoor Use. Mosaic panels to be used outdoors or in damp locations should be coated with a silicone polish which will waterproof and seal all joints and add a lustre to the surface of the tiles. Panels used in dry locations need no special preparation.

Chapter 28
Crushed glass ornaments

It is simple to make colorful transparent jewelry, mobiles, shade pulls, and other useful ornaments from crushed glass bits. Crushed glass can be used in regular stained glass design to fill in holes, crowns, or other parts . . . adding a third dimension to your work. It may also be used for high textured areas in mosaic works, and borders.

The crushed glass used for this work may be had in several ways: (1) It may be accumulated from the colored and clear chippings and broken glass resulting when working on different glass items. (2) It may be purchased by the pound from craft shops. (3) It can be made from large pieces of glass in the following manner:

A. The large pieces of glass are first cleaned, heated slowly until hot, then taken with tongs and immersed immediately in cold water where it will break up into smaller pieces.

B. These smaller pieces are then placed in a strong cloth or leather bag and pounded with a hammer until they are broken into the desired size. In place of the cloth or leather bag, the glass pieces can be placed in a clean iron skillet, wrapped with a few layers of plastic and pounded with a hammer as mentioned before.

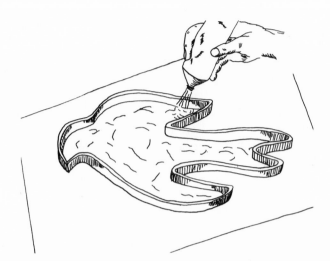

Fig. 28-2. Fill the inside of the lead design with the adhesive.

Note: The more the glass is pounded, the smaller the crushings will be.

The crushings should be carefully sorted and stored in containers, with tightly fitting caps to avoid their getting damp or dirty. If a tumbler is available, it is well to tumble the crushed glass to eliminate sharp edges.

The lead ribbon used in connection with making crushed

Fig. 28-1. Form the lead stripping around the design.

Fig. 28-3. When the crushed glass is dry, apply glass stain.

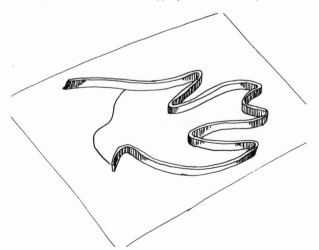

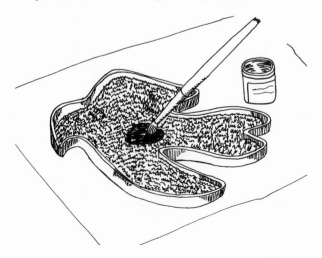

Fig. 28-4. Complete the ornament by hanging or mounting.

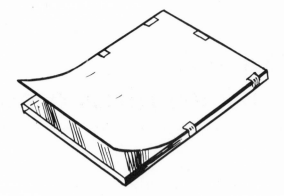

Fig. 28-6. (Courtesy: Stained Glass Products)

glass ornament may be purchased from hobby craft shops. The widths usually used are 1/8 to 1/4 inch.

Crushed glass ornaments may be made with or without a supporting panel to hold the crushings. When a support panel is used it is usually clear or transparent colored glass or plastics. The procedure for preparing a crushed glass ornament without a support panel differs from that with such a panel. Both procedures are described below.

ORNAMENTS WITHOUT SUPPORTING PANELS

1. Prepare a full size design on a piece of paper. The design should be kept rather simple and small. For beginners, cookie cutters can be used for designs. Lay the design out on a flat surface.
2. Completely cover the design with heavy wax paper (glue will not adhere to the waxed surface.)
3. Form around the outline of the design with 1/8

Fig. 28-5. (Courtesy: Stained Glass Products)

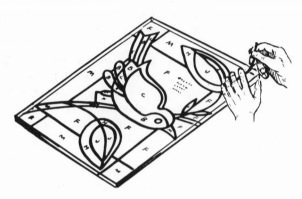

Fig. 28-7. (Courtesy: Stained Glass Products)

inch or 1/4 inch lead ribbon (Fig. 28-1). For details on bending the lead ribbon see page 167.
4. Fill inside the lead design with Duco Cement or white resin glue (Fig. 28-2).
5. Immediately fill the glued area with crushed glass bits packing them as thick or loose as desired.
6. Permit the glue to dry well. When the top side is dry turn the design over and carefully peel off the wax paper. Permit the other side to dry (this may take overnight).
7. When thoroughly dry, apply glass stain to the crushings (Fig. 28-3). The same glass stain that is used for making simulated stained glass is excellent for coloring the glass crushings (see page 167).
8. Complete the ornament by hanging or mounting (Fib. 28-4).

ORNAMENTS WITH SUPPORTING PANELS

The making of a crushed glass ornament with a support panel is very similar to making a simulated stained glass item. Up to a point, the procedures are almost identical. The designs suggestions on page 164 will work equally as well if prepared with crushed glass.
1. Prepare a full size design on a sheet of paper.
2. Cut the support panel to the size desired. This

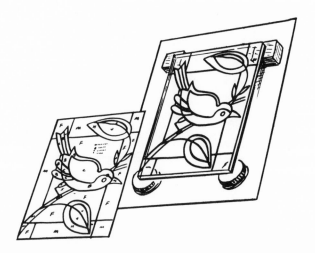

Fig. 28-8. *(Courtesy: Stained Glass Products)*

panel may be glass, plastic, or heavy acetate at least .010 inches in thickness. Thoroughly clean the support panel and then avoid handling it with bare hands.

3. If the support panel is transparent, attach the design to the back of it with masking or scotch tape (Fig. 28-6). If the panel is not transparent, draw the design directly on the panel with a china pencil.

4. Prepare the lead stripping by cutting it into lengths approximately 16 inches long then straighten and

flatten them.

5. Outline the segments of the design with the lead stripping (Fig. 28-7). This can be done two ways: First, the surface of the lead to go against the support panel can be coated with contact cement and the lead pressed permanently against the support panel as it is bent to shape. Secondly, the lead can be shaped then cemented in place with Duco Cement or white resin glue. Details on bending the lead are given on page 157.

6. Remove paper design. Prop glass up 1/2 inch over a white background, leaded side up (Fig. 28-8). Keep absolutely level.

7. Fill inside the lead design with Duco Cement or white resin glue (Fig. 28-2). A toothpick helps spread glue into small corner areas and along edges. Glue on the face of the lead can be wiped off immediately or, if already dry, cleaned off with thinner or removed by scrapping with a razor blade.

8. Immediately fill the glued area with crushed glass bits packing them as thick or loose as desired.

9. Permit to dry thoroughly (at least 12 hours).

10. When thoroughly dry, apply glass stain to crushings as indicated above.

   Note: Instead of making the design up of clear glass crushings and then coloring it with stain, the design can be composed of different colored glass crushings.

Chapter 29
Enameling with glass

Ever since the 5th century B.C. when Greek jewelers occasionally used enamel and later when the early Egyptians fused vitreous enamels (glass) to their bronze and copper armor for decoration, enameling has been an important glass activity. Its rise and fall as an industrial and art form very closely parallels the general history of glass.

Porcelain enameling is a process whereby a glass coating is baked on a metal base at a high heat of about 1250 to 1600° F. This coating is applied to many different items in common use today—kitchenware, appliances, plumbing fixtures, wall panels and tiles, store fronts, building sections, hospital and scientific equipment, jet engine linings, rocket parts, architectural murals (Fig. 29-1), jewelry, artware, etc. These items are enameled to impart to them many of the qualities found in glass itself—Permanent and attractive colors that are easily cleaned and resistant to corrosion, weather, abrasion, stains, and acids. With a minimum of care and maintenance the porcelain coating can be depended upon to last indefinitely.

Hard Enamels are Glass

Enamels are made chiefly of feldspar, quartz, silicon, borax, lead, and mineral oxides for coloring. These ingredients are mixed in various proportions depending on the use to which the enamel is to be put. When these chemicals are fused into a molten mass (Fig. 29-2) it is poured directly into water (Fig. 29-3) shattering the mass to bits (called frits—Fig. 29-4) or poured onto thick steel slabs and either left there to cool, or before being entirely cooled, dropped into water, where it breaks up into small particles before being used for enameling.

Enamels can be secured in a few different forms such as powders, lumps, threads, and liquids. The size of enamel powders is measured and indicated by the size wire mesh the fine grains will pass through. For general all around work 80 mesh is satisfactory, however, a little 60 and 100 or 150 mesh is good to have on hand. The 60 mesh can be used on large areas and the 100 or 150 for very fine lines. There are over 150 kinds and colors of lump and thread enamel available. These are used for overlay designs. The liquid enamels are used for spraying, brushing or dipping. The sprayed enamels produce a crackle effect that looks well as a backing enamel, or may be applied to a fired base or

Fig. 29-1. Mural of porcelain enamel. (Courtesy: Porcelain Enamel Institute, Inc.)

Fig. 29-2. Measuring the batch temperature and quenching the molten glass in the process of frit making. (Courtesy: Porcelain Enamel Institute, Inc.)

directly on a metal base.

The different types of enamels include transparents—those through which you can see the surface below; opaques—which are solid colors and light will not pass through; overglazes—which can be used for painting and detail work. Opalescent enamels are specially formulated to be especially effective over foils.

Ground enamels may be stored either dry or wet. Dry ground enamels should be stored in a glass jar with a screw top and kept in a warm dry atmosphere. If the grains are permitted to absorb moisture they will mat together resulting in defective firing. Ground enamels stored wet should be completely covered with water. Distilled water should be used to avoid contamination.

METALS THAT CAN BE ENAMELED

Copper is the least troublesome metal to enamel. It is also the easiest to prepare and finish. Such other copper-zinc alloys as Brass, Gilder Metal and Oreide Metal are much more difficult to enamel even though they may impart more brilliance to transparent enamels. Bronze, which is basically an alloy of copper and tin, will enamel satisfactory when not too much lead and zinc are present.

Silver enamels very satisfactorily. Although fine-silver is best, most common alloys are acceptable.

Note: Pieces of copper and silver ready-cut and shaped for enameling can be secured from many handycraft suppliers.

Gold, gold-coppery alloys, gold-plated and gold-filled metals enamel satisfactory provided they are properly prepared to accept the enameling. White gold (which is an alloy of gold, copper, zinc, and nickel) is very hard to enamel.

Aluminum and aluminum alloys are extremely difficult to enamel. It can only be accomplished with specialized equipment and processes not ordinarily available to the hobbiest.

Iron and Cast Iron and special iron alloys **especially prepared** for enameling can be coated without a great deal of difficulty. Ordinary irons and cast irons usually cause excessive boiling of the enamels.

Steel can be enameled but requires special preparation and care. Stainless steels are rather difficult to enamel.

TOOLS AND EQUIPMENT NEEDED FOR ENAMELING

Enameling requires very few and inexpensive tools and supplies (Fig. 29-5). Although the following list may appear large, a close observation will show that many of the items can be made from inexpensive odds and ends.

Fig. 29-3. Quenching the molten glass in the process of making frit. (*Courtesy: Porcelain Enamel Institute, Inc.*)

Fig. 29-4. Porcelain enamel frit. (*Courtesy: Porcelain Enamel Institute, Inc.*)

Fig. 29-5. Enameling Tools. *(Courtesy: American Art Clay Co., Inc.)*

Tweezers

Short Spatula

Long Wide Spatula

Swirl Sticks

Pancake Turner

Small Sieve (such as kitchen strainer)

Thin nose tongs with long handle

Asbestos glove

Paint Brushes—small and very small water color brushes and flat brush for adhesive

Hand anomizer or spray gun

Planche (may be made of steel wire or perforated metal)

Ceramic kiln stilts (small)

Torch

Kiln—small electric jewelry type (1400 to 1500° F.) or gas or electric chamber type.

If Shaping Own Metal Add:

Tinsnips

Round headed hammer

Jeweler's saw

Files—medium and fine, round and flat

SUPPLIES NEEDED FOR ENAMELING

Enamels

Glass lumps and glass threads

Metal to be enameled—copper, silver, gold, etc. (copper should be 16-18 gauge)

Wire to use in enamels

Alkali cleaners

Acids—Sulfuric and nitric

Steel wool

PREPARATION FOR ENAMELING

Shaping the Metal is the first thing that is done. All cutting, drilling, bending, filing, and shaping must be completed before the enameling process is begun.

Cleaning the Metal Properly is the Most Important Step in Enameling. The problem is to remove all grease and other forms of contamination from the metal and keep it that way throughout the entire enameling process. Once the metal is clean IT MUST NOT BE TOUCHED WITH THE BARE HAND. Some people wear thin cotton gloves when enameling to avoid touching the metal, others use tweezers, long handled forks, or tongs to handle the metal.

There are a few different methods that can be used for cleaning the metal to be enameled. The one to use can be determined by the effect you wish to produce, the enamel you are using (transparent enamels requires more careful cleansing than opaques), the facilities available, the degree of accuracy demanded, etc.

Commercial Cleaners are available for cleaning metals prior to enameling. When these are used, the directions on the containers should be followed without variation. These cleaners may be secured in craft shops.

To Clean Copper you can rub the surface with steel wool, followed by scouring powder. Another variation is to rub the copper vigorously with steel wool and a paste of baking soda and water, finishing with a detergent. When it is necessary to avoid scratches of any kind, a paste made of baking powder and pumice can be rubbed on with a cloth. (Very fine scratches can be removed with an ink eraser.) **All the above cleaning methods must be followed by careful rinse in clean**

water and a drying with a clean cloth or paper towel.

Some people prefer **Annealing and Pickling** (clean with acid) as a method of preparing the copper. These are done in combination—The copper is first annealed by placing it in the kiln until it turns color (approximately 2 to 3 minutes) then removed and rinsed in water. It is then placed in a warm pickle solution and left until clean and bright. After pickling, the copper is rinsed in hot clean water and dried.

The pickling solution may be made up of sulphuric acid and water or nitric acid and water in proportions of one part acid to ten parts of water to as much as one part acid to five parts of water. The more concentration of acid the faster the action. It is **important** to remember that when preparing the pickle solution **THE ACID IS ALWAYS ADDED TO THE WATER NEVER THE OPPOSITE WAY.** Mix the acid in a well ventilated area and wear a rubber apron. The solution is warmed after it is mixed.

To Clean Silver heat to a dull red color, cool, then pickle in a hot sulphuric acid solution (one part acid to ten parts water), follow with a rinse of water and dry.

To Clean Gold heat until it changes color, **cool completely**, then pickle in a warm nitric acid solution (one part acid to eight parts water), follow with a rinse of water and dry.

To Clean Cast Iron sandblast the surface until it is an even gray color.

To Clean Iron for enameling requires that the metal be subjected to a series of baths. The metal is first dipped in a hot strong alkali cleaner. Following this it is cleaned with hot clean water. It is then placed in a hot sulphuric acid pickle solution until it is an even light gray color, when it should be removed and again rinsed in hot water. Finally, it should be cleaned in a hot dilute alkaline solution and rinsed in hot water.

> **Note:** If the metal is not going to be enameled immediately after cleaning, leave it immersed in distilled water.

To Test Cleanliness of metal after cleaning, pour water on it. If water coats the metal in a thin even sheet the metal is sufficiently cleaned. If the water forms separate pools then grease or oil is most likely present and must be removed.

APPLYING THE ENAMEL

Adhesives for Enamels are **Gum Tragacanth** and **Gum Arabic.** These adhesives keep the fine grains of enamel from shifting on the metal. They are especially helpful on curved surfaces as on plates and bowls. The adhesives can be brushed or sprayed on with an atomizer or paint spray. (Sometimes brushing will leave brush marks visible under transparent enamels.) Gum Tragacanth or Arabic can be purchased in a prepared form from craft shops. Clear water is sometimes used as an adhesive. It is especially effective on small pieces

where the grains of enamel are to be held for a short time.

Some people coat all pieces to be enameled with gum tragacanth first. Others only use the adhesive on sloping surface. The answer as to whether you should use it or not will depend on how carefully you handle your work.

ENAMELING PROCEDURE

1. Paint or spray gum tragacanth on parts of metal you feel needs adhesive. Filing edge to roughen it sometimes helps in addition to gum.
2. Fold a clean piece of white paper in center and lay flat on table—this will catch excess enamel and fold will serve as funnel for returning to jar.
3. Place the metal in the center of the paper but raise it from the paper with a piece of cardboard cut smaller than the metal piece to be enameled.
4. Select the enamel to be used as a base coat. Transparent flux and white are frequently used for this purpose. Opaque colors are used by the more experienced.
5. Apply the enamel to the metal. The enamel may be dusted or sprayed on. When dusting, the 80 mesh enamel is distributed on the metal through a fine sieve or screen. To spray, use an atomizer or spray gun with a large opening in the tip and use 150-mesh enamel that has been stirred into an agar or gum solution. Try to apply the enamel in an even coat. Slight unevenness may be corrected by dropping water on the uneven spots or spraying gum tragacanth over the entire piece.
6. Pour excess enamel into proper jar and close jar.
7. Carefully slide broad knife or spatula under coated piece and transfer it to a planche or grid. (Place on planche when firing in a chamber kiln, and on

Fig. 29-6. Enamel bowel, dish and ladel. *(Courtesy: American Art Clay Co., Inc.)*

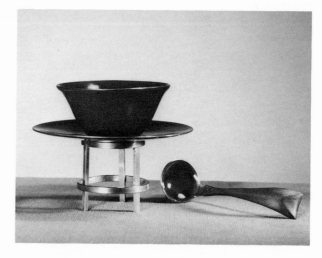

a. Enamel first turns dark and grainy
b. Then gets ripply like an orange skin.
c. Finally, turns red hot.

When the surface is entirely smooth, shiny or glazed over, remove from kiln and lay on a piece of asbestos to **cool gradually**. (Sudden cooling will cause enamel to crack.)

To Enamel Both Sides of a Dish or Bowl at the Same Time (Fig. 29-6)

For this kind of work it is best to use a slush or crackle enamel. These are very finely ground enamels mixed in a water solution which is brushed on the piece to be enameled. When fired, very special crackle effects are achieved.

1. Clean the metal dish well.
2. Place the dry dish face down on a small can or cylindrical object for ease of handling and turning.
3. Have dry enamel ready in sifter. Take a brush, (use a wider brush for larger pieces) and brush a coat of slush on the entire back of the dish. Work quickly since you are going to sift a coat of enamel on entire back before slush dries. If slush starts to dry before you can get to enamel coat . . . dip your brush in water and wet the slush evenly (Fig. 29-7).
4. Then sift a coat of enamel or counter enamel all over the slush coat. Try to cover entire slush surface making sure the slush is still wet. Let the surface dry.
5. When surface of dish is thoroughly dry pick up, turn over and hold carefully with hand under dish. Proceed to enamel the top side of the dish with the first coat (Fig. 29-8). Then place carefully on a star stilt (3 pointed stainless steel planchon) and place in kiln.
6. When dish is fired remove from kiln, knock off stilt, turn over on a piece of asbestos and place a weight or electric iron on top to help control warping (Fig. 29-9).

Fig. 29-7.

grid for a hot plate kiln.)

8. Place the coated piece to be enameled near to the kiln (**not** in it) to **thoroughly dry out gum before firing**.
9. Pre-heat kiln to 1450° or 1500°. A quick hot heat is better than a long heat.
10. Place the coated piece in the kiln in a position where it can be watched. There is not a set time that it takes to fire enamel and therefore it must be constantly watched for progress indications as follows:

Fig. 29-8.

Fig. 29-9.

Fig. 29-10. Samples of different types of enamel work. *(Courtesy: American Art Clay Co., Inc.)*

7. When cool, apply another coat of slush, this time stippling or dabbing it on. Use a brush you will reserve for this stippling process since the brush will be ruined for anything else. Stippling is a process where you dab slush all over your enameled back. When back is completely covered permit to dry thoroughly.

8. Now turn over, apply another coat of slush, carefully handling dried slush area and work on your front again—place on stilt and then into kiln and fire.

9. Cool, once again stipple the back thoroughly, dry, turn over and complete or add to design on front. After firing you can continue working on front forgetting about back since it has been sufficiently counter enameled by this time.

Remember, your back coat consists of 4 applications: (1) Coat of slush, (2) Coat of counter enamel (3) Coat of stipple slush, (4) Another coat of stipple slush. In the case of your first 2 coats the colors are not important but in the case of coats 3 and 4 a light third coat and dark fourth coat will result in a dark finish. A dark third and a light fourth coat will give a light finish. Experiment until you find the combination you prefer.

Additional Suggestions About Enameling

For best results apply at least two coats of opaque enamel, firing each time. Sometimes even a third coat may be necessary. The additional coats bring up and hold the true color. Exceptions to rule of two coats or more: Where "bleed-out" is part of the design or in case of transparents where one coat produces the desired effect.

If a number of firings are planned, underfiring early coats (removing piece from kiln when enamel is pebbly) will prevent burning out.

Do not use too much gum. The same color enamel will look different when fired over gum as compared to a dry sifting.

There is practically no limit to the number of times a piece may be enameled. The number of coats that are applied depends on the effect desired.

DESIGNING ON THE ENAMELED PIECE

The designs that can be created in enameling are limited only by the skill and imagination of the enamelist (Fig. 29-10). It is practically impossible to record all the many variations in technique that are practiced today. Following are but a few of the more popular techniques that are used. With a little practice the novice will soon be emplying variations of his own.

The Design should be carefully worked out on paper in color, if possible. To trace the design onto metal use ordinary carbon paper and then with scratch awl or scriber scratch the design into the metal carefully (Fig. 29-11). Following this the metal should be cleaned thoroughly. For tracing on to fired enamel use dark red or white carbon paper.

Stencil Designs are frequently used in enameling. Such stencils should be cut from paper toweling, which when wet, adheres tightly to the piece being enameled. The paper can be folded many different ways and cut with scissors to produce interesting design variations (Fig. 29-12). Almost anything with an interesting design can be used as a stencil—lace, leaves, doilies, screening, etc.

To Use Stencils in enameling, you first fire a base coat of enamel on the piece which could be an earring, dish, tie clasp, etc. Dampen the stencil (Fig. 29-13) and press very flat on piece where desired. (It is optional whether to brush or spray gum on stencil and piece.) Sift on a color that contrasts well with base coat (Fig. 29-14). This color should be put on thickly to be definite but not too thick or the edges will crumble when the stencil is removed. When the enamel is completely dry so it will adhere to the stencil, lift the stencil off carefully with sharp tweezers (Fig. 29-15). Then proceed to fire as explained before. After firing, the procedure may be repeated with the same stencil placed in a new position (Fig. 29-16) or another stencil used.

Scraffito (Fig. 29-17) is a process of designing in enameling. It is done by drawing lines through an un-fired coat of enamel exposing the base coat beneath. Scraffito is possible because the glass enamel fuses and

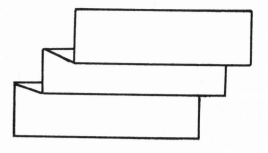

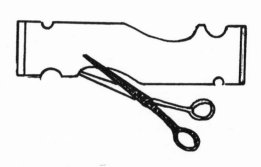

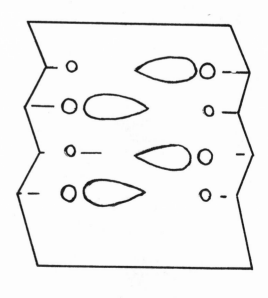

Fig. 29-11.

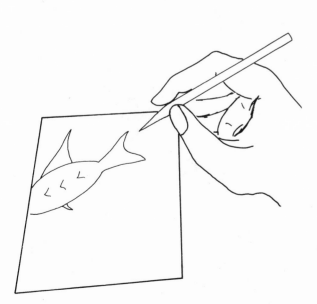

Fig. 29-12.

Fig. 29-13.

Fig. 29-15.

does not flow.

To produce the scraffito design first complete the base coats then coat the piece with gum. On the gum dust an even coat of contrasting enamel. Using the back end of a brush, a fork, the eraser end of a pencil, or any pointed object, draw lines in the unfired enamel exposing the base coat (Fig. 29-18). Dry out gum, then place carefully into kiln.

Trailing refers to the designs that can be created in enamel by sifting on contrasting colors to achieve varied effects. It is done a few different ways; You may trail enamel with your thumb and forefinger onto a prefired coat (Fig. 29-19). A second method is to brush gum in lines and spots, dust on enamel then blow (Fig. 29-20) or turn over and tap off excess. Another method is to trail on enamel using a little paper funnel. Fine dust and scattered grains that are not wanted can be removed with a brush (Fig. 29-21).

String Designs are attractive and easy to do. Wet a

string and form a line design on a prefired enamel piece (Fig. 29-22). Loop the string over itself in any design you feel is attractive. Apply gum over entire piece. Sift evenly a contrasting color and apply gum again. Pick up the loose end of the string carefully with tweezers and lift off slowly (Fig. 29-23). Fire as previously explained.

Note: Use different sizes of thread, string, or yarn for different effects.

Threads and Lumps are pieces of glass enamel in uncrushed form. They come in opaque and transparent colors. To create attractive designs using glass lumps and threads proceed as follows:

1. Wet the enameled piece with gum solution.
2. With tweezers position threads and lumps where you want them to be.
    a. Use small pieces if possible as they will offer

Fig. 29-14.

Fig. 29-16.

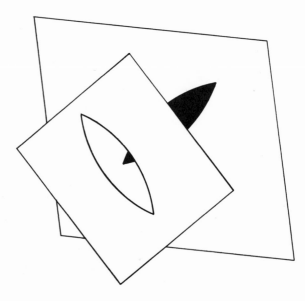

Fig. 29-17.

the least difficulty when being fired.

b. When making matching enameled pieces, break one piece of thread or lump and use half on each piece. This way you can be sure you are getting the same color and thickness.

c. The threads that stand up slightly will fire down when heated.

d. Threads and lumps can be shaped by heating them on a hot plate until they become soft.

3. Dry out gum then place in kiln.

a. Since different colors fire at different speeds it is well to partially fire the slow firing threads and lumps before adding the others. Sometimes this may avoid firing out the base coat.

Swirling (Scrolling) permits the enamellist to create

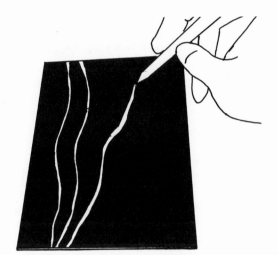

Fig. 29-18.

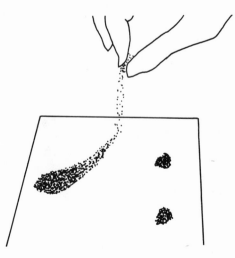

Fig. 29-19.

Fig. 29-20.

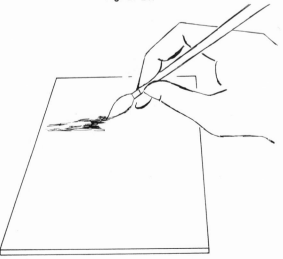

Fig. 29-21.

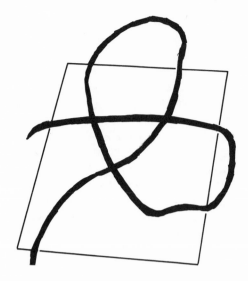

Fig. 29-22.

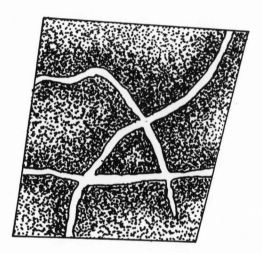

Fig. 29-23.

Fig. 29-24.

the most attractive and unusual designs (Fig. 29-24) The threads and lumps described above are used in swirling only this time in a different manner:

1. Clean the metal and dust on a fairly thin base coat of enamel.
2. Place threads and/or lumps (no thicker than 1/8 inch) on the unfired base coat and press them in with the spatula or back of a spoon.
3. Carefully place the coated piece in the preheated kiln and heat until the metal is red hot and the threads and/or lumps have melted into the base enamel.
4. Remove lid from kiln and holding the piece with a metal object so it will not move, run the point of the swirling stick through the flowing enamel (Fig. 29-25). Different effects can be produced by moving the swirling stick in small or large circles, zig zags, loops, etc. By closely watching the piece as the swirling stick is moved, the final design can be anticipated.
   a. Do not try to swirl until the enamel flows without any resistance to the tool.
   b. If enamel sticks to the tool, it means that the piece should be heated some more. Dip the tool in cold water to remove enamel that sticks.

FINISHING THE ENAMELED PIECE

When the piece is completely enameled the metal parts that are not covered with enamel will have a black oxid appearance. Most times this is the edge and the back of the object. To finish the piece, the black oxid must be properly removed and the metal surfaces

Fig. 29-25.

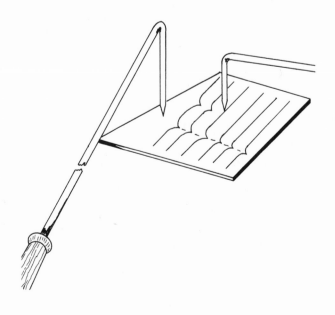

Fig. 29-26. (Courtesy: American Art Clay Co., Inc.)

Fig. 29-27. (Courtesy: American Art Clay Co., Inc.)

coated to prevent the oxide from reforming. Any metal exposed to the air will form an oxide and with copper it forms especially fast.

There are creams on the market that can be brushed on copper that will prevent it from oxidizing during firing. If the back of the dish is not to be enameled, it can be painted with "Protex" or a similar cream. After firing the enamel on top of the dish, the protective coating is peeled off the bottom leaving it clean of oxide.

The black on the edge may be removed with a very fine file or fine carborundum hand stone. Work the file or stone in a horizontal position across the edge. In this way it is less apt to chip the enamel.

The black oxide on the back of the dish (or any exposed metal part) can be removed with a medium steel wool followed by finer steel wools or a coarse scouring powder and a wet rag, followed by milder scouring powders. Fine scratches can be removed with an ink eraser.

When the exposed metal is finished to your liking, coat the exposed metal with clear metal lacquer. (Brushing lacquer if you brush, and spraying lacquer if you spray.)

Note: If you chip the enamel during the finishing process, apply gum to the chipped area, stuff in enamel and refire.

POSSIBLE DEFECTS IN ENAMELING AND CORRECTIONS

1. If enamel surface resembles an orange skin it means the piece was under-fired. The piece can be refired or the design added and then refired.

2. If opaque white pieces of enamel come out partially or mostly black after firing, it means the coat was not thick enough and the color "bled out." Give the piece another coat (and next time put more on first coat).

3. When the fired enamel is lumpy and with large separations, it indicates that there was too much enamel unevenly applied on first coat.

4. If edges "burn out" (have no enamel on them) it indicates that the piece was fired too long or there was not enough enamel around edges in the first place. Coat edge with gum, enamel, and refire.

5. Gum rings can be seen through transparent enamels when the gum was not applied evenly before

Fig. 29-28. Enamel being dip coated. (Courtesy: Porcelain Enamel Institute, Inc.)

Fig. 29-29. Enamel being spray coated. (Courtesy: Porcelain Enamel Institute, Inc.)

applying enamel.

6. Flaws (pinholes, blisters, cracks, etc.) can be corrected by cleaning out the defective enamel with a carborundum stone then refill the hole with enamel and refire.

7. Black spots are usually caused by fire scale or shreads of steel wool. This is a result of poor cleaning. It can be stoned off, refilled and refired or covered with an opaque enamel.

8. Warping of fired dishes and bowels can be prevented if they are placed bottom up, and a heavy weight is placed on them while they are cooling.

Chapter 30
Mirror making

"To see ourselves as others see us . . ." has been an urge of mankind from the beginning. Ever since the first time that man saw his image in the clear, still water of a pool, he has tried to duplicate the effect in a more convenient and permanent form. First, he polished and burnished comparatively small pieces of gold, bronze, and silver to a mirror like finish. It was not until the advent of the modern mirror, produced by depositing an extremely thin layer of silver on a sheet of glass, that mirrors became inexpensive enough for wide, universal use.

THE MIRROR MAKING PROCESS

Mirror making consists of three processes: First, a sheet of glass is polished and thoroughly cleansed on one side. Secondly, a mixture of chemicals are poured on the prepared side of the glass. When these chemicals are reduced they leave a deposit of pure silver on the glass. This process is called silvering. Thirdly, when the glass is dried, a protective coating is brushed, rolled or sprayed over the silver coating.

Today, silvering is applied to mirrors by pouring the silvering solution from a hand pitcher or by spraying from a spray gun (Fig. 30-1). The pouring method is still widely used; however, in large quantity production only the spraying method is employed. Either method may be used in the school shop, laboratory, or home workshop, however, for a small, inexpensive shop unit that will require very little space, the pouring method is recommended.

There are two methods of silvering by hand pouring. One is called the hot-table method, wherein the glass sheet is heated to 120° -130° F. on a heating table before the silvering solution is poured on the glass. The other is called the cold-table method, wherein the silvering solution is poured on the glass at room temperature. Before the advent of the much faster method of silver spraying, the hot table method was most widely used and still is quite popular today. This was true because the hot-table method was faster and easier to control than the cold-table method. However, there has recently been developed a new cold-table method that is faster and equal quality to the previous hot-table method. This new cold-table method of silvering is the most practical for school and novice use as it requires a minimum of cost, space and know-how.

COLD-TABLE HAND POURING METHOD OF MIRROR MAKING

Equipment Needed

The equipment needed for a cold process mirror making unit is extremely inexpensive and simple as indicated below:

Polishing Table. Any shop bench or sturdy table will be sufficient for this purpose. It is important that the polishing table be located as far away from the silver-

Fig. 30-1. Silvering chemicals being applied with a spray gun. *(Courtesy: Peacock Laboratories, Inc.)*

ing table as possible since the rouge used in polishing can contaminate the silvering solution. **Washing Table.** A tray, made up of wooden slats an inch wide by an inch thick and spaced an inch apart, fitted over the top of a sink will do very nicely for a wash table. If a sink is not available the silvering table described below can be duplicated and used as a washing table. In a situation where only an occasional silvering job is done, the washing and silvering can be done on the same table. **Cold Silvering Table** (Fig. 30-2). The top of the silvering table is made up of wooden slats about one inch wide by one inch thick and spaced about 2 inches apart. Below the slotted top, the table is formed into a funnel with the opening of the funnel leading into a drain or large pan. **It is important that the top of this table be** **level in all directions.** If the silvering table is to set on the floor, it should measure 30 inches from the floor to the table top. If it is to set on a bench, it need only be high enough so a drain pan can be placed beneath it. If a table model is made up to have a drain pan beneath, it is recommended that one side of the table be cut away so that the pan can be removed for dumping without disturbing the table. **Wedges** (Fig. 30-3). These are used to level up the glass on the silvering table, should the table itself be out of level. About two dozen of these should be kept at hand at all times for immediate use. **Silvering Pitchers** (Fig. 30-4) At least one porcelain finish chinaware pitcher of at least 2 quarts capacity is needed. If the Ermax cold-table method is used, a special pouring pitcher with compartments in it is needed (Fig. 30-7). **Silvering Room Glassware.** This

Fig. 30-2. Silvering Table.

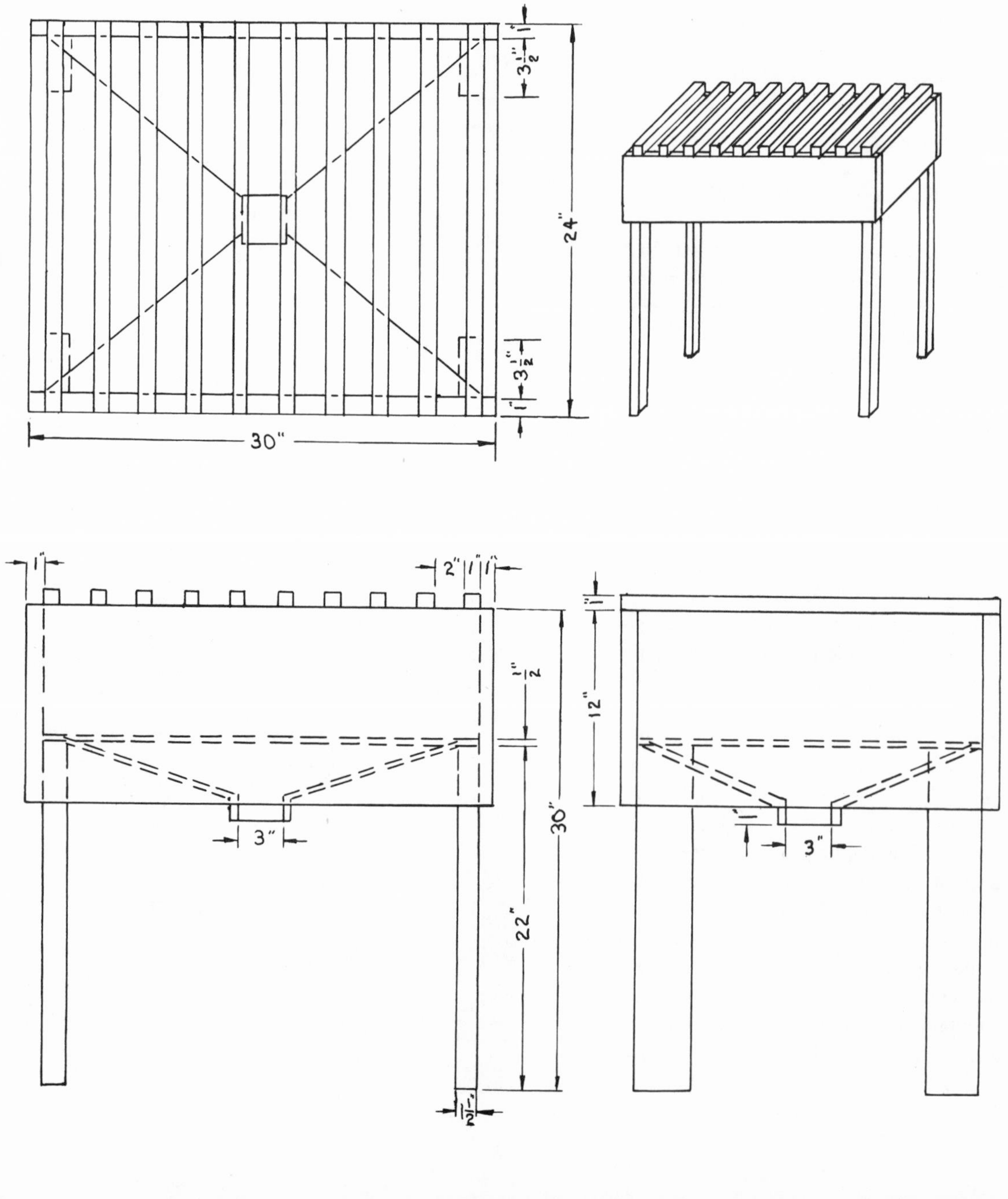

should include a graduated cylinder for measuring up to 100 ml. (1ml. - 1cc.) a second one that will measure up to 32 oz., and a glass funnel of about 7 inches diameter. **Laboratory Scale.** This scale should have facilities for measuring up to 40 grams. **Silvering Room Apron.** This apron should be made of heavy, acid proof rubber and large enough to fully cover the front of the student. **Chamois and Sponges.** At least one large, lint free chamois should be available. Only the best chamois should be used. Cellulose sponges may be used in place of the chamois. **Wash Brush.** At least one firm bristled scrub brush (Similar to a floor scrub brush) should be available for cleaning the glass. **Felt Polishing Block** (Fig. 30-5). A block of hard felt, large enough for convenient handling, should be available for polishing out scratches in the glass. **Rubber Gloves.** Use rubber gloves while performing all washing and silvering operations. Their uses protects the hands from chemicals, and equally important, protects the glass from skin contaminates.

MIXING FORMULAS FOR COLD-TABLE SILVERING

There are two leading concerns that supply the chemicals needed for the mirroring trade. One is the Peacock Laboratories Incorporated and the other is the Ermax Corporation (See "Source of Supply" for addresses). Each of these companies recommend cold-table silvering formulas of their own. The formulas differ in content, procedure for application and time needed for reaction. Both formulas and procedures for application are presented below.

Note: The following formulas may be purchased already mixed from each of the respective companies at an increased cost. The ready mixed solutions should not be purchased, however, unless they are going to be used within a reasonable length of time as they tend to deteriorate. The unmixed chemicals will last indefinitely.

Ermax Cold-Table Silvering Formulas. The Ermax formula consists of these solutions: No. 1 Activators Solution, No. 2 Silvering Solution and No. 3 Reduction Solution. The following instructions are for preparing one gallon of each solutions 1, 2, and 3. If less or more is desired the amounts should be increased or decreased in proportion. The mixed solutions will last from 4 to 6 weeks.

No. 1 Activator Solution

INGREDIENTS:
34 grams Sodium Hydroxide, reagent pellets
49cc Ammonia solution, 27% U.S.P.
60cc Metallochrome, compound (1-9 mix)
1 gal. Distilled Water

PREPARATION:
Into a thoroughly clean jug pour approximately 1/2 gallon of distilled water (This amount is not critical and can be varied a great deal at this point). Completely dissolve the sodium hydroxide pellets in the water by shaking the jug. Then add the ammonia solution and shake the mixture again. Next, add the metallochrome **compound**. (Metttallochrome compound is a 1 to 9 solution of metallochrome concentrate and distilled water. It is the **compound** and **not the concentrate** that is used to make up Solution 1.) Finally, fill the remainder of the gallon jug with distilled water.

No. 2 Silvering Solution

INGREDIENTS:
38 grams Silver Nitrate
49cc Ammonia Solution 27% U.S.P.
1 Gallon Distilled Water

PREPARATION:
Into a thoroughly clean gallon jug, pour approximately 1/2 gallon of distilled water. Add the silver nitrate to the water and completely dissolve by shaking the jug. In a graduated cylinder, measure off the 49cc of Ammonia Solution. This Ammonia Solution is to be added into the gallon jug in a few small portions and the jug shaken well after each portion is added. The first Ammonia portion added to the mixture in the jug will turn the solution black. As more portions are added a point will be reached where the mixture will suddenly turn clear once again. (Do not add anymore Ammonia Solution once the mixture becomes clear even though there still may be some left in the graduated cylinder.) Fill the remainder of the gallon jug with distilled water.

No. 3 Reducer Solution

INGREDIENTS:
33cc Super Chromic Silver Reduction

Fig. 30-3. Leveling wedge.

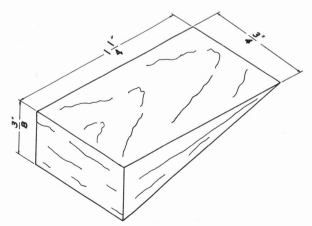

Fig. 30-4. Silvering pitchers.

7cc Chromic Silver Compound
1 gallon Distilled Water

PREPARATION:

Into a thoroughly clean gallon jar pour the Super Chromic Silver Reduction. Add to this approximately 1/2 gallon distilled water then add the Chromic Silver Compound. Fill the remainder of the gallon jug with distilled water and shake the solution.

Note: The Ermax formula requires that a Neutralizer be applied to the silvered glass following the application of the above solutions. The Neutralizer is mixed as follows:

NEUTRALIZER:

To one quart of distilled water add 5cc (1 Teaspoon) of Neutralizer Concentrate.

PEACOCK (BRASHEAR TYPE) COLD-TABLE SILVERING FORMULAS

The Peacock formulas consist of two solutions: Silvering solutions and Reducer Solutions. This cold silvering formula does involve explosion hazards but the hazard is reduced by the outlined procedure of preparation. The quantities specified are for a gallon of each solution.

Silvering Solution

INGREDIENTS:
1.6 Av. oz. Silver Nitrate
1.6 Av. oz. Sodium Hydroide, Reagent
5.8 Fl. oz. Ammonium Hydroxide, Reagent 29.4%, sp. gr. 0.89
1 gallon Distilled water

PREPARATION:

Dissolve silver nitrate in 1-1/2 pint of cold (55-60° F.) distilled water. When silver is dissolved add the ammonium hydroxide. Stir to mix. Pour into 1/2 gallon of cold distilled water in 1 gallon jug. Stir to mix. To the ammonical silver solution add the sodium hydroxide which has been previously dissolved in one pint of distilled water and cooled to 55-60°F. Dilute to 1 gallon

with cold distilled water and shake well or stir well, to mix. Maintain at 55-60° F. Discard unused solutions 'at the end of the day's work by pouring into sufficient muriatic acid solution (1 part acid to 1 part tap water) so that the resultant mixture is acid to litmus.

Reducer Solution

INGREDIENTS:
5 Av. oz. Hydrous Dextrose
1 Gal. Distilled water

PREPARATION:

Dissolve the dextrose in one pint of distilled water and then dilute the solution one gallon. Stir well to mix. Maintain temperature at 55-60° F. Discard unused solution at the end of the day's work.

MIRROR MAKING PROCEDURE

It is assumed that all cutting, bevelling and edging on the glass has been completed at this point. The mirror making procedure, as presented below, includes 4 operations. Polishing, Prewashing, Silvering, and Finishing Process. One word of caution must be injected at this time—**CLEANLINESS is of the utmost importance.** Without it, successful mirror making is impossible.

Polishing

The surface of the glass to be silvered should be polished to insure a true clean image. The polishing should be done as far from the silvering table as possible for any specks of rouge that contaminates the silvering solution will leave permanent black specks on the mirror. Before proceeding, make a crayon mark on the side of the mirror that will **not** be polished so it will be recognized as the non-polished side when the glass is later moved about.

Powdered rouge is used as the polishing agent. It can be secured in a white, red and black form. White is the one used for the general polishing before silvering. Red is used for block polishing and black for scratch polish-

Fig. 30-5. Felt polishing block.

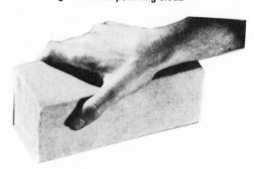

ing. For use, the rouge is mixed with a little water to make a medium thick paste. A shallow pie plate or disk is used to hold the mixture.

Place the glass on a wad of newspaper with the surface to be silvered facing up. The opposite side should be marked with a crayon. Pick up some of the paste rouge on a soft rag and rub it vigorously over the entire surface to be silvered (Fig. 30-6). Fresh supplies of paste rouge should be picked up on the rag as the polishing progresses. If there are no scratches on the glass, the polishing will be completed in a couple of minutes of rubbing with the rouge. Slight scratches in the glass can be removed by hard rubbing of the paste rouge on the scratch with the stiff felt block. Moderately deep scratches can be removed by rubbing them with black rouge on a felt block. (Separate felt blocks should be used for each kind of rouge.) Deep scratches must be removed with a polishing machine.

When the polishing is completed, wipe the rouge from the glass as completely as possible.

Pre-Washing

Transfer the glass to the wash table and scrub it **thoroughly**, including the edges, with a bristle brush and cellulose sponge. For this washing use tap water and a special wash solution such as Peacocks 37B Cleaning Agent or Ermaxs Pre-Wash Solution. When the glass is clean, rinse off the tap water with distilled water. If the room in which the silvering is going to be done is cool, preheat the distilled water that is going to be used for this final rinse.

From this point on use only distilled water and do not allow the glass surface to be silvered to dry or develop "water breaks" until the entire process is completed.

Silvering

Place the **wet** glass on the silvering table with the polished side up. **The glass must be perfectly level.** If the table is not true, the glass itself will have to be leveled with leveling wedges (Fig. 30-3). The glass is now ready to receive the silvering solutions.

Note: The actual pouring and timing of the Ermax and Peacock solutions differ enough so that they will be presented separately.

Ermax Pouring Procedure. A special Ermax pouring pitcher is needed. It is one that has three compartments each numbered 1, 2, and 3. Pour the prepared solution 1 into compartment 1, solution 2 in compartment 2 and solution 3 in compartment 3. Hold the pitcher about 10 inches above the glass and start pouring slowly and evenly **next** to the glass then, while you continue to pour, carry the pitcher onto the glass (Fig. 30-7). (This procedure eliminates the possibility of bubbles forming.) When the glass is completely covered to a depth of

about 3/32" stop pouring and let the glass stand undisturbed for about 1-1/2 to 2 minutes. At the end of this time dump the solution off the glass and pour on a second coat the same as the first. Let this second coat remain on the glass for approximately 6 minutes. (The 6 minute time limit is recommended for a room temperature of 75-80 degrees F. In colder temperature give it more time; in hot temperature give it less.) When the specified time has passed, dump the spent solutions, sponge clean with the neutralizing solution and rinse with distilled water.

Peacock Pouring Procedure. Mix equal volumes of Silvering solution and Reducer Solution in a porcelain or glass pitcher and **pour promptly**, onto the glass until it is covered to a depth of about 3/32". The room and glass temperature should not be over 60-65 degrees F. Allow silver to deposit from 8 to 10 minutes. After specified time, dump spent solutions rinse with distilled water under pressure, and drag wetted chamois over deposited silver film to remove sludge.

(If a second coat of silvering is desired to produce a mirror of finer quality, repeat the above making sure to dump, rinse, and chamois between coats as well as at completion of silvering.)

Finishing Process

Dry the mirror. This may be done in several ways: 1. The mirror may be left on the silvering table and air-wiped with clean warm air as may be had from a home hair dryer (Fig. 30-8). 2. The mirror may be air-wiped with clean compressed air. 3. The mirror may be placed in a drying rack and left to air dry. When the third method is used it is recommended that the mirror be stood on one of the corners for most efficient drying.

Fig. 30-6. Polishing the glass with paste rouge in preparation for silvering.

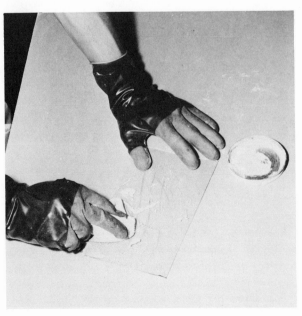

Fig. 30-7. The silvering chemicals are poured on the prepared glass.

Fig. 30-8. A home hair dryer is good for quick drying the mirror.

When the mirror is thoroughly dry, apply the protective backing on the silvering. For a backing apply one coat of mirror shellac; allow to dry 15-20 minutes, then apply a final coat of mirror paint. These coatings may be applied with a brush, roller or spray gun. Allow the final coating to dry in a well ventilated place for 24 hours.

To remove the silver that has flowed around onto the face of the mirror, take **dry** 3F or 4F pumice on the hand and rub it over the silver you want removed. Dry pumicing will not scratch the glass much if at all; rubbing with wet pumice will scratch glass fairly readily.

HOT-TABLE HAND POURING METHOD OF MIRROR MAKING

Equipment Needed

The equipment needed for the hot-table silvering process is identical with that needed for the cold-table process (page 239) with only one exception—the silvering table.

Hot Silvering Table. This silvering table consists of an angle iron frame on which are placed steam pipes running parallel about 4″ apart, covered with soapstone or slate. The final covering consists of 5 layers of heavy cotton cloth. The table must be perfectly level in all

directions. The table must have thermostatic facilities to keep the temperatures between 120-130 degrees F.

Formula for Hot-Table Silvering

The following non-explosive type hot-table silvering formula was used extensively before silver spraying became popular:

Solution No. 1. Dissolve 16 av. oz. of Silver Nitrate in 32 fl. oz. of distilled water, then add 12-1/2 fl. oz. of Ammonia Water. (This solution must **not** be clear.) Filter clear, then add 160 fl. oz. of distilled water. Shake well to mix thoroughly.

Solution No. 2. Dissolve 56 av. oz. of Rochelle Salts in 128 fl. oz. of distilled water. Then add 4 av. oz of Epsom Salts and dissolve. Mix thoroughly and filter clear.

Important Note: Do the following just before you are ready to prepare the mirror.

One-Coat Process. To every gallon of distilled water add and mix well:
9 fl. oz. of Solution No. 1
9 fl. oz. of Solution No. 2

Two-Coat Process. To every gallon of distilled water add and mix well:
4-1/2 fl. oz. of Solution No. 1
4-1/2 fl. oz. of Solution No. 2

Whenever possible, it is recommended that this formula be used without any Tartaric Acid. However, in extremely warm weather, when it is necessary to retard the reduction of the silver and to prevent turning in the pitcher, a speed control may be used. In such cases Solution No. 3 shown below can be added to the pouring mixture in quantities not exceeding 6 fl. oz. per gallon for the Single-Coat Process or 3 ft. oz. per gallon in the Two-Coat Process. Use the minimum amount needed to give the desired results.

Solution No. 3. Dissolve 2 av. oz. of Tartaric Acid in 112 fl. oz. of distilled water; filter clear.

HOT-TABLE MIRROR MAKING PROCEDURE

All procedures for hot-table silvering are identical with that described above for cold-table silvering with only two exceptions: 1. The glass should be at a temperature of 120-130 degrees F. when the hot-table silvering solution is applied. 2. The hot-table silver should be allowed to deposit 20-30 minutes.

ONE-WAY-VISION MIRRORS

One-way-vision mirrors are made by coating one side of a sheet of plate glass with a very thin film of silver, such a thin film that it may actually appear either transparent or reflecting, depending upon the conditions of illumination. Looking from a darker room into a brightly lighted room, the glass appears transparent; from the opposite side it appears as a mirror.

Chapter 31
Glazing windows and panels

Webster defines glazing as the "art, or trade of setting glass." Until quite recently this was a rather simple art but with the introduction of new types of glasses, new materials used for sashes, new sealants, (materials used to seal around edges), and new construction practices, glazing has become a rather complex art that can only be performed by those familiar with all the facts.

GENERAL CONDITIONS GOVERNING GLAZING

The theory of proper installation of glass is basically a very simple one. It can be stated as follows: 1st—Do not allow glass to come into contact with metal or other hard or sharp materials. 2nd—Use a resilient type of sealant. 3rd—Use a stop of sufficient size to get a good grip on the glass. The larger the glass the larger the grip. 4th—Install glass in openings that are rigid, plumb and square. 5th—Glazing with putty or mastic shall not be done in temperature below 40 degrees F. 6th—Allow sufficient clearance at edge of glass to compensate for expansion of glass or some settlement of the building.

Glass that absorbs heat will expand to a greater degree than clear glass. Therefore, heat-absorbing glass (and this includes the newer reflective types of glass and spandrel glass) should have a greater clearance around the outer edge of the glass and a resilient sealant between the face of the glass and rabbet or stop.

Movement between sash and glass is the most common cause of early seal failure. This may involve temporary or permanent displacement of the sealing material or subsequent vibration of the glass. If the extent of this movement is kept small, satisfactory service may be expected. The use of stiff sash, stops, and shims, and the generous provision of expansion joints and "isolation" connections to the structure, help by keeping the glass-sash movement within tolerable dimensions.

GLASSES USED IN GLAZING

Window Glass and Heavy Sheet Glass. This glass is manufactured by the horizontal or vertical drawn process. It is supplied in a number of thicknesses but only two thicknesses are commonly used in buildings and they are termed single strength and double strength. Four different qualities are supplied: "AA," "A" and

"B" for single and double strength, and greenhouse in double strength only. Heavy sheet glass manufactured by the same flat or vertical drawn process is supplied in several thicknesses of 3/16", 7/32" and 1/4" in sizes up to 76" x 120" and in "AA," "A," and "B" quality material. Heavy sheet is also made in 3/8" and 7/16" thicknesses and is most commonly used for shelving where greater strength is required.

The draw or wave distortion incurred in its manufacture runs in one direction. For best appearance glazing should be with the draw horizontal ,or parallel to the ground, where sizes do not exceed manufacturer's ability to furnish. Thus to insure this, when specifying, the width dimension should be listed first.

Plate Glass. This type glass is manufactured in a continuous ribbon and is cut in large sheets. Both surfaces are then ground, producing uniformity of thickness. This glass is then polished. Polished plate glass is furnished in thicknesses of 1/8" to 1-1/4". Thicknesses of 5/16" and over are termed heavy polished plate. One-quarter inch polished plate glass is available in four qualities: silvering, mirror glazing, selected for mirrors and glazing. The glazing quality is that used generally where ordinary glazing is required clear. Plate glass is

Fig. 31-1. Glass used as a curtain wall. (Courtesy: American-Saint Gobain Corp.)

made in thicknesses of 1/8", 1/4", 5/16", 3/8", 1/2", 5/8", 3/4", 1" and 1-1/4". IN BLUE GREEN (Heat Absorbing) in 1/4", 3/8" and 1/2". IN GRAY AND BRONZE (also Heat Absorbing) in 13/64", 1/4", 3/8", and 1/2".

Float Glass. A new method of manufacturing glass has recently been developed on a commercial basis. Developed in England, it is now being made in the United States by domestic producers. Basically, it is a very simple process. The molten glass flows on to a bed of molten metal where it is first reheated to produce a perfectly flat fire finished parallel surface and while still on the molten metal is cooled sufficiently to allow it to be withdrawn on to a series of rollers in the lehr where the temperature is gradually reduced so that the glass will be properly annealed.

Tempered Glass. This type is a glass which has been reheated to just below its melting point and suddenly cooled. As a result, fully tempered glass is 3 to 5 times stronger than regular plate from both impact and from temperature variances. When fully tempered glass shatters, it disintegrates into small pieces. It cannot be cut or drilled after tempering. The exact size desired and details must be specified when ordering. The qualities of tempered glass are determined from the characteristics of glass used in its processing. Tempered glass is not recommended as a fire-retardant glass.

Fig. 31-2. Glare-reducing, patterned glass made by acid etching or sand blasting. *(Courtesy: American-Saint Gobain Corp.)*

Spandrel Glass, Heat-Treated, Color Fired On. Principally, this type of glass is for use in the curtain wall design as spandrel glazing (Fig. 31-1) It is usually made of polished plate glass or patterned glass on one surface of which is fused a colored ceramic frit. Other textures are available.

This reheating for fusing of the ceramic and subsequent cooling also strengthens the glass. It may be supplied in a variety of surface textures and an almost unlimited color range. Some manufacturers fully temper their spandrel glass making it 3 to 5 times stronger than plate glass of equal thickness, while other manufacturers heat-strengthen their glass and this type is about twice as strong as regular glass.

Heat-Absorbing Glass. The composition of this glass is such that the ferrous add-mixture absorbs much of the energy of the sun. A considerable amount of this

Fig. 31-3. Patterned glass. *(Courtesy: American-Saint Gobain Corp.)*

heat is dissipated externally, thus reducing the amount of heat which enters the building. This heat-absorbing quality is such that the glass retains a portion of this heat. Thus, it is important that installations be designed minimizing cold edge effects and temperature differences over the plate. Further precautions must be heeded to prevent chipping or edge damage of any sort. This type of glass should be glazed with a permanently elastic glazing compound.

Glare-Reducing, Patterned glass with heat-absorbing qualities is made by acid etching one or two surfaces or by sandblasting and then etched with acid (Fig. 31-2).

Heat-Absorbing Glass is made in Sheet Glass, Plate Glass and Patterned Glass in various thicknesses.

Patterned Glass. This is a rolled flat glass with an impressed design on one or both sides accomplished during the rolling process (Fig. 31-3). It is available in

Fig. 31-4. Wired glass. *(Courtesy: American-Saint Gobain Corp.)*

a variety of designs, and finishes. Maximum width some patterns 48"—other patterns 60" and 96".

Wire Glass. This is a regular rolled flat glass with either a hexagonal twisted, diamond shaped or square welded wire mesh placed as near as possible in the center of the sheet cross section (Fig. 31-4). The surfaces may be either patterned or polished. Maximum size 72".

Insulating Glass. Insulation glass units comprise two or more sheets of glass separated by either 3/16", 1/4" or 1/2" air space. These units are factory-sealed and the captive air is dehydrated at atmospheric pressure. Typical units are made up of either window glass or polished plate glass. However, special units may be obtained of varying combinations of heat-absorbing, laminated, patterned or tempered glass (Fig. 31-5).

Because of the likelihood of breakage from stresses caused by temperature differences, care should be taken to avoid covering or painting areas of this glass, thus causing a heat trap. High insulating value and condensation are qualities which render insulating glass extremely useful for a variety of installations.

Glare-Reducing Glass. Glare-reducing glass is plate or sheet glass, which, by composition has a blue green, gray or bronze tint. This allows for "clear vision" as opposed to translucence in the case of surface-treated glass and patterned glass.

Patterned and wire glass may have one or both surfaces acid-etched to diffuse or soften the light (Fig. 31-6)(See Heat Absorbing Glass).

Mirrors. Mirrors can be made of Plate Glass or Sheet Glass. FHA requirements are that all mirrors shall be made of Plate Glass. The best mirrors are made of mirror glazing Plate Glass or Float Glass selected for this purpose.

Fig. 31-5. Laminated glass helps keep the room at a moderate temperature. *(Courtesy: American-Saint Gobain Corp.)*

Low Transmission Glass. Gray glass with a daylight transmission of less than 20% is available in 1/8" and 7/32" thick sheet glass. Also in laminated glass with a tinted vinyl interlayer.

Laminated Glass. This type consists of two or more sheets of glass with a layer of transparent vinyl plastic sandwiched between. The glass and plastic are bonded into one unit.

The elasticity of the plastic cushions any blow against the glass, and if the glass cracks, it holds firmly to the plastic preventing sharp pieces from flying.

There is also laminated glare-reducing glass where the pigment in the vinyl plastic laminate provides the glare-control quality.

A laminated glass with an increased thickness of plastic film has sound-attenuating properties.

Bullet-resistant glass is made by laminating a number of sheets of glass, building the unit up to various thicknesses according to the requirements of the job.

> **Note:** Manufacturers warranty this type of glass for a period of 15 months against film separation and discoloration. Not recommended for fire-retardant glass.

Heat-Reducing Glass. New methods are constantly being developed to produce a glass or unit that can be glazed in an opening to reduce the transmission of solar energy into a building by reflecting the heat.

Some of the newest are—

1. A double glazed unit with a coating on the inside of the outer panel that reflects the heat.
2. A glass sandwich that has a louvered screen of coated steel or stamped aluminum between two pieces of glass.
3. A single light of heat-resistant glass with a permanent reflecting coating on both sides of the glass that cannot be removed by the ordinary cleaning solutions or methods.
4. A laminated glass with a thin coating of metal on one of the inner surfaces of the glass to reflect the heat.

Art Glass. Art Glass, or at times called Cathedral Glass, is made in many colors and patterns. Usually 1/8" thick. Maximum domestic size is 32" x 84".

Quality Standards. Federal Specification DD-G-451-a cover the major quality characteristics of flat glass products as well as the dimensional tolerances, and is the standard of the industry.

DDM411 is a Federal Specification covering mirrors of plate glass quality.

GLAZING MATERIALS

I. Bulk Compounds

Bulk compounds are glazing materials supplied in containers or cartridges capable of being extruded in place.

A. Mastics

Mastics are compositions which do not change appreciably in viscosity. They remain in a semi-fluid or pliable state. Mastics may or may not form a skin on the surface exposed to the atmosphere, and are essentially non-curing.

1. Elastic Glazing Compounds

Elastic glazing compounds are formulated from selected processed oils and pigments which will remain plastic and resilient over longer periods of time than conventional hard putties. They may be used for either interior or exterior glazing, alone or in combination glazing as a topping bead over curing-type sealants. They should form a skin on the surface, yet remain soft and plastic underneath for considerable periods of time. This soft or plastic condition is desirable where windows or doors are subject to twisting and vibration. It should cling tenaciously to the sash and glass. Compound (if paintable variety) should be painted immediately after a thin skin has formed on the surface. Certain compounds are available in a range of colors.

2. Non-Skinning Compounds

Non-skinning compounds are part polybutene or polyisobutylene base. These materials are composed of viscous liquids rendered immobile by the addition of fibers and fillers. The vehicle may be a non-drying oil or polybutene. Such compounds are suggested for use only when little movement is anticipated and where dirt pick-up is not objectionable. Since most types are of an oily nature that would impair adhesion, they are not suggested for combination glazing with curing-type bulk sealants such as polysulfides,

Fig. 31-6. Linex patterned glass. *(Courtesy: American-Saint Gobain Corp.)*

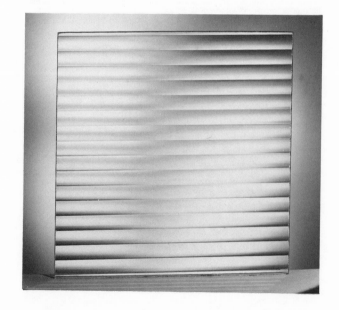

silicones or urethanes.

B. Putties

Putties are oil-base compounds which tend to harden with age.

1. Wood Sash Putty

Wood sash putty is generally a mixture of pigment and linseed oil. They may contain other drying oils such as soybean and perilla. A good grade of wood sash putty should not adhere with too great a tenacity to the putty knife or to the glazier's hand, yet it should not be too dry to apply to the sash. Application of a suitable primer, such as a priming paint or boiled linseed oil, to wood sash before applying putty is a necessary practice.

Putty should not be painted until it is thoroughly set. Premature painting may form an air-tight film, retarding its drying, and later may cause the surface of the paint to wrinkle or crack.

2. Metal Sash Putty

Metal sash putty differs from wood sash putty in that it is formulated to adhere to a non-porous surface. It is used for glazing aluminum or steel sash, either inside or outside. It should be applied as recommended by the manufacturer. Metal sash putty should be painted within two weeks after application. Some types used on aluminum sash require no paintings; if they do, they should be thoroughly set and hard before painting commences. As a rule, there are two grades, one for interior and one for exterior glazing.

C. Sealants—One Part Synthetic Polymer Base

1. Curing Type (Polysulfide, Silicone, Urethane Base)

These one part sealants undergo a chemical reaction at room temperature to form a firm, resilient seal. Chemically curing sealants are capable of tolerating more movement over wider temperature ranges than other bulk compounds. Positive adhesion may or may not require the use of a sealant primer or other surface conditioner. They generally exhibit flexibility, elongation, resilience and other rubber-like characteristics. Listed are the various base materials and some of their properties.

Polysulfides exhibit resistance to a wide variety of solvents, fuels and chemicals. They bond to themselves, facilitating field repair of damaged areas where necessary. Primers usually are not required for adhesion to metal and glass.

Silicones exhibit rubbery properties over a wide temperature range; also, the ability to withstand high temperatures. Sealant primers or surface conditioners are generally suggested for maximum adhesion to all surfaces.

2. Solvent Release Type (Acrylic, Butyl, Hypalon, Neoprene, Vinyl Base)

Sealants of this type are essentially non-curing in the sense that no extensive chemical change takes place to alter their form from a gunnable sealant to a more firmly set semi-resilient state. Although some forms may achieve a partial state of polymerization, these materials generally rely upon solvent release which allows them to "set" on exposure to the atmosphere. Certain of these may remain soft and pliable. Others are subject to volume change and increase in hardness with subsequent reduction in serviceability. Performance properties of the various types vary considerably, ranging from similarity to elastic glazing compounds to properties of the curing-type sealants.

D. Sealants—Two Part Synthetic Polymer Base (Polysulfide or Urethane Base)

Chemically curing, resilient two part sealant systems offer numerous cure characteristics (usually in a matter of hours) as an advantage for field installation.

1. Mixing and "Work Life"

Proper mixing technique requires that both parts be thoroughly blended together, preferably with a mechanical device. Attention is called to the possibility of reducing the sealant "work life" by excessive mixing, or of mixing at too high a speed (100 RPM is the suggested maximum). The "work life" of two part sealants is further influenced by temperature and humidity once they have been mixed. Warm, humid weather shortens work life, while cooler, less humid weather prolongs work life. Storage, mixing and application of two part sealants must be coordinated at the job site to obtain the most favorable working conditions.

2. Frozen Cartridges

Many two part polysulfide and urethane base sealants may be pre-mixed and quick frozen and stored until needed for use.

Proper utilization of this technique requires:

a. Mixed sealant be immediately loaded into cartridges and quick frozen at temperatures below -40°.

b. Storage of cartridges be maintained at temperatures below -20° F.

c. Before application, cartridges be allowed to thaw (at job site) at room temperature. Heat should not be used for thawing, since it may substantially reduce work life.

3. Hardness Ranges

Two part curing sealants are available in several hardness ranges for specific applications.

Shore A 20, or less (soft set): May be desirable where maximum flexibility and minimum strain on the adhesive bond are required by the sealant.

Shore A 20 to 35: Used for the most standard glazing and sealing applications, reamin serviceable under conditions of vibration and movement.

Shore A 35 to 50: For semi-structural applications, or where high resistance to abrasion is required.

4. Consistency, Adhesion, Non-Staining Characteristics

Two part sealants are generally available in both gun and pourable grades. Gun grade should be specified for use in vertical or overhead joints. Pourable grades are intended for use on horizontal areas, decks, paving, etc. (Not normally suggested for glazing applications.)

Specific adhesive qualities may be "built into" certain two part sealants at time of manufacturing which provide position adhesion to unusual types of glass and metal finishes as well as to other surfaces. In most cases polysulfide base sealants will not require a sealant primer for adhesion. Many urethane base sealants require a sealant primer.

Where two part sealants are employed for glazing areas adjacent to masonry or stone the sealant should be designated "non-straining."

II. Preformed Sealants

Flexible, pre-molded joint sealants may be made of rubber, natural or synthetic, as well as plastics such as polyvinyl chloride. Glazing materials are supplied in preformed tapes, beads, ribbons, or as mastics, resilient compositions or impregnated cellular materials.

A. Preformed Synthetic Polymer Base Materials

1. Resilient Type (Tapes, Ribbons and Beads)

These resilient, preformed materials may be used for many glazing applications, including curtain wall and store front construction. Tapes may be used either as a primary seal for interior or exterior use, or as a stop, spacer, shim or secondary seal in combination with the curing-type compounds, one or two part, such as polysulfide base sealants. Resilient tapes are most frequently based on butyl or polyisobutylene polymers. They are furnished in various widths and thicknesses, in roll form, either in a pre-vulcanized or modified raw material, rubbery state. Many extruded shapes are supplied either plain or reinforced with scrim, twine, rubber or other materials. Reinforced varieties resist deformation and contribute some shape-holding ability to non-vulcanized compositions. Most resilient tapes are available with a pressure-sensitive surface. They possess good resistance to ultra-violet radiation, water and water vapor. They are generally non-staining, non-bleeding and exhibit good low temperature properties.

2. Non-Resilient Type (Tapes and Beads)

These products are usually based on poly-butene polymers in combination with asbestos fibers and other inert fillers. They may be high in solids content. They are available in a variety of extruded widths and thicknesses; are easily deformed; have low cohesive strength; flow under pressure; and remain permanently soft and tacky. Exposed surfaces collect dust and dirt. When bedding glass in this type of compound,

spacers should be used. They are not suggested for use in combination glazing with curing-type sealants (such as polysulfides).

B. Preformed Gaskets—Compression and Structural Types

1. Gasket, Compression Type

These products may be made of natural or synthetic rubber, or plastic and are supplied in a solid or sponge-rubber form. They may be extruded to specific shapes for field installation or furnished as an integral part of a metal angle, channel, or tube section. They may be used as primary seals where they are maintained under positive compression, or supplemented with field-applied sealants of all types.

2. Gaskets, Structural Type

These products are usually made from factory-vulcanized neoprene rubber compounds. Vulcanized neoprene rubber has good weather and oil resistance, as well as excellent resistance to permanent deformation; thereby allowing it to be used for compression-type sealing. Many structural gasket shapes and sizes are available as stock items, others are custom-extruded for each opening.

Installation should be limited to temperature conditions at which the rubber gasket can be easily handled. Cold-weather applications may require warming of gasketing for ease of installation.

C. Preformed Impregnated Type

Flexible urethane foams impregnated with asphalt or polyisobutylene materials have recently been introduced as sealants. To function as a weather-tight seal, these materials must be maintained under constant compression. Asphalt impregnated foam exhibits recovery at temperature above 40° F. and limited recovery from compression at low temperatures. Applications of this material for glazing should include the use of a compatible, non-curing bulk-type sealant. Use of asphalt impregnated foams with the curing-type sealants such as polysulfides is not suggested.

PREPARATION BEFORE GLAZING

Movable sash shall be fastened and remain stationary until putty or glazing compound has "set".

Priming

Wood. Application of a priming paint is essential to good results when glazing in a wood surround.

Priming fills the pores of the wood lessening the extraction of the oil from the putty which would leave it dry and brittle. Good grades of wood preservatives do not interfere with the performance of glazing material. However, a good priming paint should be applied to a clean, dry surface before glazing is started. The use of shellac or varnish as a primer should be avoided because most glazing materials will not adhere to them.

Absorbent hardwood frames that are not to be painted

may be primed with varnish to be completely sealed when glazed with a non-setting compound. Some hardwoods such as teak may be almost completely non-absorbent and metal sash putty should be used.

Steel. In the absence of a bonderized shop paint coat or zinc coating, steel sash should be treated to inhibit rust and be prime painted.

A hot dip galvanized smooth surface is usually suitably conditioned by normal weathering at the site.

Since there is no absorption of vehicle in the glazing of steel sash, a different type of priming paint is used. A satisfactory steel primer retards the formation of rust that would loosen the bond between the putty and the steel and also provides the proper "tooth" or roughened surface for adhesion, giving a long-lasting bond between metal and glazing material.

The priming should be allowed to dry thoroughly in all cases before glazing material is applied over it.

Aluminum. Be sure to remove completely any protective coating in the glazing rabbet area with a methyl ethyl ketone (MEK) or similar solvent material that will not etch or mar the finish.

Where two dissimilar metals are in contact, the contacting surfaces should be treated to retard electrolysis.

Dry Surfaces

At outdoor temperatures of 40° F. or below, glazing surfaces are often covered with a film of moisture. For this reason many manufacturers of glazing compounds caution against applying their materials at temperatures below 40° F.

If glazing schedules call for continuation of work during cold periods, provision shall be made to warm the area sash to avoid seal failures related to the trapped moisture film.

GLAZING DETAILS FOR SINGLE GLASS IN WOOD AND METAL SASH

Preparation. Glazing rabbets and grooves shall be rigid and true. Wood sash shall be well seasoned and dry. Metal sash shall be filled and adjusted before glazing. Wood, and steel glazing surfaces of sash shall be properly primed. Moisture or condensation should be prevented during glazing installation.

General Glazing Conditions. Glaze sash from outside (or inside for industrial). Glaze when temperature is above 40° F. Thoroughly mix glazing compound to a uniform consistency. The entire perimeter of the glass should be bedded. Glaze all sash in a closed position. Do not handle or operate the sash until the glazing compound has set.

Application (Face Glazed with Glazing Compound). Clean sash and glass free from dirt, moisture and oil. Apply ample sealant to rabbet in sash so it will ooze out after pressing glass into it, covering completely glass in rabbet allowing proper back sealant. Press glass into

sash maintaining equal tolerance on four edges. Each light hold in place by application of glazing points (in wood) and spring wire clips (in metal). Points spaced not to exceed 24" o.c. around perimeter of frame and spring wire clips spaced 18" o.c. around perimeter of frame. (Minimum of 4 points or clips per sash.) Strip surplus back putty at an angle (not undercut). Fill front pane edge in rabbet with putty to form a triangle fillet stopping 1/16" short of sight line. Form a neat trim line. Corners should be treated with care for neat workman like finish. Strip surplus putty from front bed.

Application (Stop Glazed with Bead). Clean sash and glass free from dirt, moisture and oil. Apply ample glazing compound to rabbet. Place setting blocks if required. Press glass centered into rabbet so that compound covers glass completely in rabbet allowing proper depth of back sealant. Each light hold in place by application of stop bead all around. Bed bead against glass and bottom of rabbet with compound. Leave proper bed of compound between glass and bead. Strip surplus compound from both sides of glass at angle (not undercut).

Note: Patterned glass should be glazed with smooth side to face putty.

After Application. When putty has hardened sufficiently, within 4 weeks from glazing, paint putty and sash at least two coats. Paint should overlap glass slightly above sight line. When painting sash with bead, take care not to paint glazing material if such is not recommended.

The Following Illustrations indicate the recommended tolerances for the different methods of glazing. A, B, C, and D are equal to the following:

A & B. Dimensions for these are shown on Chart I. See page 107.

C. Dimension equal to 1/16" for glass up to 5 sq. ft. 1/8" minimum on all over 5 sq. ft. Except for Method 6 which is 0.

D. Dimension for Methods 1 and 3, minimum 3/8". For Methods 5; 6, and 11, dimension is 0. Other Methods, dimension is 1/8".

GLAZING DETAILS FOR SINGLE GLASS

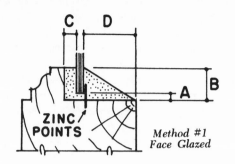

Method #1
Face Glazed

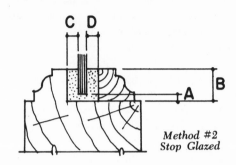

Method #2
Stop Glazed

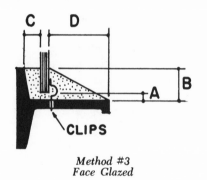

Method #3
Face Glazed

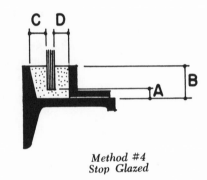

Method #4
Stop Glazed

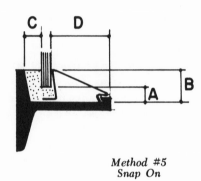

Method #5
Snap On

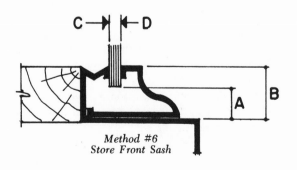

Method #6
Store Front Sash

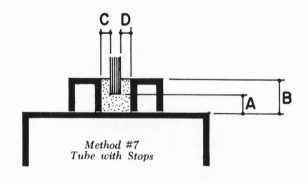

Method #7
Tube with Stops

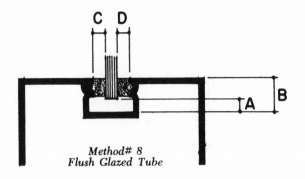

Method# 8
Flush Glazed Tube

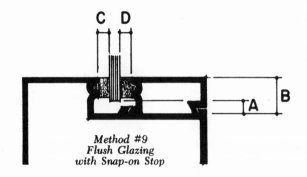

Method #9
Flush Glazing
with Snap-on Stop

METAL SLIDING SASH AND DOORS (Often Pre-glazed)

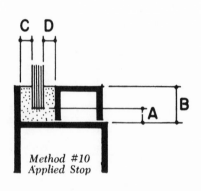

Method #10
Applied Stop

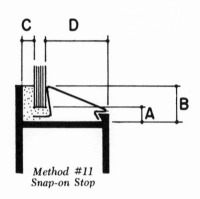

Method #11
Snap-on Stop

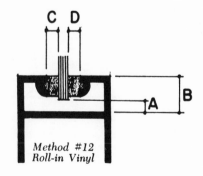

Method #12
Roll-in Vinyl

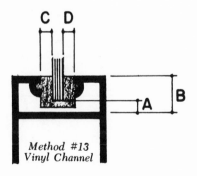

Method #13
Vinyl Channel

A & B Dimensions shown on Chart — see Page 16.

C Dimension ¹⁄₁₆ for glass up to 5 sq. ft. ⅛ minimum on all over 5 sq. ft. Except for Method #6 which is 0.

D Dimension for Methods 1 and 3, Minimum ⅜″.
For Methods 5-6 and 11, Dimension is 0. Other Methods, dimension is ⅛″.

Fig. 31-7. Chart I *(Courtesy: Flat Glass Jobbers Assoc.)*

VARIOUS EDGE CLEARANCES AND TOLERANCES FOR SINGLE LIGHTS OF GLASS
Methods 1 thru 13

GLASS TYPE AND THICKNESS	GLASS SIZE		TOLERANCES		Clearance at Head, Sill & Jambs (allow)	Rabbet Depth (min.)	Block Setting Height (range)
	Area (sq. ft.)	Width or Height (in.)	Glass Cutting Size (in.)	Sash Daylight Opening (in.)			
SHEET GLASS							
SS	5	40	±1/32	±1/16	1/16	3/8	None Req.
SS	14	50	±1/32	±3/32	1/8	7/16	1/16-3/16
DS	5	40	±1/32	±1/16	1/16	3/8	None Req.
DS	25	80	±1/32	±3/32	1/8	7/16	1/16-3/16
3/16	25	120	±1/16	±3/32	11/64	1/2	3/32-1/4
3/16	70	120	±1/16	±7/32	15/64	5/8	3/32-3/8
7/32	25	120	±1/16	±3/32	11/64	1/2	3/32-1/4
7/32	70	120	±1/16	±7/32	15/64	5/8	3/32-3/8
POLISHED PLATE (CLEAR AND TINTED)							
1/8	25	128	±1/16	±3/32	11/64	1/2	3/32-1/4
1/8	67	128	±1/16	±7/32	15/64	5/8	3/32-3/8
1/4	100	120	±1/16	±3/32	11/64	1/2	3/32-1/4
1/4	140	156	±1/16	±3/16	1/4	5/8	1/8-3/8
1/4	207	229	±3/32	±7/32	11/32	3/4	3/16-1/2
5/16	207	229	±3/32	±7/32	11/32	3/4	3/16-1/2
3/8	258	286	±3/32	±5/32	3/8	3/4	1/4-1/2
3/8	258	286	±3/32	±17/32	7/16	7/8	1/4-5/8
1/2	258	286	±1/8	±1/4	7/16	7/8	1/4-5/8
3/4	258	286	±1/8	±1/4	7/16	7/8	1/4-5/8
1	76	148	±5/32	±7/32	7/16	7/8	1/4-5/8
1	76	148	±5/32	±11/32	1/2	1	1/4-3/4
1 1/4	76	148	±3/16	±5/16	1/2	1	1/4-3/4
SPANDREL GLASS AND HEAT-STRENGTHENED GLASS							
1/4	25	80	+0—3/16	+0—3/16	5/32	1/2	1/16-1/4
1/4	84	168	+0—3/16	+1/16—1/4	1/4	5/8	1/8-3/8
3/8	25	80	+0—3/16	+0—3/16	5/32	1/2	1/16-1/4
3/8	84	168	+0—3/16	+1/16—1/4	1/4	5/8	1/8-3/8

Chart II (Courtesy: Flat Glass Jobbers Assoc.)

GLAZING DETAILS FOR INSULATING GLASS IN WOOD AND METAL SASH

A good knife grade of glazing compound should be used. Do not use putty.

Use 2 Neoprene or metal setting blocks at bottom. Place them at 1/4 points. All units over 10 sq. ft. should have Neoprene spacer strips on side and top of unit to assure uniform setting.

If the window is glazed with the fixed stop to the outside—weep hole should be provided from below edge of glass to outside for drainage.

The recommended tolerances indicated by A, B, C, and D on the following drawings can be found on the following chart:

GLAZING DETAILS FOR INSULATING GLASS IN WOOD AND METAL SASH

Methods 14 through 21

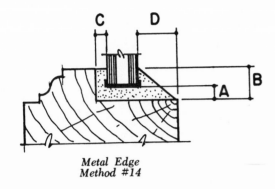

Metal Edge
Method #14

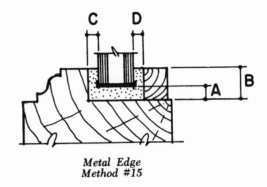

Metal Edge
Method #15

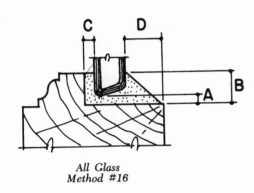

All Glass
Method #16

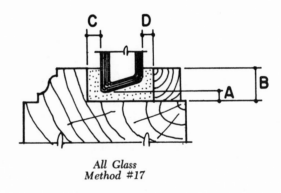

All Glass
Method #17

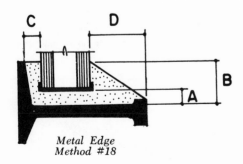

Metal Edge
Method #18

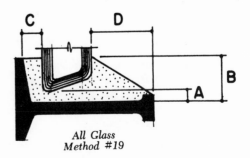

All Glass
Method #19

Fig. 31-8. Chart III *(Courtesy: Flat Glass Jobbers Assoc.)*

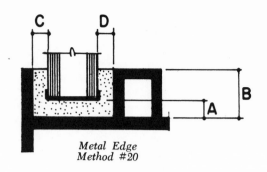

Metal Edge
Method #20

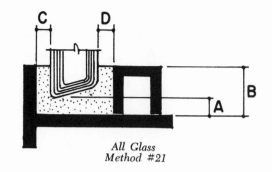

All Glass
Method #21

Type	Thickness	Max. Sq. Ft.	Minimum Dimensions				
			A	B	C	D 14-16-18-19	D 15-17-20-21
Glass Edge S.S.	⅜	to 10	⅛	⅝	⅛	½	⅛
Glass Edge D.S.	⁷⁄₁₆	to 24	⅛	⅝	⅛	½	⅛
Metal Edge	⁹⁄₁₆ to ¹³⁄₁₆	12	⅛	⅝	⅛	½	⅛
Metal Edge	¹¹⁄₁₆ to ¹⁵⁄₁₆	27	¼	¾	⅛	⅝	⅛
Metal Edge	¹³⁄₁₆ to 1¹⁄₁₆	70	¼	¾	⅛	⅝	⅛

SUSPENDED GLAZING

Suspended Glazing allows the use of very large lights of plate glass without distortion with or without metal mullions. The glass hangs distortion-free and in perfect plane from its own weight and sight lines are virtually non-existent when metal mullions are eliminated and glass stabilizers are used. The possibility of breakage is greatly reduced because the suspended glass is less subject to damage caused by movement of the building and the stresses normally occurring where setting blocks are used. Should breakage occur, the glass is held at the top and does not fall with a guillotine effect. Also, a full cantilever roof may be used as the glass wall will readily adjust to the expansion, contraction and wind movements of such a roof.

Suspended Glazing is a method of glazing by the use of a lateral head suspension device attached to the structural steel above the plate glass and thus suspending the glass without the use of any blocking at the sides or bottom. The glass hangs in a channel below without conventional blocking. The entire weight of the glass is carried by a structural member above. Wind load requirements are met by metal mullions or by using glass stabilizers at right angles at the vertical joint formed by the main light when all glass construction is desired.

SPECIAL GLAZING CONSIDERATIONS

1. Protection of Glass

DO NOT mark installed glass with an "X" or other symbol with any material whatsoever. Tapes or banners may be fastened to the sash bead and suspended over the glass. The practice of placing an "X" or other materials on the surface of the glass has at times caused the surface of the glass to become damaged and has lead to costly replacements.

2. Mirrors

a. The use of mastic is not recommended in the installation of mirrors.

b. If, however, it is not feasible to install with mechanical fasteners, the mirror should be given an extra

Fig. 31-9. GLAZIERS' HAMMER. A manually operated point driver for sinking triangle glaziers points into the toughest wood. *(Courtesy: Red Devil Tools)*

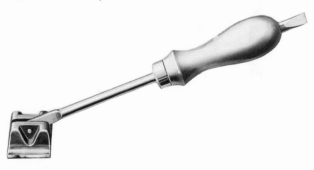

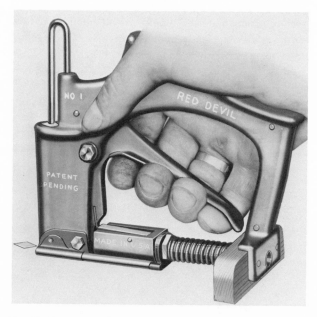

Fig. 31-10. AUTOMATIC POINT DRIVER. This gun allows one man to glaze more sash in an hour than five men by the old hand method with triangle points. *(Courtesy: Red Devil Tools)*

Fig. 31-11. ELECTRICAL PUTTY SOFTENER. When heat is transmitted by this tool to old hardened putty on a window sash it will, in a few minutes, soften the putty to a consistency which enables it to be easily removed with an ordinary putty knife. *(Courtesy: The Fletcher-Terry Co.)*

coat of moisture-resistant paint and allowed to dry thoroughly before installing.

c. Mirrors, either in a frame or fastened to the wall with clips or metal moulding, constitute the best installation.

d. An air space should be left in back of the mirror if design requirements permit.

e. The use of felt or felt paper should be discouraged as those products have a tendency to trap and hold moisture which may damage the silver coating of the mirror.

f. Mirrors should not be mounted against unpainted plywood or other wood as the resin in the wood may cause silver spoilage. Spoilage may also result from damp plaster or damp masonry walls.

g. Care should be taken to have the mirror blocked up on the bottom with setting blocks at the quarter points.

h. Mirrors may be installed by drilling holes in the mirror and fastening the mirror to the wall with a screw and rosettes. The hole should be 1/8" larger than the screw and a felt or rubber sleeve should be placed around the screw so that the metal will not come in contact with the mirror.

i. Glass is weakened by drilling the holes and is more likely to crack at the screw hole than is the solid sheet with no holes drilled in it.

j. All exposed edges of mirrors should be polished.

k. Two-way or transparent mirrors should be glazed with the coated side to the darker area. When glazed in exterior openings, the coated side should be glazed inside.

3. Tempered and Heat-Strengthened Glass

a. May be glazed the same as ordinary plate glass EXCEPT that due to the manufacturing process warpage of the lights is common. This warpage may be as much as 1/2" in large sizes so that it may be impossible to glaze in a "flush-glazed" type of setting. A double-stop installation, that may be adjusted to this warpage, is recommended.

b. Some spandrel glass (heat-strengthened colored glass) will contain pin holes that can be seen when glazed in an opening such as a transom light.

4. Heat-Absorbing Glass

a. Special attention must be given to the installation of all types of heat-absorbing glass, because of its ability to absorb heat. Partial shading, painted signs, large interior labels, tight draperies or blinds, heavy masonry structure and heating-cooling outlets directing air against glass may increase edge tension stresses.

b. The ability of heat-absorbing glass to resist solar energy breakage is primarily related to its edge strength. THEREFORE:

(1) EDGES MUST BE CLEAN CUT.
(2) DO NOT INSTALL GLASS WITH FLARED EDGES AT BOTTOM.
(3) DO NOT SEAM EDGES.
(4) DO NOT NIP EDGES NOR SCARF CORNERS.
(5) BE CAREFUL NOT TO BUMP OR BRUSH EGES AGAINST METAL OR OTHER HARD OBJECTS.
(6) AVOID THE USE OF POCKETED FLUSH GLAZING.
(7) RADIUS CUTTING SHOULD BE REVIEWED BY MANUFACTURER.

5. Patterned Glass

a. In glazing glass with one surface patterned, the smooth side of the glass should be to the weather side.

6. Laminated Glass (with Vinyl Plastic Center)

a. Avoid using any glazing material that will bring oil in contact with the plastic encased in the glass. Such oils will cause discoloration and occasionally separation or shrinkage.

   b. Protect laminated glass with a piece of cellulose tape stretched along the edges and slightly overlapping the faces of the glass.

   c. Glazing material should extend against the face of the glass sufficiently above the edges of the sealing tape to provide an adequate and satisfactory weather seal.

Chapter 32
Miscellaneous Information

THE PAINTING OF DISPLAY WINDOWS

In many instances, glass which has been painted has broken. The effects of paintings have been extensively investigated, with the following conclusions.

The most dangerous exposure is the southern exposure, the next most dangerous is the western exposure, the least most dangerous is the eastern exposure where direct sunlight is concerned. Since there is no direct sunlight, there should seldom, if ever, be any breakage from the painting of windows on northern exposures.

The effects are emphasized where the glass is held rigidly in a frame and cannot expand or contract with changing conditions.

Windows which are painted with any type of opaque paint, and particularly with opaque black paint, are especially subject to breakage under direct solar exposure.

It is dangerous to paint any appreciable area of plate glass or other glass window with any completely opaque paint, quite irrespective of the amount of area of the window which such painting may cover. It is particularly dangerous to paint valances across the top of a pane since this is the second most dangerous condition which can exist. The most dangerous condition is to paint across the top of a pane a valance which does not reach entirely to the edges, while extending reasonably into the window. An area which approximates one-quarter of the total area of the window, or more, and extending in from the edge of the window to approximately half of the depth of the window, is quite the most dangerous type of painting which can be done (Fig. 32-1).

It is least dangerous to paint the entire area of the window or the central area of the window, not in excess of 25 per cent of the total area.

Most of the plate glass insurance companies are advised of these conditions and most of them will impose a larger premium, if they will accept the risk at all.

White paint, or very light-colored paints, are less likely to cause breakage than are the darker and more absorptive paints, but the application of valances, borders, or similar painted areas, to glass may be expected to cause breakage, especially on southern exposures.

A very thin coat of lamp black suspended in shellac, which permits the greater portion of the solar light and heat to penetrate, while at the same time preventing the possibility of seeing into the building through the glass, seems safe to use. It is essential that the greater portion of the solar light and heat shall be able to penetrate through any coating which is put over the glass, if the glass is to be maintained without breakage.

Fig. 32-1. *(Courtesy: Pittsburgh Plate Glass Co.)*

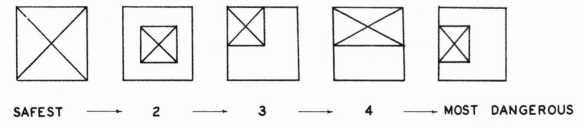

SAFEST ⟶ 2 ⟶ 3 ⟶ 4 ⟶ MOST DANGEROUS

PAINTED AREA
HEATED AREA

WINDOW PAINTING

WHY IS GLASS TRANSPARENT?

An Explanation of the Phenomenon of Transparency

Light can be explained in many respects by the fact that space has the property of carrying electric and magnetic forces that vary periodically with time. The whole phenomenon is known as an electromagnetic light wave. Matter influences this periodically varying field through the charged particles of which matter is composed. The particles are moved by the electrical forces of the field.

Visible light has a wave length of from 0.4 to 0.7 The formula $c=\lambda v$ ($c-3\times10^{10}$ cm./sec., λ=wave length, and v=frequency) shows that the vibration frequency is extremely high, namely of the order of 10^{14} per sec. Now electrons are light enough to follow these extremely quick varying fields if the electrons are free to move and not bound by atomic forces to the heavy positively charged nuclei of the atoms. The nuclei are some 2000 times heavier than the electrons. Such free electrons are found in metals. They move inside a metal like gas particles with high velocities that are due to the heat motion and cause the high electric conductivity of metals. If light strikes a metal these free electrons start to swing in phase with the light vibration and convert the electrical energy into mechanical energy of motion and finally into heat energy. Therefore metals are not transparent—at least in thick layers.

From this picture it follows that substances having high electrical conductivity, that is, having free electrons, should be opaque, whereas insulators should be transparent. Examples are the metals that are opaque. Another example is diamond and graphite. Diamond is transparent and an insulator. Graphite is opaque and a conductor. All organic liquids are transparent and insulators. Water is transparent and an insulator. Of course, liquids can acquire some small conductivity and yet be transparent. But this conductivity is ionic in origin. This explains why glass in heating does not become opaque, as the conductivity of glass acquired by heating is ionic and not electronic in character. If a metal becomes part of a compound, its free electrons are bound to the radical with which the metal is compounded and the compound is transparent. For example $CuSO_4$ (copper sulfate) or lead glass. In the latter the lead is present as a lead compound.

On the other hand, there are many insulators that are opaque, as for instance, ceramics. This is explained by the grain structure of these substances that are composed of grains of different indices of reflections at the grain boundaries prevent light from passing through. Another example is sintered glass.

The explanation given here for opacity as due to absorption by free electrons applies only to the visible part of the spectrum.

In the infra-red region light waves are capable of inciting the whole molecular structure, including the nuclei, to vibration and rotation. The variety of these motions is very great and therefore many absorption bands may occur, or even a continuous absorption may take place.

In the ultra-violet region the electrons in the electronic shells are lifted into higher energy levels. The whole subject is governed by the laws of quantum mechanics.

In the X-ray region the particle properties of light are more apparent and absorption is chiefly a function of the nuclei contained in the glass. The heavier these nuclei are the more absorption takes place. Lead glass, for instance, absorbs more than common silica glass.

Conclusion. Glass is transparent for visible light because it is a homogenous insulator and has no free electrons.

POLARIZED GLASS

Polarized glass, an integral product, is nonexistent. There is not known any material which will retain its properties of polarizing light after being incorporated into a glass melt. However, within the last few years, there has been developed a process for producing reasonably large plastic sheets incorporating a polarizing material and which may be laminated between two pieces of glass by the usual safety glass lamination processes to produce a very satisfactory polarizing medium. This material, known as Polaroid, consists of very fine crystals of a polarizing substance widely dispersed in a plastic film in such a way that the crystals are all parallel, one to another. The resultant product is a polarizer of great merit. When two sheets of this material are properly placed together, complete interception of light is achieved and consequent darkness results on that side of the system away from the light source. Now, if a sheet of cellophane is placed between the two sheets of Polaroid, and turned slowly, relative to the two plates of Polaroid, the system may be opened or closed at will. This same opening or closing of the system may be achieved by rotating the plates themselves relative to each other. This material will transmit somewhat less than 50% of normal white light and somewhat less than 90 per cent of plane polarized light.

MARKINGS ON WINDOWS

Unfortunately, there is absolutely nothing which can be done to restore the finish on a window on which soap has been allowed to remain even for several days under humid conditions.

The alkali in some soaps definitely attacks the surface

of glass and etches it sufficiently so that its original brilliance cannot be restored by any amount of rubbing or other work which may be undertaken in the field.

This difficulty is particularly noticeable around the period of Halloween, but is also frequent when the opening of buildings is long delayed and the windows are marked up with soap to make them definitely noticeable to workmen.

Every possible effort should be made to discourage the use of soap as a marking material on windows, and if by any means possible, no markings should be resorted to.

In general, surgeon's adhesive tape, or other pressure-sensitive adhesive tpaes, appear to be safe for use. Also, very thin films or ordinary white paint offer a satisfactory means for indication of the presence of glass, if not left too long.

An oil-type glass-frosting paint, which upon drying becomes translucent to opaque, seems to be a satisfactory obscuring or masking paint. Applied with a brush, it dries to the touch in about 4 hours, and dries hard in 12 to 24 hours. Aside from brush marks, it gives the glass a very satisfactory frosted appearance, comparable to a matt etch. A satisfactory marking paint may be prepared by mixing powdered chalk or whiting plus a small amount of wheat flour in water, and using as a watercolor paint on the windows.

Never use any marking paint which contains glue or sodium silicate.

Sources of supply

Important Note: Almost every community has an excellent source of supply for FREE MATERIALS. Your local glass shops disposes of large quantities of glass that is not satisfactory for their use because of size but is excellent for items customarily made in schools, studios, home workshops, craft classes, etc. A friendly request of the glass shop owner will probably get you, at no cost, enough glass sections to meet your needs for a long time.

A search for supplies should start with the yellow pages in your local phone book. The phone books for neighboring communities can be borrowed from the central office of the telephone company, and some-times, from the local library. The supplies and equipment needed can be secured from dealers in most large communities.

Listed below are suppliers of some of the more specialized materials that might be needed. In some cases the listed are dealers who will take care of your needs. On the other hand, many of the listed are prime manufacturers who will not supply small quantities but will send you the address of the nearest dealer who can take care of your needs.

The following list is by no means complete. There are many fine companies making identical or similar products. Inquiries at the local stores may reveal companies that supply the material you seek right in your vicinity.

Companies That Supply Glass Material

1. American Art Clay Co., 4717 W. Sixteenth St., Indianapolis, IN 46222
2. American Crayon Co., 1706 Hayes St., Sandusky, OH 44870
3. A. Bendheim Co. Inc., 122 Hudson St., New York, NY 10013
4. Bellco Glass Inc., Vineland, NJ 08360
5. Berton Plastics, Inc., 170 Wesley St., So. Hackensack, NJ 07606
6. Bethlehem Apparatus Co. Inc., Hellertown, PA 18055
7. Blenko Glass Co. Inc., P.O. Box 67, Milton, WV 25541
8. Blum Ornamental Glass Co. Inc., 314 Jacob St., Louisville, KY 40202
9. C.R. Laurence Co. Inc., P.O. Box 21345, Los Angeles, CA 90021
10. C.S.C. Scientific Co., 2600 Koftner Ave., Chicago, IL 60613
11. Cadillac Plastics & Chemical Co., 1924 N. Paulina, Chicago, IL 60622
12. Chase Brass & Copper Co., (See address in local phone book)
13. Corning Glass Works, Corning, NY 14830
14. Crystal International Corp., 1 World Trade Center, New York, NY 10017
15. Dow Corning Corp., Midland, MI 48640
16. Eastern Smelting & Refining Corp., Lynn, MA 01903
17. Ferro Corp., 20 Culvert Ave., Nashville, TN 37210
18. Fletcher-Terry Co., Spring Lane, Farmington, CT 06032
19. Fulton Glass Co. Inc., West Washington, Hartford City, IN 47348
20. Glazing Products Corp., 78th St. & 68th Rd., Middle Village, NY 11379
21. High Point Glass & Decorative Co., 624A Breenboro Rd., High Point, NC 27261
22. Kaufman Glass Co., 1303 Northeast Blvd., Wilmington, DE 19899
23. L. Levin Co., 100 Fountain St., Providence, RI 12903
24. L. Reusche & Co., 2 Lister Ave., Newark, NJ 07105
25. Laminated Glass Corp., 9797 Erwin, Detroit, MI 48213
26. Libby-Owens-Ford Glass Co., 811 Madison Ave., Toledo, OH 43602
27. Litton Engineering Laboratories, P.O. Box 949, Grass Valley, CA 95945
28. Lunzer Industrial Diamond, Inc., 48 W. 48th St., New York, NY 10036
29. Maas & Waldstein Co., 2121 McCarter Hwy., Newark, NJ 07104
30. Newall Mfg. Co., 139 N. Wabash Ave., Chicago, IL 60602
31. O. Hommel Co., P.O. Box 475 Pittsburgh, PA 15230
32. Owens-Illinois Glass Co., Ohio Building, Toledo, OH 43601
33. Pangborn Corp., Hagerstown, MD 21740
34. Parallel MFG. Co., 423 W. 43rd St., New York, NY 10003
35. Peacock Laboratories, Inc., 54th St. & Paschall Ave., Philadelphia, PA 19143
36. Pittsburgh Plate Glass Co., Gateway Center, Pittsburgh, PA 15222
37. Plaid Enterprises Inc., 6553 Warren Dr., Norcross, GA 30091
38. Plastic Molders Supply Co. Inc., 109 N. Brunswick Rd., Sumerset, NJ 08873
39. Pressure Blast Mfg. Co., 41 Chapel St., Manchester, CO 06040
40. Red Devil Tools, 2400 Vauxhall Rd., Union, NJ 07083
41. Ruemelin Mfg. Co., 3860 N. Palmer St., Milwaukee, WI 53212
42. S.S. White Co., 3 Parkway, Philadelphia, PA 19102

43. Shatter Proof Glass Corp., 4815 Cabot Ave., Detroit, MI 48210
44. Sommer & Maca Glass Machinery Co., 5501 W. Ogden Ave., Chicago, IL 60650
45. Standard Ceramic Supply Co.: 9 Sandhurst Dr., Pittsburgh, PA 15241
46. T.E. Conklin Co., 345 Hudson St., New York, NY 10013
47. The Paul Wissmach Glass Co. Inc., Paden City, WV 26159

48. Thomas C. Thompson, 1205 Deerfield Rd., Highland Park, IL 60035
49. Wepra Co., 49 W. 37th St., New York, NY 10018
50. White Metal Rolling & Stamping Corp., 80-84 Moultrie St., Brooklyn, NY 11222
51. William Dixon Inc., Carlstadt, NJ 07072
52. Atlantic Art Glass Inc., 3601 Clearview Place NE, Atlantic, GA 30340

Subject Classification

The number following the product refers to the number in front of the company listed above.

PRODUCT

Glasses

Bent 26,36
Bulb Edge 14
Ground, Chipped Glass 14,36
Heat Absorbing 13
Jalousie, Edged 19
Laminated 14,25,26
Picture 14,26,36
Sheet, Colored 7,36
Sheet, Gray 14,36
Sheet, Heavy 14,26,36
Sheet, Extra Heavy 14,36
Silvering Quality 14,26,36
Window, Double 14,26,36
Window, Single 14,26,36

Machinery Equipment

Beveling Machines 9,44
Boards, Cutting 9,18,40,44
Compressors 44
Cut-Off Machines 9,18,44
Cutting Machines, Glass 18,40,44
Drilling Machines, Glass 9,20,44
Handling Equipment, Glass 9,20,44
Heat Forming Equipment 6
Grinding Equipment 9,44
Handy Grinders 9,44
Kilns 1
Laths, Glass 6,27
Lehrs 39
Pliers, Glass 9,44
Sand Blasting Equipment 33,39,42,44
Tables, Cutting 9,44
Torches 10
Wheels, Diamond 9,44

Supplies & Accessories

Belts, Abrasive 9,20,44
Blades, Carbide 9,44
Block, Polishing 9,44

Chisels, Glaziers 9,20,44
Clips, Mirror 9,44
Cutters, Glass 9,24,44
Drills 9,44
Felt 9,44
Glass Enamels 24,45
Glass Lifting Tools 9,20,44
Hammers, Glaziers 9,20,44
Heat Forming Supplies 6
Silvering Supplies 35
Point Drivers 9,18,20,40,44
Sand Blasting Supplies 44
Tools, General 9,44
Tubing, Glass 4,10,13
Vitrified Colors 24,45

Enameling Supplies 1,29,48,51

Engraving Tools 23,28,9,44

Fiber Glass Supplies 5,11,15,17

Glazing Supplies 9,15,44

Mirror Making Supplies 35

Mosaics, Glass, Supplies 49

Separators 17,31

Simulated Stained Glass Supplies 37,44

Stained Glass

Glasses 3,7,47,52
Cames 9,44,50,52
Channels, Zinc 9,50,52
Cements 9,44,52
Colors, Glass 2,31,45,52
Compounds, Setting 44,52
Copper Foil 9,44,52
Paint, Glass 2,31,45,52
Solder, Tape & Wire 9,44,52
Stains, Glass 2,31,45,52

Tools, General 9,44

Plate Glass

Bent 23,26,36
Bullet Resisting 23,26,36,43
Electricity Conducting 13,23,26,36
Heat Absorbing 23,26,36
Heat, Reflecting Laminated 23,25,43
Mirror Quality 26,36
Parallel Twin, Ground 26,36
Plate, Glazing 26,36
Plate, Gray 23,26,36
Plate, Polished, Heavy 23,26,36
Plate, Polished, Extra Heavy 23,36
Plate, Polished, Tinted 26,36
Safety, Laminated 25,26,36,43
Spandrel 26,36

Float Glass 26,36

Rolled Glass

Colored 14,21,26,27
Corrugated 14
Decorative 14,21,25,26,36
Glare Reducing Finish 14,25,26,36
Glass Block 32
Heat Absorbing 13,14,26,36
Pattern 14,26,36
Sand Blast Finish 14,26,36
Soft Finish 36
Translucent 26,36
Wire, Corrugated 14
Wire Pattern 14,26

Specialty Glass

Acid Etched 23,36
Antique 7,8,34
Insulating 26,36
Low Transmission 23,25,36
Opal 8,34
Optical 13,23,36
Sheet, Colored 7,7,34

Glossary

GENERAL TERMS

ACID POLISHING—method of polishing glass by immersion in a mixture of concentrated sulphuric and hydrochloric acids.

ACID PROOF—unaffected by acid contact.

ACID RESISTING—resistant to a large degree of acid attack.

AGATE BURNISHER—an agate-tipped tool used in rubbing down and polishing silver and gold decorations.

AIR BRUSH—see spray gun.

ALKALI RESISTING—resistant to a large degree of alkali attack.

ALKALINE—having properties of basic compounds such as neutralizing acids.

ALUMINA HYDRATE—a fine white powder used as a separator between glass and other materials.

ANNEALING—the toughening treatment of glass and metals by controlled heating and cooling so as to remove brittleness and strains; tempering.

ANTIQUE GLASS—glass made in imitation of the color and quality of old glass.

AQUA REGIA—mixture of three parts hydrochloric acid and one part nitric acid which will dissolve gold and platinum.

AQUEOUS—solution made with water.

ASBESTOS—fibers obtained from chrysolite, a hydrated magnesium silicate rock which is unaffected by fire, and hence is used for fireproofing.

BAIT—the tool dipped into molten glass to start any drawing (pulling) operation.

BALL MILL—a revolving porcelain-grinding cylinder half filled with flint pebbles in which enamels, glazes and colors are ground.

BALSAM OF COPAIBA—a natural resinous liquid used as a color vehicle in decorating work.

BAND—to decorate with bands or stripes (applied while object rotates on a banding wheel)

BANDING WHEEL—a small revolving disc on which a decorator sets his ware while it is banded.

BAT—a flat slab of material, usually plaster; used as a base for construction of clay forms.

BATCH—the raw materials needed to make glass, properly proportioned and mixed, for delivery to the furnace.

BATCHING BIN—container in which raw ingredients for glass are mixed.

BATH—a term applied to developing, fixing or other photographic solutions.

BEAD—to form beads or beadlike bubbles.

BENT GLASS—sheet glass shaped into three-dimensional form by firing over a form.

BEVEL—angle of slanted finish of glass edge; angle at which cut is made; to cut or shape to a bevel slant.

BEZEL—grooved rim or flange in which a watch crystal or ring gem is set.

BISQUE—term applied to pottery which has had one fire, and is unglazed.

BLANCHON—shelf or tray for holding glass pieces in the kiln during firing.

BLANK—a piece of glass cut to a predetermined measurement.

BLISTER—a large gas bubble which is an imperfection in the glass.

BLOOD STONE BURNISHER—a tool used in rubbing down and polishing metal decorations, as burnished gold.

BLOW-OVER—a thin-walled bubble formed above a hand-blown mold to make it easier to crack off.

BLOWPIPE—(blowing iron)—a long segment of tubing on which molten glass is gathered and blown into shape.

BOLTING CLOTH—a silk cloth of even mesh used for screening purposes.

BOND—to bind glass to a material by means of firing.

BORE—interior diameter of tube, as in a gun barrel.

BRIGHT GOLD—a gold solution which furnishes a bright surface of gold at once when fired on the ware.

BRUSH, RUBBING—a stiff bristle brush used in making decal and steel plate transfers.

BURN—to heat intensely in order to perfect or condition.

BURNISHED GOLD—a gold paste preparation which must be polished after firing.

BURNISHED SAND—a special sand of round grains used in polishing burnished silver and gold.

CALCIUM CARBONATE (whiting)—a finely ground white powder, used as a separator in glass firing.

CAPILLARY TUBING—tubing having a very small bore so that capillary action takes place (liquid is drawn along it as in a thermometer).

CARBON TISSUE—a pigmentized gelatin-coated paper.

CARBORUNDUM—carbide of silicon; an extremely hard abrasive.

CARRY-IN BOY—the person who places the completed blown objects in the annealing lehr.

CHECK—an imperfection such as a surface crack.

CHUCK—contrivance for holding and guiding work in a machine, especially lathe.

CLEAVAGE—severance of the glass.

COEFFICIENT—a number indicating degree of change, as in expansion and contraction.

COHESION—the state or act of sticking together.

COEFFICIENT OF EXPANSION—the fractional part of its length that a rod elongates when raised one degree in temperature.

CONES—Seger—see pyrometric cones.

COPPER PLATES—used in transfer printing in the same manner as steel plates.

CORROSION—the process of being worn or eaten away, as in oxidation of metals through chemicals and/or heat.

CRAZING—the name given to an enamel defect in which fine cracks appear in the enamel surface.

CULLET—small pieces of glass left over from previous batches that are added to each new batch of glass.

DECAL—see decalcomania.

DECALCOMANIA—a specially prepared paper from which designs or pictures which have been printed with ceramic colors may be later transferred to ceramic ware.

DECALCOMANIA VARNISH—a special varnish used in making decalcomania transfers.

DRAGONS BLOOD—a dark red, resinous substance often mixed with stamping oils so that the stamped print can be seen easily.

DUPLEX PAPER—a special paper used in preparing decalcomanias. Consists of two sheets stuck together by a patented process, the top sheet being thinnest and having on it a special coating consisting chiefly of water soluble gum and albumin.

ELECTRODE—either terminal of an electric source.

ENAMEL—a smooth, tough substance used in coating surfaces of glass, pottery, or metal for ornament or protection.

EUTECTIC—of maximum fusibility; lowest melting point consistent with components of a mixture.

FIRE—to subject to intense heat, either from a direct source or in an oven or kiln.

FLUX—a pulverized low-melting glass used to soften or lower the melting point of decorating colors. A substance that promotes fusion.

FORMULATE—to express or as in a formula.

FRACTURE—to break or crack.

FRAGMENTATION—sections or shards of glass fired together.

FRIT—a smelted mixture of soluble and insoluble materials forming a glass, which, when quenched in cold water, is shattered into small friable pieces.

FRIT FURNACE—a smelter used for making frits and fluxes.

FRITTED GLAZE—glaze which has been made by melting together all constituent materials.

FUSE—to liquify by means of heat; to unite or blend, as by melting together.

GAFFER—the master craftsman in charge of a shop of offhand glass blowers.

GATHER—a gob, or blob, of glass collected on the end of a blowtorch.

GATHERER—craftsman who first takes the proper amount of molten glass from the furnace and starts to blow the form.

GELATIN SHEET—a transparent gelatin film used in making photographic screen stencils.

GLASS—an amorphous substance usually hard, brittle and transparent consisting ordinarily of a mixture of silicates, but in some of borates, phosphates, etc.

GLAZE—a thin vitreous coat that attaches itself firmly to the body of the ware and imparts a gloss and smoothness to the surface.

GLAZER—glazier; also one who works with and applies glaze.

GLAZIER—one who sets glass in window frames.

GLORY HOLE—an opening in a small furnace used to reheat glass in handworking.

GLOST FIRE—the second or decorating fire—pottery term.

GROG—fired clay, firebrick, etc., which has been ground up and screened.

GROUND GLASS SLAB—a plate glass square, with sand blasted surface, on which decorators mix small quantities of color for banding.

GROUND LAYING—a method of decorating with powdered color, which is dusted onto large areas of ware coated with tacky oils.

GROZE—to finish a raw edge of glass by controlled chipping with a hooked tool or grozing iron.

GUM ARABIC—a water-soluble vegetable gum somewhat similar to gum tragacanth in its uses.

GUM TRAGACANTH—an original substance of vegetable origin used as a binder for glazes and enamels.

ICE—a fusible, fine, granular material used as a decorating medium.

INCISE—to engrave.

INFRARED—denoting hot rays outside the red end of the spectrum; employed as a means of drying.

INTAGLIO—incised design-elements below surface level of material.

ION—electrically charges atom or group of atoms.

IRIDESCENCE—a rainbowlike surface effect produced by spraying hot glass with solutions containing chlorides or nitrates of such elements as tin, bismuth, iron, and antimony.

JEWEL—small ornamental piece of glass.

JIG—contrivance to guide a tool or to form a shield or template.

KILN—a furnace or chamber in which clay or glass objects are heated or fired. Sometimes also used for firing decorating colors on glass.

LAMINATE—to construct of superimposed layers bonded together.

LATHE—machine in which work is held and rotated while being worked with a tool.

LAWN—a silk or brass sieve for screening or lawning color.

LEADLESS—material containing no lead, such as lead-free glaze.

LEHR (LEER)—a name given to an oven used in the annealing and decorating of glass.

LIGHT—segment of glass—medium through which light is admitted, as a window or pane.

LUSTERS—solutions of metallic resinates that produce mother-of-pearl and colored irridescent effects when fired on glass.

MANDREL—axle or spindle inserted in work to support it while it is machined.

MARVEL—a flat plate, usually metal, on which a gather of glass is rolled, shaped, and cooled.

MAT FINISH—a dull, satiny finish, not glossy.

MERCURY OXIDE, YELLOW—often used to dilute burnished gold.

MICA—a mineral silicate that crystallizes into forms readily separating into thin layers or scales.

MOLD—a hollow form which determines the contour of a glass blank fired upon it.

MUFFLE—an enclosed furnace in which the contents may be heated without exposing them directly to the fire.

MUDDING—an operation usually necessary to make liquid bright gold adhere to glass. A thin paste mixture of equal parts ochre and copper sulphate is brushed onto the glass and fired to about 850° F. After firing, the ware is washed, dried and decorated with the gold application.

MULLER—a flat-bottomed glass pestle for grinding color.

NEGATIVE—a photographic image on a film in which the dark portions of the subject appear light, and the light portions dark.

OBSCURE—an agent added to an enamel to produce opaqueness.

OPAL GLASS—a glass with a milky or resinous appearance.

OVERGLAZE—a low-firing pigment requiring a glossy surface as a base coating.

OXIDE—a compound of oxygen and a base substance.

OXIDIZE—to combine with oxygen, forming an oxide, as iron oxide.

OXIDE COLORS—inorganic base pigments used in porcelain and glass enamel colors.

PAINT MILL—small mixing unit used to incorporate an oil vehicle with ceramic color.

PALETTE KNIFE—a special tool for mixing color.

PEBBLE MILLS—see ball mills.

PIN-HOLES—name given to a defect consisting of small holes which may occur in any ceramic glaze.

PONTIL, OR PUNTY—a solid iron rod, about the size of a blowpipe, used to carry and manipulate small amounts of glass during fire polishing and finishing.

POSITIVE—opposite to a negative; the true picture.

PROFILM—trade name for a specially prepared film paper used in making silk screen stencils.

PYROMETRIC CONES—small triangular pyramids with flat based, used to measure the effect of heat upon clay products.

REFRACTORY—a material used to resist the action of heat.

REFRACTIVE INDEX—angle at which glass appears distorted when immersed in liquid.

RESISTANCE WIRE—electric wire which resists the passage through it of current and consequently becomes hot (used in toasters and appliances)

RESISTS—waxy or resinous substances used to resist the action of acids in etching.

RESPIRATOR—device covering mouth and nose to prevent inhalation of noxious vapor.

RHEOSTAT—an instrument regulating the strength of an electric current by controlling the amount of resistance.

ROUGE—red powder (ferrice oxide) used in polishing.

RUN—to exert steady pressure following along the scored line of an intended cut.

RUNNING—to become fluid or flowing, as melting glass.

SAGGED—see BENT.

SAGGERS—fire clay boxes in which ceramic ware is fired. Also used as containers for the manufacturing process of sintering and fusing color oxides.

SAND BLAST—a process of abrading glass by means of sharp sand driven by a jet of air.

SCORE—to mark with grooves.

SEGER CONES—see PYROMETRIC CONES.

SEGMENT—a separate part.

SEPARATOR—any of several materials to which molten glass will not adhere; used usually as powder to prevent the glass sticking inadvertently to objects such as the supports used during firing.

SERVITOR—the craftsman in the glass blowing process who takes the molten glass from the gatherer and continues the forming.

SHARD—a fragment of broken glass or clay.

SILICA—a raw material used in manufacturing glass.

SILICATE OF SODA (sodium silicate)—"water glass"; sometimes used as an adhesive.

SILK SCREEN PLATE—the stencil used in making silk screen prints.

SLURRY—a thin, watery mixture.

SMELT—to melt or fuse, usually to cause one component to separate out from the others in a mixture as one.

SOLDER—a material used when melted to join surfaces.

SPAUDLE—type of chuck for holding tubing in alignment while being joined.

SPRAY GUN (air brush)—an apparatus by means of which enamel is sprayed onto the surface of the ware.

SQUEEGEE—rubber blade for forcing color through stencil in silk screen printing.

SQUEEGEE OIL—a mixing medium for silk screen stencil colors.

STEEL PLATES—engraved plates from which color is transferred to ware being decorated.

STENCIL—pieces of thin material (paper, metal, etc.) so perforated that when laid on a surface and pigment applied a desired design appears.

STENCIL SILK—a special silk cloth of even mesh used in making silk screen stencils.

STICK-UP BOY—the assistant to the servitor.

STONE—any non-glassy material imbedded in the glass. It is an imperfection.

SUMP—drain or receptacle for fluids.

TEMPLATE—gauge or pattern used as a guide to the form of the work—commonly a thin plate or board.

TERRA COTTA—a composition of unfired and finely ground fire clay.

TEXTURE—discernibly uneven surface.

THERMAL—pertaining to heat.

TRAMMEL—an instrument for drawing large circles or ellipses.

TRANSLUCENT—partially transmitting light.

TRANSPARENT—permitting the passage of light through a substance so that objects may clearly be seen.

TUNGSTEN CARBIDE—very hard metal alloy used for wheels of glass cutters.

TURRET—a pivoted, revolving head for holding a part of a tool, as on the turret head glass cutter.

UNDERGLAZE—an opaque colorant applied prior to glazing, and requiring glaze or vitreous material as a surface coating.

VEHICLE—oil, spirits, etc. used as the liquid medium with which colors are applied.

VENT—to allow escape of air or fumes through an opening.

VISCOSITY—resistance to flow, as in molten glass.

SPECIFIC PROCESSES

ENAMELING

Agar Solution—A material used for wetting metal surfaces for suspending enamels. (Available in drug stores.)

Alkali Cleaners—Strong alkaline cleaners such as trisodium phosphate.

Annealing—The treatment of glass and metals by controlled heating and cooling in such a manner as to remove strains.

Bright Dip—Strong acid dipping solution that gives metal a bright finish.

Centigrade—A temperature scale used for recording temperature (universally used).

Counter Enameling—Applying enamel on both sides of a piece of metal.

Crazing—(See Fiber Glass.)

Electrolyte—A material that is added to a clay, enamel and water mixture that aids in suspension of enamel.

Enamel—A smooth, tough substance used in coating surfaces of glass or metal.

Fahrenheit—A temperature scale ordinarily used in the United States for thermometer readings.

Flux—A pulverized low melting glass.

Frit—A smelted mixture of soluble and insoluble materials forming a glass, which, when quenched in cold water, is shattered into small friable pieces.

Gum Arabic—A water soluble vegetable gum used somewhat similar to gum tragacanth.

Gum Tragacanth—An original substance of vegetable origin used as a binder for glazes and enamels.

Kiln—A furnace or chamber in which clay products are burned. Sometimes used for firing decorating colors on glass.

Opaque Enamel—An enamel of solid color which completely covers the metal or glass background.

Overglaze—A low-firing pigment requiring a glossy surface as a base coating.

Oxide Scale—A scale formed of a compound of oxygen and a base substance.

Oxidize—To combine with oxygen.

Pickle—An acid and water mixture used for cleaning metal pieces.

VISCOUS—sticky.

VITREOUS—consisting of or resembling glass.

VITRIFIED—converted into glass.

VOLATILE—easily changed into gas (vaporized).

WHITING—see CALCIUM CARBONATE.

Placque—A flat piece of screening, or perforated metal on which the work is rested when being placed in the kiln or lehr.

Planche—(Same as Placque.)

Pyrometer—An instrument to measure high degrees of heat.

Reducing Atmosphere—An atmosphere which contains a comparatively small amount of oxygen.

Refractory—A material used to resist the action of heat.

Scratch-Brushing—A scratch type of finish applied with a wire brush.

Spatula—A flexible knife with a long, flat blade.

Star Stilts—Usually triangular shaped pieces of highly fired clay material and having metal wires with sharp points located at the corners of the triangle.

Stoning—Gringing away material with a hand or power grinding stone.

Transparent Enamels—Glassy enamels that allows the metal backing to show through.

Vitreous—Consisting of or resembling glass.

FIBER GLASS

Accelerator—Used in conjunction with a catalyst to produce internal heat in a liquid plastic to cure it.

Acetone—A chemical used as a cleaning fluid to remove uncured plastic resin.

Activator—A material added to the catalyst to alter its action.

Barcol Hardness—A measurement of the hardness of the part.

Catalyst—In FRP terminology, a material which is utilized to activate resins to cause them to harden.

Color Pigments—Ground coloring materials supported in a thick liquid.

Crazing—Tiny cracks which appear in a material. Usually caused by internal stresses.

Cross-Linking—Can be compared to two straight chains which are joined together by means of links. The more of these links that are present, the more rigid the material.

Cure—The transformation of the resin from the liquid to the solid state.

Cure Time—The time required for the liquid resin to

reach a cured or polymerized state after the catalyst has been added.

Draft—The relief angle allowed on the sides of a mold to allow the formed laminate to be removed.

Duplication Mold—A mold made by casting over another article.

Exotherm—Signifies the heat given off as a result of the action of a catalyst on the resin.

Fiber Glass—A material made from strands or threads drawn from glass.

Fillers—Materials added to the plastic resin to extend its volume and lower the cost of the article being produced.

Finish—The treatment applied to the glass fibers.

Flash—The excess of material that builds up around the edge of a FRP part which must be removed.

Gel—The state of the resin prior to its becoming a hard solid.

Gel Coat—This is a specially formulated polyester resin, which is pigmented and contains filler materials. It provides a smooth, pore free surface for an FRP part.

Hardener—An additive for the catalyst that alters the curing period.

Isotropic—Refers to a material whose properties are the same in every direction, e.g. metals, glass mats.

Lamination—The laying on of layers of glass material and resin, much like the layers of plywood.

Lay-Up—Placing fiber glass material onto the mold and applying resin to it.

Mold Release—A substance used to coat the mold to prevent the molded laminate from sticking to the mold.

Monomer—Refers to a reactive material which is compatible with the base resin. It serves to make the resin less viscous, thus easier to handle.

Polyester-Epoxy—Generic names for a series of resin materials.

Polymerization—The name for the reaction that takes place when resin is activated.

Pot Life—The available time, once a resin is catalyzed, until the material is no longer able to be used.

Preform—A glass mat which is formed into a shape which is similar to the mold in which the mat will be used.

Pre-Mix—Plastic resin and glass materials mixed together before placing in the mold.

Promoted Resin—Resin with accelerator added but not catalyst.

Resin—A highly reactive material which, in its initial stage, is pourable liquid. Upon activation, it is transformed into a solid.

Roller—Consists of a serrated piece of aluminum which is used to work on FRP laminate. Its function is to compact a laminate and to break up large pockets to permit easier release of entrapped air.

Roving—The name given to a bundle of continuous, untwisted glass fibers. These glass fibers are wound on a roll which is called a "roving package".

Sandwich Lay-Up—A type of fiber glass lamination that consists of two outside layers of glass mat and an inside layer or layers of glass cloth, or other material.

Shelf Life—The length of time an uncatalyzed resin remains workable while stored in a tightly sealed container.

Staple Fiber—A glass fiber of short length, formed by blowing molten glass through holes.

Squeegee—A tool similar to the ones employed by window washers to remove water. It is utilized to move excess resin and entrapped air out of a laminate.

Surfacing Agent—An oily material which rises to the surface of polyester resin during curing.

Thermoplastic—The reverse of thermoset. These materials can be reprocessed by means of heat.

Thermoset—Generic name for all plastic materials which harden with heat and cannot be reliquified.

Thinner—Material added to plastic resins to thin or lower the viscosity of the resins.

Tracer Yarn—A strand of glass fiber which is colored differently from the remainder of the roving package.

Thixotropy—The property of resin or gel-coat which reduces draining. It permits a material to remain on a vertical surface without sagging or running.

Undercut—Negative or reverse draft on the mold.

Yarn—A twisted strand or strands of glass fiber used to form glass cloth.

GLAZING

Back Putty—See Bed or Bedding.

Back-Up—A material placed into a joint, primarily to control the depth of the sealant.

Bead—A sealant or compound after application in a joint irrespective of the method of application, such as caulking bead, glazing bead, etc. Also a molding or stop used to hold glass or panels in position.

Bed or Bedding—The bead of compound applied between light of glass or panel and the stationary stop or sight bar of the sash or frame, and usually the first bead of compound to be applied when setting glass or panels.

Bedding of Stop—The application of compound at base of channel, just before the stop is placed in position, or buttered on inside face of stop.

Bevel of Compound Bead—Bead of compound applied so as to have a slanted top surface so that water will drain away from the glass or panel.

Bite—Amount of overlap between the stop and the panel of light.

Block—A small piece of wood, lead, neoprene or other suitable material used to position the glass in the frame.

Buttering—Application of putty or sealant compound to the flat surface of some member before placing the member in position, such as the buttering of a removable stop before fastening the stop in place.

Channel—A three-sided, U-shaped opening in sash or frame to receive light or panel, as with sash or frame units in which the light or panel is retained by a removable stop. Contrasted to a rabbet, which is a two-sided L-shaped opening, as with face glazed window sash.

Channel Depth—The measurement from the bottom of the channel to the top of the stop, or measurement from sight line to base of channel.

Channel Glazing—The sealing of the joints around lights or panels set in a U-shaped channel employing removable stops.

Channel Width—The measurement between stationary and removable stops in a U-shaped channel at its widest point.

Clips—Wire spring devices to hold glass in rabbeted sash, without stops, and face glazed.

Compound—A formulation of ingredients, usually grouped as vehicle or polymer pigment and fillers to produce caulking compound, elastomeric joint sealant, etc.

Compression—Pressure exerted on a compound in a joint, as by placing a light or panel in place against bedding, or placing a stop in position against a bead of compound.

Concave Bead—Bead of compound with a concave exposed surface.

Consistency—Degree of softness or firmness of a compound as supplied in the container, and varying according to method of application, such as gun, knife, tool, etc.

Convex Bead—Bead of compound with a convex exposed surface.

Curing Agent—One part of a two-part compound which when added to the base will cause the base compound to set up by chemical reaction between the two parts.

Durometer—A machine to measure shore hardness. (See shore hardness.)

Elastomer—An elastic, rubber-like substance, as natural or synthetic rubber.

Exterior Glazed—Glass set from the exterior of the building.

Exterior Stop—The removable molding or bead that holds the light or panel in place when it is on the exterior side of the light or panel, as contrasted to an interior stop located on the interior side of the light.

Face Glazing—On rabbeted sash without stops, the triangular bead of compound applied with a glazing knife after bedding, setting, and clipping the light in place.

Face Lock—An extruded piece of metal of a variety of cross-sections which slides along the interior of store sash face members. A portion of this piece is so positioned that when the set screw on the opposite side is

tightened the wedging action draws the sash face member tight against the glass and gutter or flashing members.

Front Putty—The putty forming a triangular fillet between the surface of the glass and the front edge of the rabbet.

Gasket—Pre-formed shapes, such as stripes, grommets, etc., or rubber or rubber-like composition, used to fill and seal a joint or opening either alone or in conjunction with a supplemental application of a sealant.

Glazing—The securing of glass in prepared openings in windows, door panels, screens, partitions, etc.

Gun Consistency—Compound formulated in a degree of softness suitable for application through the nozzle of a caulking gun.

Heel Bead—Compound applied at the base of channel, after setting light or panel and before the removable stop is installed, one of its purposes being to prevent leakage past the stop.

Interior Glazed—Glass set from the interior of the building.

Interior Stop—The removable molding or bead that holds the light in place, when it is on the interior side of the light, as contrasted to an exterior stop which is located on the exterior side of a light or panel.

Knife Consistency—Compound formulated in a degree of firmness suitable for application with a glazing knife such as used for face glazing and other sealant applications.

Light—Another term for a pane of glass used in a window.

Lock—See Face Lock.

Mastic—Descriptive of heavy-consistency compounds that may remain adhesive and pliable with age.

Non-Drying—Descriptive of a compound that does not form a surface skin after application.

Points—Thin, flat, triangular or diamond shaped pieces of zinc used to hold glass in wood sash by driving them into the wood.

Polymer—A high molecular weight chemical structure consisting of a long chain of small molecular units.

Polybutene Base—Compounds made from polybutene polymers.

Polysulfide Base—Compounds made from polysulfide synthetic rubber.

Priming—Sealing of a porous surface so that compound will not stain, lose elasticity, shrink excessively, etc., because of loss of a curing type sealant to certain surfaces.

Rabbet—A two-sided L-shaped recess in sash or frame to receive lights or panels. When no stop or molding is added, such rabbets are face glazed. Addition of a removable stop produces a three-sided U-shaped channel.

Sash—The frame including mounting bars when used, and including the rabbets to receive lights of glass, either with or without removable stops, and designed either for face glazing or channel glazing.

Screw-On Bead or Stop—Stop, molding or bead fastened by machine screws as compared with those that snap into position without additional fastening.

Sealant—Compound used to fill and seal a joint or opening, as contrasted to a sealer which is a liquid used to seal a porous surface.

Setting—Placement of lights or panels in sash or frames. Also action of a compound as it becomes more firm after application.

Shims—Small blocks of composition, lead, neoprene, etc., placed under bottom edge of light or panel to prevent its settling down onto bottom rabbet or channel after setting, thus distorting the sealant.

Shore Hardness—Measure of firmness of a compound by means of a Durometer Hardness Gauge (Range of 20-25 is about the firmness of an art gum eraser. Range of 90 is about the firmness of a rubber heel.)

Sight Line—Imaginary line along perimeter of lights or panels corresponding to the top edge of stationary and removable stops, and the line to which sealants contacting the lights or panels are sometimes finished off.

Size of Bead—Normally refers to the width of the bead, but there are many situations in which both the width and depth should be taken into account in design, specification, and application.

Spacers—Small blocks of composition, wood, neoprene, etc., placed on each side of lights or panels to center them in the channel and maintain uniform width of sealant beads. Prevent excessive sealant distortion.

Spacer Shims—Devices that are U-shaped in cross section and an inch or more in length, placed on the edges of lights or panels to serve both as shims to keep frames, and as spacers to keep the lights or panels centered in the channels and maintain uniform width of sealant beads.

Stationary Stop—The permanent stop or lip of a rabbet on the side away from the side on which lights or panels are set.

Stop—Either the stationary lip at the back of a rabbet, or the removable molding at the front of the rabbet, either or both serving to hold light or panel in sash or frame, with the help of spacers.

Striking Off—The operation of smoothing of excess compound at sight line when applying compound around lights or panels.

Thinning—Addition of a slight amount of unleaded

gasoline to an oleo-resinous glazing compound by the glazier to soften its consistency.

Unit—Term normally used to refer to one single light of insulation glass.

United Inches—Total of one width and one height in inches.

Vegetable Oil Base—Formulated with a vehicle of vegetable oils usually proceed with resins by application of heat.

Vehicle—The liquid portion of a compound.

Vinyl Glazing—Holding glass in place with extruded vinyl channel or roll-in type.

Work Life—The time during which a curing sealant (usually 2 components) remains suitable for use after being mixed with a catalyst.

STAINED GLASS

Antique Glass—Glass made in imitation of the color and quality of old glass.

Came—Lead or zinc strips shaped like an "H" into which the separate pieces of a stained glass panel are fit.

Cartoon—A full size drawing of the stained glass panel.

Crazing—(See Fiber Glass above.)

Cutline—An exact tracing of the leadlines of the cartoon drawn on heavy paper.

Firing—The process of heating glass.

Flashed Glass—A glass formed with color on one side and white glass on the other.

Fuse—To liquify by means of heat.

Glaze—A thin vitreous coat that attaches itself firmly to the surface and imparts a gloss.

Leading Up—Joining the pieces of glass in a stained glass panel by means of lead or zinc strips called "came".

Lehr—An oven used in the annealing and decorating of glass.

Plasticine—A non-hardening clay.

Silver Staining—A coloring process used to obtain pale yellows and deep oranges.

Solder—An alloy of tin and lead used for joining metals.

Stained Glass Work—A decorative composition constructed of many pieces of irregularly cut colored glass, bound together by strips of grooved lead or zinc.

Waxing Up—Positioning of the colored pieces of glass on a sheet of clear glass. The colored pieces are held in place by drops of beeswax at the corners.

Related reading

1. Almy, Ruth C. *Simulated Stained Glass For Amateurs*. Harper & Bros., 1949.
2. Amaya, Mario. *Tiffany Glass*. Walker & Co., 1966.
3. Anderson, Harriet. *Kiln-fired Glass*. Chilton Book Co., 1970.
4. Armitage, E.L. *Stained Glass*. Charles T. Branford Co., 1957.
5. Barr, W.E. and Anhorn, V.J. *Scientific and Industrial Glass Blowing and Laboratory Techniques*. Instrument Publishing Co., 1959.
6. Bates, Kenneth F. *Enameling: Principles and Practice*. T.Y. Crowell Co., 1951.
7. Bell, Charles. *How to Build Fiberglass Boats*. Coward-McCann, Inc., 1957.
8. Burton, John. *Glass: Hand Blown, Sculptured, Colored*. Chilton Book Co., 1967.
9. Connick, Charles J. *Adventure in Light and Color*. Random House, 1937.
10. Davidson, D.M.J. *Glass and Glazing*. Crosby Lockwood & Son, Ltd. (London), 1947.
11. Divine, J.A. and Bachford, G. *Stained Glass Craft*. Warner Publishing Co., 1972.
12. Drepperd, Carl W. *The ABC's of Old Glass*. Doubleday & Co., 1972.
13. Haynes, E.B. *Glass Through the Ages*. Penguin Books, 1968.
14. Henson, Catherine M. *How to Enamel on Copper*. Walter T. Foster Co., 1950.
15. Hutton, Helen. *Mosaic Making*. Batsford, 1966.
16. Isenberg, Anita and Seymour. *How to Work in Stained Glass*. Chilton Book Co., 1972.
17. Jenkins, Loisa and Mills, Barbara. *The Art of Making Mosaics*. D. Van Nostrand, 1957.
18. Kinney, Kay. *Glass Crafts*. Chilton Book Co., 1962.
19. Lloyd, John G. *Stained Glass in America*. Agora Books, 1963.
20. Morey, G.W. *Properties of Glass*. Reinhold Publishing Corp., 1954.
21. Norman, Barbara. *Engraving and Decorating Glass*. McGraw Hill Book Co., 1981.
22. Pact, Greta. *Jewelry and Enameling*. D. Van Nostrand, 1961.
23. Sonneborn, Ralph E. *Fiberglass Reinforced Plastics*. Reinhold Publishing Corp., 1954.
24. Steele, Gerald L. *Fiber Glass*. McKnight & McKnight Publishing Co., 1962.
25. Twining, E.W. *The Art and Craft of Stained Glass*. Sir Isaac Pitman & Son, 1928.

Index

Abrasives, 14, 66-67, 87-91, 93, 102, 240, 242-243, 244
  use of, 75, 77
Accelerators, **See** Activators
Acetate film, 78
Acid, 18, 249
  hydrofluoric, 61-64, 249
  muriatic, 218, 221, 242
  nitric, 64, 230
  sulphuric, 61-62, 64, 230
  tartaric, 245
Activators, 183, 194, 197-198, 201
Additives, 183, 194, 203
Adhesive(s), 216, 219
  for enameling, 230
  mastic, 216, 217
  non-fired glass, 137-139, 146-148, 208
  pressure sensitive, 163, 166-168
Alkali, 262-263
A. Ludwig Klein and Co., 139
Aluminum, 228
American Art Clay Company, 99
American Crayon Company, 84
Annealing, 12, 96, 98, 101-102, 117-118, 136, 140, 248
  methods of, 125-126
Annealing oven, **See** Lehr
Annealing point, 96
Antique glass, 149
Art glass, 4, 250
Asia, 2

Bacon, Roger, 3
Bas-relief, 76
Batch, 8
Batcher, 131
Batching kiln, 131
Bent glass, cutting, 32
Bethlehem Apparatus Company, 115, 127
Beveling, 87, 91-92
Bit-gatherer, 135
Blow-and-blow machine, 11
Blowing, **See** glass blowing
Blowpipe, 2, 9, 131, 133, 135
Bonding materials, **See** adhesives
Bondley, Ralph J., 144
Boring holes in glass, 55-59
Borosilicate glass, 16, 17, 18, 118,
  characteristics and uses of, 15
  viscosity reference point of, 96, 98
Boston and Sandwich Glass Company, 4
Burners, 97, 103, 113
  bunsen, 97-98, 118
  liquified petroleum, 97
  meker, 97-98, 118
  Universal Blast, 98
Byzantium, 2

Carbide tips, 55
Carborundum powder, 53, 66
Carborundum stone, 53, 87, 90, 237, 238
Calcium carbonate, **See** whiting
Cartoon, 152, 153-155, 166, 167
  for slab glass work, 172
Catalysts, 183, 194, 197, 200, 201, 203, 204, 206, 210, 211, 212, 213
Cathedral glass, 250
Cellular glass, 16

Cement, 161, 220, 221
  clear plastic, 216, 217
  contact, 225
  silvercrete, 215
  special for slab glass panels, 169, 171, 172, 173-174
  See also adhesive(s)
Chipped surface glass, 69-70
Coefficients of expansion, 13, 15, 18, 116, 118, 127, 140, 142-143
Coefficients of thermal expansion, 146
Color, use of, 77, 78, 151
Colored glass, 16
Cooling agents, 56
Copper, 143, 227, 228, 229-230, 237
Copper enamels, 86
  See also enameling, metal
Copper wheel(s), 75-76
  See also engraving, copper wheel
Corning Glass Center, 131
Corning Glass Works, 4, 6, 14, 15, 19, 127
Corning Glass Works Laboratory Catalogs, 124
Corning ribbon machine, 11
Corrosion resistance, 18
Corrosive gases, 102
C.R. Laurence Company, Inc. 68, 159, 160, 161
Crushed glass ornaments, 223
  making, 224-225
Cullet, 8, 131
Cutline drawing, 152-153, 157, 172
Cut-off machine, wet cut, 53
Cutter, glass, 25-32, 41-43, 45, 50, 150, 153, 215
  using the, 30-34, 35-37
Cutters, glass disk, 41, 42-43
Cutting gauge, 29, 49-50
Cutting glass, 32-33, 153, 166, 172
  equipment for, 25-28
  problems of, 32-33
  procedure for, 30-33
  unusual cuts, 45-47
  use of heat in, 51-52
  See also cutting glass disks
Cutting glass disks, 41-43
Cutting machine, 29
Cutting table, 25, 30, 38
Cylinders, glass
  cutting, 51-53
  methods of scoring, 49-51
  polishing edges of, 53

Decalcomanias, 83
Decorating glass, 101
  and firing, 102-103
  materials used in, 77, 81, 83-86, 189, 194-195, 196
    permanent, 84-85
    semi-permanent, 83-84
    vitrified, 85-86
  preparing designs for, 77-79
  techniques of, 79-83
Decorating media, 107-108, 109-110
Dek-All, 84
de Nehou, Louis Lucas, 3
Design(s), 71-76, 77-79
  metal enameling, 232-236
  mosaic, 217, 219-220
  overlay, 227

Design(s) **(Continued)**
  painting on, 79-80
  preparing, 77-79
  stencil, 233
Disk-cutting machine, 42
Disks, glass, **See** Glass disks, cutting
Display windows, painting, 261
Doerr Glass Company, 131
Dow Corning Glass and Ceramic Adhesive, 137, 146, 208
Drawing
  hand, 8, 9
  machine, 11-12
Drills, Glass, 55, 56-57, 58
Drilling glass, procedures for, 55-58
Duco cement, 137-138, 146, 208, 216, 224, 225
Duplex Paper, 83

EC 826 (adhesive), 138
EC 1103 (adhesive), 139
Edging glass, methods of, 87-92
Edison, Thomas A., 4
Egypt, 2
Electric heating coil locators, 37
Electrically-conducting glass, 16
Elmer's Glue, 138
Enamels, vitreous (glass), 85-86, 227, 233-234
  applying, 230-232, 237-238
  types of, 228-231
Enameling
  metal, 227
    preparation for, 229-230
    procedure for, 230-232, 236-238
  porcelain, 227
England, 2, 3
Engraving, 71-76
  copper wheel, 74-76
Epoxy glue, 138, 208
Epoxy resins, **See** Resins, epoxy
Ermax cold-table method for mirror making, **See** Mirror making, Ermax cold-table method of
Ermax Corporation, 241
Etching, 61-68, 126, 249
  materials for, 117-118
  procedure for, 64-66
  solutions for, 62
    commercially prepared, 64
  See also Frosting
Etching box, 61-62
Exacto knife, 59, 68
Exothermic reaction, 182, 194, 196
Expansion coefficient, 96

Fabric, fiber glass, 179
Faceted glass, **See** Slab glass panel(s)
Fernico-General Electric Company, 144
Fiber glass, 16
  physical properties of, 176, 178, 179-180, 182-183, 196
  repairing metal with, 212-213
  repairs, 209-212
  See also Fiber glass reinforced plastics (FRP); Fiber glass materials
Fiber glass laminated plastics, **See** Fiber glass reinforced plastics (FRP)
Fiber glass laminates, **See** Laminates, fiber glass
Fiber glass materials, 180-183

Fiber glass materials, **(Continued)**
  manufacture of, 177-180
  use of, 159
Fiber glass paintings, 203-204
Fiber glass reinforced plastics (FRP), 175-177, 188-190
  decorating, 201
  manufacture of, 154-157
  as a protective covering, 199-201
  repair of, 198-199
  resin mix for, 180-183
  Selection of molding processes, 183-188
  shop procedures for, 190-199
Fiber glass sculpture, 205-208
  plastic form, 205-208
  solid form, 208
File, three-cornered, 49
Fillers, inorganic, 182-183
Firing devices, 97
  direct, 97-98
  indirect, 97, 98-99
Firing glass, 22, 95-103, 107-109, 156-157
  temperatures for, 84, 85-86
Flake glas (flake glass), 201
Flashed glass, 150
Flat glass, 3, 4-5
Float glass, 248, 249
Flux, 161
Foam glass, **See** cellular glass
Foil, aluminum design, 65-66
Ford, John B., 4
Forming machines, 8
Formulas, silvering, **See** solutions, silvering
Fotoceram, 15
Fotoform glass, 15
Frabel, Hans Godo, 141, 127
France, 2, 3
Franklin, Benjamin, 4
Frits (ceramic), 227, 248
Frosted glass, 61-70
Frosting glass
  chemical methods of, 61-66
  physical methods of, 66-67
  by sandblasting, 67-68
Frosting paint, 263
Frosting solutions, commercial, 64
FRP, **See** Fiber glass reinforced plastics
Frying, 155-157
Fulgurites, 1
Fumes, poisonous, 61, 64
Furnace glass, 8-9, 129-136
Fused silica glass, 15
Fusing, 139
  See also laminated glass

Galileo, 3
Gaffer, 9, 135-136
Gather, 9, 131
Gatherer, 131
Gauge, glass, **See** cutting gauge
Gel, **See** resin; resin mix
General Electric Company, 144
Germany, 2
Glare-reducing glass, 249
Glass, 21-22, 87, 129, 146-148
  boring holes in, 55-59
  cutting, **See** cutting glass
  definition of, 8
  designs on, 67-70
  firing, **See** firing glass
  history of, 1-7
  identification of, 116-117
  methods of forming, 8-14, 19-20
  properties of, 7, 17-19
  types of, 14-17, 149-150, 247-250
  uses of, 6-7, 19-20
Glass blowing, laboratory, 111-128

Glass Blowing **(Continued)**
  equipment for, 111-115
  procedure for, 119-121
  machine, 10-11
  off-hand, 8-9
    American system of, 133
    German system of, 133
    materials for, 131
    methods and tools of, 129-136
Glass blowers table, 111-113
Glass chips, 86
Glass cleaners, 83
Glass crushings, 86, 223-225
Glass cutter, **See** cutter, glass
Glass disks, cutting, **See** cutting glass disks
Glass jewels, 86
  firing, 103
  preparation of, 108-109
Glass joints
  fired, 139-146
    glass to glass, 139-140
    glass to metal, 140-142
    metalizing, 144-146
  non-fired, 137-139
  See also seals
Glass knife, 49, 114
Glass lift tool, 22
Glassmakers, 2-4, 8-9, 17, 131-136
Glassmaking, 2-4, 12-14
  hand methods of, 8-9
  machine methods of, 9-12
Glass mosaics, **See** Mosaics, glass
Glass pliers, **See** glazier's pliers
Glass shards, 86, 108-109
Glass sling, 22
Glassware, laboratory, 118
Glassworks, 3-6, 8-12
Glazier's pliers, 29, 36, 41, 46, 47
Glazing, 247-260
  preparation for, 252-253
  procedures for, 253-260
  theory of, 247
Glazing compounds, 249, 250-251, 253
Glory hole, **See** Furnace, glass
Glue
  animal or noodle, 69
  white resin, 216, 217, 224, 225
Gold, 228, 230
Grinding, 14, 38, 53, 126-127
  edge, 87, 89-90
  See also polishing
Grinding equipment, 50-51, 72-73, 89, 90-91
Grog clay, 106-107
Grout, 215, 216, 218-219, 220

Hale telescope, 9, 15
Handmade glass, **See** glass blowing, offhand
Heat-absorbing glass, 248-249, 259
Heat, absorption of, 18
  effect on glass, 95-96, 108, 110, 139-145
  use in glass blowing, 118-126
Heating coil, 37
Heaton, Maurice, 137, 138
Heat-reducing glass, 250
Herschel, William, 5
Hogue, Lawrence J., 144
Holland, 3
Houghton, Amory, 4
Hydrofluoric acid, **See** acid, hydrofluoric

Indices of reflection, 262
Inhibitors, 183
Insulating glass, 249
Intaglio, 52
Internal squares, cutting, 45-47
Ion exchange, 85
Iron, 228-23

Jack Frost (frosting solution), 64
Jansen, Zacharias, 3
Jars, cutting, 52-53
Jarves, Deming, 4
Jasper, 1
Jewels, glass, **See** Glass jewels
Johnson, Samuel, 7

Kelly, Floyd C., 144
Kepler, Johann, 3
Kiln(s), 84, 86, 95, 97, 98-99, 139-140, 156-157
  ceramic, 98, 107
  metal enameling, 98-99, 102, 103, 107-109, 230-231, 232, 234, 235
  hot-plate type of, 109
  procedures for using, 78-80
  See also lehr(s)
Kimble Glass Company, 140
Kleins #200 Cement, 139, 146
Kovar-Stupakoff Ceramic and Manufacturing Company, 144

Laminated glass, 139-140, 249, 250, 259-260
  cutting, 35-36
  decorating, 86
Laminates, fiber glass, 179, 196, 197, 198, 199, 203, 204, 205-206
  curing, 196-198, 199, 200
  repairing, 209-210, 211-212
  See also Fiber glass reinforced plastics (FRP)
Laminating, 101-102, 107-108, 109, 140-142
Lamination process, 262
Lampworking, 13
Lathes, glassworking, 127
Lead borosilicate glass, 85
Lead glass, 16, 17, 18, 118, 262
  characteristics and uses of, 14
  viscosity reference point of, 96, 98
Lead oxide glass, 3
Lead ribbon, 223-225
Leads, 171
  came, 157-158
  soldering the, 160-161
  use in stained glass work, 151-153, 161
Lead strips, 163
  applying, 164-167, 168
Lead tape, **See** lead strips
Lehr(s), 84, 95, 97, 101, 102, 125, 136, 156-157, 248
Libby, Edward Drummond, 4
Lifting devices for glass, 22
Lime-soda glass
  properties of, 118
  viscosity reference point of, 96, 98
Litton Engineering Laboratories, 127
Low transmission glass, 250
Lusters, 84-85, 86

Mandrel, 11, 125
Marveling table, 133
Masking, 58-59, 68, 74-75
Masking tape, 67, 68
Mastic, 247, 250-251, 258
Metal enameling, **See** enameling, metal
Metals for enameling, 228
  cleaning, 229-230
Metalizing, 13
Microscope, 3, 5
Minnesota Mining and Manufacturing Company, 138-139
Mirror(s), 3, 249, 258-259
  one-way vision, 245
Mirror making
  cold table method of, 239-244
  Ermax, 240
  hot table method of, 244-245
  spraying method of, 239, 244

Mississippi Glass Company, 4
Mold release, 192-194
Molds, 190-198, 204-208
    choice of, 183-188
    closed, 184, 185
    glass sagging, 105-109
    open, 183-184, 185, 189-190
    use of, 8, 9, 11, 136
Molten glass
    hand forming of, 8-9
    machine forming of, 9-12
Monce, Samuel, 25
Monomer(s), 180, 181, 182, 183, 203
Mosaics, glass, 2
    materials used in, 215-217, 223
    procedure for making, 217-221
Multiform glass, 16
Museum of Contemporary Crafts, 129

Natron (soda), 1
New England Glass Works, 4

Obsidian, 1
Onyx, 1
Opal glass, 16, 64
Optical glass, 17
    properties of, 18-19
Optics, science of, 3
Owens-Illinois Company, 140
Owens, Michael J., 4
Oxides, 16, 154

Paint, 261, 263
    vitrifiable, 154-155, 156
Parker-Garrick, Inc., 71, 92
Pattern drawing, 152-153, 172
Patterned glass, 249, 253
    cutting, 32
Patterns, cutting, 35-36
Peacock Laboratories, Inc., 241
Pellet box, 66
Photosensitive glass, 15, 16
Pitcoirn, John, 4
Pittsburgh Plate Glass Company, 4
Plastic, cutting the (in safety glass), 37-39
Plate glass, 27, 35, 245, 261
    beveling edges of, 91-92
    heat absorbing, 248
    manufacture of, 247-248, 249, 258
Plate Glass Company, 4
Platinum, 143, 145
Pliers, glazier's, **See** glazier's pliers
Pliny the Elder, 1
Points, 119-120, 121, 124
Polarized glass, 262
Polaroid, 262
Polisher, portable scratch, 93-94
Polishing, 14, 53, 76, 87, 90, 91-92, 242-243
    felt blocks for, 241, 243
    removing scratches by, 93-94
Polishing agents, 87, 90, 91, 93, 240, 242
    See also abrasives
Polishing wheels, 76, 91
Polyester resins, **See** resins, polyester
Polymerization, 180, 183
Polysulfides, 251, 252
Pontil rod, 9, 135-136
Porcelain enameling, **See** enameling, porcelain
Portman, John, 128, 202
Pot metal glass, **See** Antique glass
Prang Products (American Crayon Company), 84
Press-and-blow machine, 11
Pressing
    hand, 8, 9
    machine, 11
Punty iron, **See** Pontil rod
Turret-chain machine, 11
Tweezers, glass blowing, 114, 116, 120, 125

Putty(ies), 247, 250
    and aluminum sash, 253
    and metal sash, 251, 253
    and wood sash, 251, 252-253
Pyrex Brand Glass, 111, 113, 115, 125-126, 127, 140, 142-144
    etching of, 126
    flasks of, 114
    identification, 116-117
    properties of, 118
Pyrometric cones, 99, 102

Quartz, 1

Radiation-absorbing glass, 16, 18
Radiation-sensitive glass, 16, 18
Ravenscraft, George, 3
Reamers, 55, 114, 121
Refraction, index of, 14, 16, 17, 19, 116-117
Refractories, 8, 20
Reinforcements, fiber glass, 195-197
    forms of, 175, 176, 178-180, 184, 188
    See also Fiber glass materials
Rendzunas, Paul, 109, 110
Resin(s), 183-188, 189, 192-194, 197, 198-200, 201, 206-208
    epoxy, 175, 209-213
    as a painting medium, 203-204
    performance qualities of, 175, 176, 180-182
    polyester, 175, 176-177, 180-183, 203
    thermoplastic, 180, 182
    thermosetting, 180, 182
Resin mix, 180-182, 188, 210, 211
    application of, 194-198, 199-201
    fillers in, 182-183
    processing of, 185-188
Rome, 2

Safety glass, laminated, 90-91, 262
    cutting, 35-39
    See also laminated glass
Safety precautions, 21-22, 62, 64, 68, 70, 76, 87, 90, 114, 115, 116, 126, 139, 140, 168, 194, 201, 204, 213, 218, 230, 242
Sagging, 96, 101, 102, 105-110
Sandblasting, 55, 58-59, 67-68, 74-75, 249
Sand box, 147, 208
Sander
    portable hand, 87
    upright belt, 89
Sanding disk, 89-90
Sandwich pressed glass, 4
Sanereisen Cement Company, 173
Sanereisen No. 54 Pour-lay Cement, 173
Scratches, removing, 93-94
Sealant(s), 247, 250-253
Seals, 12, 124-125
    glass to glass, 127
    graded glass, 13
    ring, 122-123
    straight, 121-122
    T, 122, 123
    See also glass joints
Separators, glass, 96-97, 102, 108
Serigraph, **See** decorating, glass
Servitor, 135
Sheet (flat) glass, 33, 34, 105, 108
    ground edge on, 87
    manufacture of, 247, 249
Sidon, 2
Silica, crystallized forms of, 1
Silica glass, 262
Silica glass, 96%, 18
    characteristics and uses of, 15
Silicon-carbide, 53, 57
Silicones, 251
Silk screening, **See** decorating, glass

Silver, 228, 230
Silvering, 239, 243-244
    See also mirror making
Silver nitrate, 241, 242-244
Slab glass panels, 171-174, 175
Soda-lime glass, 16, 17, 18, 115
    characteristics and uses of, 14
    properties of, 118
    viscosity reference point of, 96, 98
Softening point, 96
Solder, 144-145
    glass, 140
Soldering iron, 150
Soldering stained glass panels, 158-161
Solution(s)
    Ermax pre-wash, 243
    Peacock 37B Cleaning Agent, 243
    Silvering, 239
        Ermax Cold-Table, 241-242
        hot-table, 244-245
        Peacock (Brashear Type) Cold-Table, 242
Spandrel glass, 248, 259
Spectacles
    bifocal glass, 4
    glass-lensed, 3
Stained glass panels and windows, 2, 150-153, 169, 215, 223
    assembling, 153-154
    copper (Tiffany) foil method, 158-160
    cutting glass for, 153
    firing, 156-157
    leading, 157-158
    pointing, 154-155
    types of glass used on, 149-150
Stained glass panels, simulated
    construction of, 166-168
    materials for, 163-166
Stains, glass, 85, 155-156, 163-166, 224, 225
    applying, 167-168
Steel, 228
Stencil(s), 65-66
    silk screen, 82-83, 84
Steuben Glass (Div. of Corning Glass Works), 14
Stick-up boy, 135
Stiegel, Henry William, 4
Straight edge, 25-26, 30
Strain point, 96
Strains, 111, 125, 136
    avoiding, 95, 111
Stresses, 9, 140
    avoiding, 95
    removing, 96
Sulphuric acid, **See** acid, sulphuric
Syria, 1

Telescope, 3, 5-6
Tempered glass, 248, 249, 259
Tempering, 12, 17, 248, 249
Template, 45
    See also patterns, cutting
Textured glass, 163, 168
Thermometer tubing updraw machine, 9
Third man sling, 21, 22
Tiles, setting, 217, 219-220
Titanium hydride, 144
Titanium hydride metalizing process, 144-145
Translucent glass, 61-69
Transparency, 2, 262
Trimming, 12
Tubing, 116, 118
    cutting, 51, 52, 53, 117
    making, 9, 11-12
    methods of marking, 117-118
    procedure for working, 119-127
Tungsten, 142-143
Turpentine, 56

United States, 3-5
Vacuum cup holders, 22
Van Leeuwenhold, Anthony, 3
Velvet Glass Frosting Base No. 77, 64
Venetian glass, 215
Venice, 2-3
Ventilation, need for, 61, 62, 64, 65, 67, 70, 82
Vinyl plastic, transparent, 250, 259-260
    See also laminated glass
Viscosity, 241
Viscosity reference points, 95-96
Vitreous enamels, **See** enamels, vitreous

Wax
    use in etching, 61, 62, 64, 65, 70, 82
    use in stained glass work, 153-154
Weldwood Glue, 138
West Coast Display Mannequin, 204, 206, 207
Wheel cutter, steel, 25
    type of, 26-28
    care of, 27, 29, 30
    using the, 30-34
    See also cutter, glass
White acid, 64
Whiting, 96-97, 108, 140, 263
Window glass
    manufacture of, 247, 249
    marking, 262-263
    paint on, 261
Windshields, automobile, 93
    cutting, 37
Wire glass, 249
Wistar, Casper, 4
Working point, 96

Yarn (fiber glass), 179
Yellowstone National Park, 1